Big Data Technologies and Applications

Borko Furht · Flavio Villanustre

Big Data Technologies and Applications

 Springer

Borko Furht
Department of Computer and Electrical
    Engineering and Computer Science
Florida Atlantic University
Boca Raton, FL
USA

Flavio Villanustre
LexisNexis Risk Solutions
Alpharetta, GA
USA

ISBN 978-3-319-83077-3   ISBN 978-3-319-44550-2   (eBook)
DOI 10.1007/978-3-319-44550-2

Printed on acid-free paper

This Springer imprint is published by Springer Nature
The registered company is Springer International Publishing AG
The registered company address is: Gewerbestrasse 11, 6330 Cham, Switzerland

Preface

The scope of this book includes leading edge in big data systems, architectures, and applications. Big data computing refers to capturing, managing, analyzing, and understanding the data at volumes and rates that push the frontiers of current technologies. The challenge of big data computing is to provide the hardware architectures and related software systems and techniques which are capable of transforming ultra large data into valuable knowledge. Big data and data-intensive computing demand a fundamentally different set of principles than mainstream computing. Big data applications typically are well suited for large-scale parallelism over the data and also require extremely high degree of fault tolerance, reliability, and availability. In addition, most big data applications require relatively fast response. The objective of this book is to introduce the basic concepts of big data computing and then to describe the total solution to big data problems developed by LexisNexis Risk Solutions.

This book comprises of three parts, which consists of 15 chapters. Part I on *Big Data Technologies* includes the chapters dealing with introduction to big data concepts and techniques, big data analytics and relating platforms, and visualization techniques and deep learning techniques for big data. Part II on *LexisNexis Risk Solution to Big Data* focuses on specific technologies and techniques developed at LexisNexis to solve critical problems that use big data analytics. It covers the open source high performance computing cluster (HPCC Systems®) platform and its architecture, as well as, parallel data languages ECL and KEL, developed to effectively solve big data problems. Part III on *Big Data Applications* describes various data-intensive applications solved on HPCC Systems. It includes applications such as cyber security, social network analytics, including insurance fraud, fraud in prescription drugs, and fraud in Medicaid, and others. Other HPCC Systems applications described include Ebola spread modeling using big data analytics and unsupervised learning and image classification.

With the dramatic growth of data-intensive computing and systems and big data analytics, this book can be the definitive resource for persons working in this field as researchers, scientists, programmers, engineers, and users. This book is intended for a wide variety of people including academicians, designers, developers, educators, engineers, practitioners, and researchers and graduate students. This book can also be beneficial for business managers, entrepreneurs, and investors.

The main features of this book can be summarized as follows:

1. This book describes and evaluates the current state of the art in the field of big data and data-intensive computing.
2. This book focuses on LexisNexis' platform and its solutions to big data.
3. This book describes the real-life solutions to big data analytics.

Boca Raton, FL, USA   Borko Furht
Alpharetta, GA, USA   Flavio Villanustre
2016

Acknowledgments

We would like to thank a number of contributors to this book. The LexisNexis contributors include David Bayliss, Gavin Halliday, Anthony M. Middleton, Edin Muharemagic, Jesse Shaw, Bob Foreman, Arjuna Chala, and Flavio Villanustre. The Florida Atlantic University contributors include Ankur Agarwal, Taghi Khoshgoftaar, DingDing Wang, Maryam M. Najafabadi, Abhishek Jain, Karl Weiss, Naeem Seliva, Randal Wald, and Borko Furht. The other contributors include I. Itauma, M.S. Aslan, and X.W Chen from Wayne State University; Chun-Wei Tsai, Chin-Feng Lai, Han-Chieh Chao, and Athanasios V. Vasilakos from Lulea University of Technology in Sweden; and Akaterina Olshannikova, Aleksandr Ometov, Yevgeni Koucheryavy, and Thomas Olsson from Tampere University of Technology in Finland.

Without their expertise and effort, this book would never come to fruition. Springer editors and staffs also deserve our sincere recognition for their support throughout the project.

Contents

About the Authors

 **Borko Furht** is a professor in the Department of Electrical and Computer Engineering and Computer Science at Florida Atlantic University (FAU) in Boca Raton, Florida. He is also the director of the NSF Industry/University Cooperative Research Center for Advanced Knowledge Enablement. Before joining FAU, he was a vice president of research and a senior director of development at Modcomp (Ft. Lauderdale), a computer company of Daimler Benz, Germany; a professor at University of Miami in Coral Gables, Florida; and a senior researcher in the Institute Boris Kidric-Vinca, Yugoslavia. Professor Furht received his Ph.D. degree in electrical and computer engineering from the University of Belgrade. His current research is in multimedia systems, multimedia big data and its applications, 3-D video and image systems, wireless multimedia, and Internet and cloud computing. He is presently the Principal Investigator and Co-PI of several projects sponsored by NSF and various high-tech companies. He is the author of numerous books and articles in the areas of multimedia, data-intensive applications, computer architecture, real-time computing, and operating systems. He is a founder and an editor-in-chief of two journals: *Journal of Big Data* and *Journal of Multimedia Tools and Applications*. He has received several technical and publishing awards and has been a consultant for many high-tech companies including IBM, Hewlett-Packard, Adobe, Xerox, General Electric, JPL, NASA, Honeywell, and RCA. He has also served as a consultant to various colleges and universities. He has given many invited talks, keynote lectures, seminars, and tutorials. He served on the board of directors of several high-tech companies.

Dr. Flavio Villanustre leads HPCC Systems® and is also VP, Technology for LexisNexis Risk Solutions®. In this position, he is responsible for information and physical security, overall platform strategy, and new product development. He is also involved in a number of projects involving Big Data integration, analytics, and

Business Intelligence. Previously, he was the director of Infrastructure for Seisint. Prior to 2001, he served in a variety of roles at different companies including infrastructure, information security, and information technology. In addition to this, he has been involved with the open source community for over 15 years through multiple initiatives. Some of these include founding the first Linux User Group in Buenos Aires (BALUG) in 1994, releasing several pieces of software under different open source licenses, and evangelizing open source to different audiences through conferences, training, and education. Prior to his technology career, he was a neurosurgeon.

Part I
Big Data Technologies

Chapter 1
Introduction to Big Data

Borko Furht and Flavio Villanustre

Concept of Big Data

In this chapter we present the basic terms and concepts in Big Data computing. Big data is a large and complex collection of data sets, which is difficult to process using on-hand database management tools and traditional data processing applications. Big Data topics include the following activities:

- Capture
- Storage
- Search
- Sharing
- Transfer
- Analysis
- Visualization

Big Data can be also defined using three Vs: Volume, Velocity, and Variety.

Volume refers to size of the data from Terabytes (TB) to Petabytes (PB), and related big data structures including records, transactions, files, and tables. Data volumes are expected to grow 50 times by 2020.

Velocity refers to ways of transferring big data including batch, near time, real time, and streams. Velocity also includes time and latency characteristics of data handling. The data can be analyzed, processed, stored, and managed in a fast rate, or with a lag time between events.

Variety of big data refers to different formats of data including structured, unstructured, semi-structured data, and the combination of these. The data format can be in the forms of documents, emails, text messages, audio, images, video, graphics data, and others.

In addition to these three main characteristics of big data, there are two additional features: Value, and Veracity [1]. Value refers to benefits/value obtained by the user from the big data. Veracity refers to the quality of big data.

© Springer International Publishing Switzerland 2016   3
B. Furht and F. Villanustre, *Big Data Technologies and Applications*,
DOI 10.1007/978-3-319-44550-2_1

Table 1.1 Comparison between traditional and big data (adopted from [2])

	Traditional data	Big data
Volume	In GBs	TBs and PBs
Data generation rate	Per hour; per day	More rapid
Data structure	Structured	Semi-structured or Unstructured
Data source	Centralized	Fully distributed
Data integration	Easy	Difficult
Data store	RDBMS	HDFS, NoSQL
Data access	Interactive	Batch or near real-time

Sources of big data can be classified to: (1) various transactions, (2) enterprise data, (3) public data, (4) social media, and (5) sensor data. Table 1.1 illustrates the difference between traditional data and big data.

Big Data Workflow

Big data workflow consists of the following steps, as illustrated in Fig. 1.1.
These steps are defined as:

Collection—Structured, unstructured and semi-structured data from multiple sources
Ingestion—loading vast amounts of data onto a single data store
Discovery and Cleansing—understanding format and content; clean up and formatting
Integration—linking, entity extraction, entity resolution, indexing and data fusion
Analysis—Intelligence, statistics, predictive and text analytics, machine learning
Delivery—querying, visualization, real time delivery on enterprise-class availability

Fig. 1.1 Big data workflow

Big Data Technologies

Big Data technologies is a new generation of technologies and architectures designed to economically extract value from very large volumes of a wide variety of data by enabling high-velocity capture, discovery, and analysis. Big Data technologies include:

- Massively Parallel Processing (MPP)
- Data mining tools and techniques
- Distributed file systems and databases
- Cloud computing platforms
- Scalable storage systems

Big Data Layered Architecture

As proposed in [2], the big data system can be represented using a layered architecture, as shown in Fig. 1.2. The big data layered architecture consists of three levels: (1) infrastructure layer, (2) computing layer, and (3) application layer.

The infrastructure layer consists of a pool of computing and storage resources including cloud computer infrastructure. They must meet the big data demand in terms of maximizing system utilization and storage requirements.

The computing layer is a middleware layer and includes various big data tools for data integration, data management, and the programming model.

The application layer provides interfaces by the programming models to implement various data analysis functions including statistical analyses, clustering, classification, data mining, and others and build various big data applications.

Fig. 1.2 Layered architecture
of big data (adopted from [2]

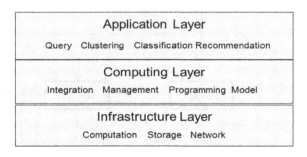

Big Data Software

Hadoop (Apache Foundation)

Hadoop is open source software framework for storage and large scale data processing on clusters computers. It is used for processing, storing and analyzing large
amount of distributed unstructured data Hadoop consists of two components:
HDFS, distributive file system, and Map Reduce, which is programming framework. In Map Reduce programming component large task is divided into two
phases: Map and Reduce, as shown in Fig. 1.3. The Map phase divides the large
task into smaller pieces and dispatches each small piece onto one active node in the
cluster. The Reduce phase collects the results from the Map phase and processes the
results to get the final result. More details can be found in [3].

Splunk

Captures, indexes and correlates real-time data in a searchable repository from
which it can generate graphs, reports, alerts, dashboards and visualizations.

LexisNexis' High-Performance Computer Cluster (HPCC)

HPCC system and software are developed by LexisNexis Risk Solutions.
A software architecture, shown in Fig. 1.4, implemented on computing clusters

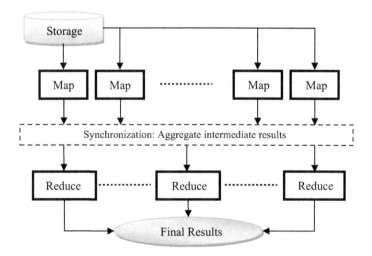

Fig. 1.3 MapReduce framework

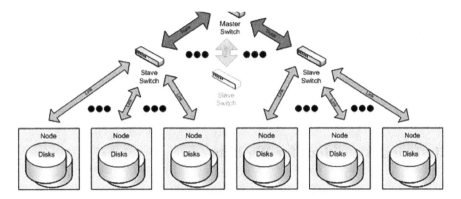

Fig. 1.4 The architecture of the HPCC system

provides data parallel processing for applications with Big Data. Includes a data-centric programming language for parallel data processing—ECL. The part II of the book is focused on details of the HPCC system and Part III describes various HPCC applications.

Big Data Analytics Techniques

We classify big data analytics in the following five categories [4]:

- Text analytics
- Audio analytics
- Video analytics
- Social media analytics
- Predictive analytics.

Text analytics or text mining refers to the process of analyzing unstructured text to extract relevant information. Text analytics techniques use statistical analysis, computational linguistics, and machine learning. Typical applications include extracting textual information from social network feeds, emails, blogs, online forums, survey responses, and news.

Audio analytics or speech analytic techniques are used to analyze and extract information from unstructured audio data. Typical applications of audio analytics are customer call centers and healthcare companies.

Video analytics or video content analysis deals with analyzing and extracting meaningful information from video streams. Video analytics can be used in various video surveillance applications.

Social media analytics includes the analysis of structured and unstructured data from various social media sources including Facebook, Linkedin, Twitter, YouTube, Instagram, Wikipedia, and others.

Predictive analytics includes techniques for predicting future outcomes based on past and current data. The popular predictive analytic techniques include NNs, SVMs, decision trees, linear and logistic regression, association rules, and scorecards.

More details about big data analytics techniques can be found in [2, 4] as well as in the chapter in this book on "Big Data Analytics."

Clustering Algorithms for Big Data

Clustering algorithms are developed to analyze large volume of data with the main objective to categorize data into clusters based on the specific metrics. An excellent survey of clustering algorithms for big data is presented in [5]. The authors proposed the categorization of the clustering algorithms into the following five categories:

- Partitioning-based algorithms
- Hierarchical-based algorithms
- Density-based algorithms
- Grid-based algorithms, and
- Model-based clustering algorithms.

The clustering algorithms were evaluated for big data applications with respect to three Vs defied earlier and the results of evaluation are given in [5] and the authors proposed the candidate clustering algorithms for big data that meet the criteria relating to three V.

In the case of clustering algorithms, Volume refers to the ability of a clustering algorithm to deal with a large amount of data. Variety refers to the ability of a clustering algorithm to handle different types of data, and Velocity refers to the speed of a clustering algorithm on big data. In [5] the authors selected the following five clustering algorithms as the most appropriate for big data:

- Fuzzy-CMeans (FCM) clustering algorithm
- The BIRCH clustering algorithm
- The DENCLUE clustering algorithm
- Optimal Grid (OPTIGRID) clustering algorithm, and
- Expectation-Maximization (EM) clustering algorithm.

Authors also performed experimental evaluation of these algorithms on real data [5].

Big Data Growth

Figure 1.5 shows the forecast in big data growth by Reuter (2012) that today there are less than 10 zettabytes of data. They estimate that by 2020 there will be more than 30 Zettabyte of data, with the big data market growth of 45 % annually.

Big Data Industries

Media and entertainment applications include digital recording, production, and media delivery. Also, it includes collection of large amounts of rich content and user viewing behaviors.

Healthcare applications include electronic medical records and images, public health monitoring programs, and long-term epidemiological research programs.

Life science applications include low-cost gene sequencing that generates tens of terabytes of information that must be analyzed for genetic variations.

Video surveillance applications include big data analysis received from cameras and recording systems.

Applications in transportation, logistics, retails, utilities and telecommunications include sensor data generated from GPS transceivers, RFID tag readers, smart

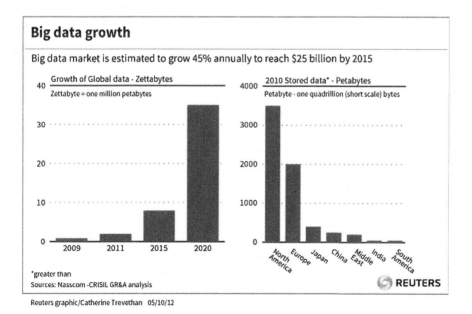

Fig. 1.5 Big data growth (*Source* Reuter 2012)

meters, and cell phones. Data is analyzed and used to optimize operations and drive operational business intelligence.

Challenges and Opportunities with Big Data

In 2012, a group of prominent researchers from leading US universities including UC Santa Barbara, UC Berkeley, MIT, Cornell University, University of Michigan, Columbia University, Stanford University and a few others, as well as researchers from leading companies including Microsoft, HP, Google, IBM, and Yahoo!, created a white paper on this topic [6]. Here we present some conclusions from this paper.

One of the conclusions is that Big Data has the potential to revolutionize research; however it has also potential to revolutionize education. The prediction is that big database of every student's academic performance can be created and this data can be then used to design the most effective approaches to education, starting from reading, writing, and math, to advanced college-level courses [6].

The analysis of big data consists of various phases as shown in Fig. 1.6, and each phase introduces challenges, which are discussed in detail in [6]. Here we summarize the main challenges.

In the Data Acquisition and Recording phase the main challenge is to select data filters, which will extract the useful data. Another challenge is to automatically generate the right metadata to describe what data is recorded and measured.

In the Information Extraction and Clustering phase the main challenge is to convert the original data in a structured form, which is suitable for analysis.

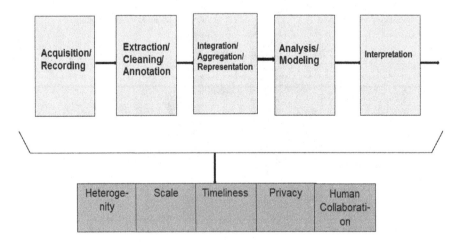

Fig. 1.6 The big data analysis pipeline [6]

Methods for querying and mining Big Data are fundamentally different from traditional statistical analysis on small data samples. The characteristics of Big Data is that it is often noisy, dynamic, heterogeneous, inter-related, and untrustworthy. These is another challenge.

The interpretation of the obtained results from big data analysis is another challenge. Usually, the interpretation involves examining all the assumptions made and retracting the analysis.

References

1. Sharma S, Mangat V. Technology and trends to handle big data: a survey. In: 2015 fifth international conference on advanced computing and communication technologies. p. 266–71.
2. Hu H, Wen Y, Chua T-S, Li X. Toward scalable systems for big data analytics: a technology tutorial. IEEE Access. 2014;2(14):652–87.
3. Menon SP, Hegde NP. A survey of tools and applications in big data. In: IEEE 9th international conference on intelligence systems and control; 2015. p. 1–7.
4. Vashisht P, Gupta V. Big data analytics: a survey. In: 2015 international conference on green computing and internet of things; 2015.
5. Fahad A, et al. A survey of clustering algorithms for big data: taxonomy and empirical analysis. IEEE Trans Emerg Top Comput. 2014;2(3):267–79.
6. Challenges and opportunities with big data. White paper; 2012.
7. Fang H et al. A survey of big data research. In: IEEE network; 2015. p. 6–9.

Chapter 2
Big Data Analytics

Chun-Wei Tsai, Chin-Feng Lai, Han-Chieh Chao and Athanasios V. Vasilakos

Abbreviations

PCA	Principal components analysis
3Vs	Volume, velocity, and variety
IDC	International Data Corporation
KDD	Knowledge discovery in databases
SVM	Support vector machine
SSE	Sum of squared errors
GLADE	Generalized linear aggregates distributed engine
BDAF	Big data architecture framework
CBDMASP	Cloud-based big data mining and analyzing services platform
SODSS	Service-oriented decision support system
HPCC	High performance computing cluster system
BI&I	Business intelligence and analytics
DBMS	Database management system
MSF	Multiple species flocking
GA	Genetic algorithm
SOM	Self-organizing map
MBP	Multiple back-propagation
YCSB	Yahoo cloud serving benchmark
HPC	High performance computing
EEG	Electroencephalography

This chapter has been adopted from the Journal of Big Data, Borko Furht and Taghi Khoshgoftaar, Editors-in-Chief. Springer, Vol. 2, No. 21, October 2015.

13

Introduction

As the information technology spreads fast, most of the data were born digital as well as exchanged on internet today. According to the estimation of Lyman and Varian [1], the new data stored in digital media devices have already been more than 92 % in 2002, while the size of these new data was also more than five exabytes. In fact, the problems of analyzing the large scale data were not suddenly occurred but have been there for several years because the creation of data is usually much easier than finding useful things from the data. Even though computer systems today are much faster than those in the 1930s, the large scale data is a strain to analyze by the computers we have today.

In response to the problems of analyzing large-scale data, quite a few efficient methods [2], such as sampling, data condensation, density-based approaches, grid-based approaches, divide and conquer, incremental learning, and distributed computing, have been presented. Of course, these methods are constantly used to improve the performance of the operators of data analytics process.[1] The results of these methods illustrate that with the efficient methods at hand, we may be able to analyze the large-scale data in a reasonable time. The dimensional reduction method (e.g., principal components analysis; PCA [3]) is a typical example that is aimed at reducing the input data volume to accelerate the process of data analytics. Another reduction method that reduces the data computations of data clustering is sampling [4], which can also be used to speed up the computation time of data analytics.

Although the advances of computer systems and internet technologies have witnessed the development of computing hardware following the Moore's law for several decades, the problems of handling the large-scale data still exist when we are entering the age of big data. That is why Fisher et al. [5] pointed out that big data means that the data is unable to be handled and processed by most current information systems or methods because data in the big data era will not only become too big to be loaded into a single machine, it also implies that most traditional data mining methods or data analytics developed for a centralized data analysis process may not be able to be applied directly to big data. In addition to the issues of data size, Laney [6] presented a well-known definition (also called 3Vs) to explain what is the "big" data: volume, velocity, and variety. The definition of 3Vs implies that the data size is large, the data will be created rapidly, and the data will be existed in multiple types and captured from different sources, respectively. Later studies [7, 8] pointed out that the definition of 3Vs is insufficient to explain the big data we face now. Thus, veracity, validity, value, variability, venue, vocabulary, and vagueness were added to make some complement explanation of big data [8].

---

[1]In this chapter, by the data analytics, we mean the whole KDD process, while by the data analysis, we mean the part of data analytics that is aimed at finding the hidden information in the data, such as data mining.

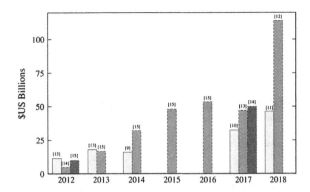

Fig. 2.1 Expected trend of the marketing of big data between 2012 and 2018. Note that *yellow*, *red*, and *blue* of different *colored box* represent the order of appearance of reference in this paper for particular year

The report of IDC [9] indicates that the marketing of big data is about $16.1 billion in 2014. Another report of IDC [10] forecasts that it will grow up to $32.4 billion by 2017. The reports of [11] and [12] further pointed out that the marketing of big data will be $46.34 billion and $114 billion by 2018, respectively. As shown in Fig. 2.1, even though the marketing values of big data in these researches and technology reports [9–15] are different, these forecasts usually indicate that the scope of big data will be grown rapidly in the forthcoming future.

In addition to marketing, from the results of disease control and prevention [16], business intelligence [17], and smart city [18], we can easily understand that big data is of vital importance everywhere. A numerous researches are therefore focusing on developing effective technologies to analyze the big data. To discuss in deep the big data analytics, this paper gives not only a systematic description of traditional large-scale data analytics but also a detailed discussion about the differences between data and big data analytics framework for the data scientists or researchers to focus on the big data analytics.

Moreover, although several data analytics and frameworks have been presented in recent years, with their pros and cons being discussed in different studies, a complete discussion from the perspective of data mining and knowledge discovery in databases still is needed. As a result, this paper is aimed at providing a brief review for the researchers on the data mining and distributed computing domains to have a basic idea to use or develop data analytics for big data.

Figure 2.2 shows the roadmap of this paper, and the remainder of the paper is organized as follows. "Data analytics" begins with a brief introduction to the data analytics, and then "Big data analytics" will turn to the discussion of big data analytics as well as stateof-the-art data analytics algorithms and frameworks. The open issues are discussed in "The open issues" while the conclusions and future trends are drawn in "Conclusions".

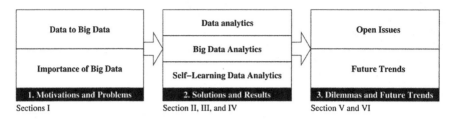

Fig. 2.2 Roadmap of this paper

Data Analytics

To make the whole process of knowledge discovery in databases (KDD) more clear, Fayyad and his colleagues summarized the KDD process by a few operations in [19], which are selection, preprocessing, transformation, data mining, and interpretation/evaluation. As shown in Fig. 2.3, with these operators at hand we will be able to build a complete data analytics system to gather data first and then find information from the data and display the knowledge to the user. According to our observation, the number of research articles and technical reports that focus on data mining is typically more than the number focusing on other operators, but it does not mean that the other operators of KDD are unimportant. The other operators also play the vital roles in KDD process because they will strongly impact the final result of KDD. To make the discussions on the main operators of KDD process more concise, the following sections will focus on those depicted in Fig. 2.3, which were simplified to three parts (input, data analytics, and output) and seven operators (gathering, selection, preprocessing, transformation, data mining, evaluation, and interpretation).

Fig. 2.3 The process of
knowledge discovery in
databases

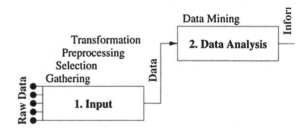

Data Input

As shown in Fig. 2.3, the gathering, selection, preprocessing, and transformation operators are in the input part. The selection operator usually plays the role of knowing which kind of data was required for data analysis and select the relevant information from the gathered data or databases; thus, these gathered data from different data resources will need to be integrated to the target data. The preprocessing operator plays a different role in dealing with the input data which is aimed at detecting, cleaning, and filtering the unnecessary, inconsistent, and incomplete data to make them the useful data. After the selection and preprocessing operators, the characteristics of the secondary data still may be in a number of different data formats; therefore, the KDD process needs to transform them into a data-mining-capable format which is performed by the transformation operator. The methods for reducing the complexity and downsizing the data scale to make the data useful for data analysis part are usually employed in the transformation, such as dimensional reduction, sampling, coding, or transformation.

The data extraction, data cleaning, data integration, data transformation, and data reduction operators can be regarded as the preprocessing processes of data analysis [20] which attempts to extract useful data from the raw data (also called the primary data) and refine them so that they can be used by the following data analyses. If the data are a duplicate copy, incomplete, inconsistent, noisy, or outliers, then these operators have to clean them up. If the data are too complex or too large to be handled, these operators will also try to reduce them. If the raw data have errors or omissions, the roles of these operators are to identify them and make them consistent. It can be expected that these operators may affect the analytics result of KDD, be it positive or negative. In summary, the systematic solutions are usually to reduce the complexity of data to accelerate the computation time of KDD and to improve the accuracy of the analytics result.

Data Analysis

Since the data analysis (as shown in Fig. 2.3) in KDD is responsible for finding the hidden patterns/rules/information from the data, most researchers in this field use the term data mining to describe how they refine the "ground" (i.e., raw data) into "gold nugget" (i.e., information or knowledge). The data mining methods [20] are not limited to data problem specific methods. In fact, other technologies (e.g., statistical or machine learning technologies) have also been used to analyze the data for many years. In the early stages of data analysis, the statistical methods were used for analyzing the data to help us understand the situation we are facing, such as public opinion poll or TV programme rating. Like the statistical analysis, the problem specific methods for data mining also attempted to understand the meaning from the collected data.

Fig. 2.4 Data mining
algorithm

1 Input data D
2 Initialize candidate solutions r
3 **While** the termination criterion is not met
4      $d = Scan(D)$
5      $v = Construct(d, r, o)$
6      $r = Update(v)$
7 **End**
8 Output rules r

After the data mining problem was presented, some of the domain specific algorithms are also developed. An example is the apriori algorithm [21] which is one of the useful algorithms designed for the association rules problem. Although most definitions of data mining problems are simple, the computation costs are quite high. To speed up the response time of a data mining operator, machine learning [22], metaheuristic algorithms [23], and distributed computing [24] were used alone or combined with the traditional data mining algorithms to provide more efficient ways for solving the data mining problem. One of the well-known combinations can be found in [25], Krishna and Murty attempted to combine genetic algorithm and k-means to get better clustering result than k-means alone does. As Fig. 2.4 shows, most data mining algorithms contain the initialization, data input and output, data scan, rules construction, and rules update operators [26]. In Fig. 2.4, D represents the raw data, d the data from the scan operator, r the rules, o the predefined measurement, and v the candidate rules. The scan, construct, and update operators will be performed repeatedly until the termination criterion is met. The timing to employ the scan operator depends on the design of the data mining algorithm; thus, it can be considered as an optional operator. Most of the data algorithms can be described by Fig. 2.4 in which it also shows that the representative algorithms—clustering, classification, association rules, and sequential patterns—will apply these operators to find the hidden information from the raw data. Thus, modifying these operators will be one of the possible ways for enhancing the performance of the data analysis.

Clustering is one of the well-known data mining problems because it can be used to understand the "new" input data. The basic idea of this problem [27] is to separate a set of unlabeled input data[2] to k different groups, e.g., such as k-means [28]. Classification [20] is the opposite of clustering because it relies on a set of labeled input data to construct a set of classifiers (i.e., groups) which will then be used to classify the unlabeled input data to the groups to which they belong. To solve the classification problem, the decision tree-based algorithm [29], naïve Bayesian classification [30], and support vector machine (SVM) [31] are widely used in recent years.

---

[2]In this chapter, by an unlabeled input data, we mean that it is unknown to which group the input data belongs. If all the input data are unlabeled, it means that the distribution of the input data is unknown.

Unlike clustering and classification that attempt to classify the input data to k groups, association rules and sequential patterns are focused on finding out the "relationships" between the input data. The basic idea of association rules [21] is find all the co-occurrence relationships between the input data. For the association rules problem, the apriori algorithm [21] is one of the most popular methods. Nevertheless, because it is computationally very expensive, later studies [32] have attempted to use different approaches to reducing the cost of the apriori algorithm, such as applying the genetic algorithm to this problem [33]. In addition to considering the relationships between the input data, if we also consider the sequence or time series of the input data, then it will be referred to as the sequential pattern mining problem [34]. Several apriori-like algorithms were presented for solving it, such as generalized sequential pattern [34] and sequential pattern discovery using equivalence classes [35].

Output the Result

Evaluation and interpretation are two vital operators of the output. Evaluation typically plays the role of measuring the results. It can also be one of the operators for the data mining algorithm, such as the sum of squared errors which was used by the selection operator of the genetic algorithm for the clustering problem [25].

To solve the data mining problems that attempt to classify the input data, two of the major goals are: (1) cohesion—the distance between each data and the centroid (mean) of its cluster should be as small as possible, and (2) coupling—the distance between data which belong to different clusters should be as large as possible. In most studies of data clustering or classification problems, the sum of squared errors (SSE), which was used to measure the cohesion of the data mining results, can be defined as

$$\text{SSE} = \sum_{i=1}^{k} \sum_{j=1}^{n_i} D(x_{ij} - c_i), \qquad (2.1)$$

$$c_i = \frac{1}{n_i} \sum_{j=1}^{n_i} x_{ij}, \qquad (2.2)$$

where k is the number of clusters which is typically given by the user; n_i the number of data in the ith cluster; x_{ij} the jth datum in the ith cluster; ci is the mean of the ith cluster; and $n = \sum_{i=1}^{k} n_i$ is the number of data. The most commonly used distance measure for the data mining problem is the Euclidean distance, which is defined as

$$D(p_i, p_j) = \left(\sum_{l=1}^{d} |p_{il}, p_{jl}|^2 \right)^{1/2}, \qquad (2.3)$$

where p_i and p_j are the positions of two different data. For solving different data mining problems, the distance measurement $D(p_i, p_j)$ can be the Manhattan distance, the Minkowski distance, or even the cosine similarity [36] between two different documents.

Accuracy (ACC) is another well-known measurement [37] which is defined as

$$ACC = \frac{\text{Number of cases correctly classified}}{\text{Total number of test cases}}. \qquad (2.4)$$

To evaluate the classification results, precision (p), recall (r), and F-measure can be used to measure how many data that do not belong to group A are incorrectly classified into group A; and how many data that belong to group A are not classified into group A. A simple confusion matrix of a classifier [37] as given in Table 2.1 can be used to cover all the situations of the classification results. In Table 2.1, TP and TN indicate the numbers of positive examples and negative examples that are correctly classified, respectively; FN and FP indicate the numbers of positive examples and negative examples that are incorrectly classified, respectively. With the confusion matrix at hand, it is much easier to describe the meaning of precision (p), which is defined as

$$p = \frac{TP}{TP + FP}, \qquad (2.5)$$

and the meaning of recall (r), which is defined as

$$r = \frac{TP}{TP + FN}. \qquad (2.6)$$

The F-measure can then be computed as

$$F = \frac{2pr}{p + r}. \qquad (2.7)$$

In addition to the above-mentioned measurements for evaluating the data mining results, the computation cost and response time are another two well-known measurements. When two different mining algorithms can find the same or similar

Table 2.1 Confusion matrix of a classifier [37]

	Classified positive	Classified negative
Actual positive	TP	FN
Actual negative	FP	TN

results, of course, how fast they can get the final mining results will become the most important research topic.

After something (e.g., classification rules) is found by data mining methods, the two essential research topics are: (1) the work to navigate and explore the meaning of the results from the data analysis to further support the user to do the applicable decision can be regarded as the interpretation operator [38], which in most cases, gives useful interface to display the information [39] and (2) a meaningful summarization of the mining results [40] can be made to make it easier for the user to understand the information from the data analysis. The data summarization is generally expected to be one of the simple ways to provide a concise piece of information to the user because human has trouble of understanding vast amounts of complicated information. A simple data summarization can be found in the clustering search engine, when a query "oasis" is sent to Carrot2 (http://search. carrot2.org/stable/search), it will return some keywords to represent each group of the clustering results for web links to help us recognize which category needed by the user, as shown in the left side of Fig. 2.5.

A useful graphical user interface is another way to provide the meaningful information to an user. As explained by Shneiderman in [39], we need "overview first, zoom and filter, then retrieve the details on demand". The useful graphical user interface [38, 41] also makes it easier for the user to comprehend the meaning of the results when the number of dimensions is higher than three. How to display the results of data mining will affect the user's perspective to make the decision. For instance, data mining can help us find "type A influenza" at a particular region, but without the time series and flu virus infected information of patients, the government could not recognize what situation (pandemic or controlled) we are facing now so as to make appropriate responses to that. For this reason, a better solution to

Fig. 2.5 Screenshot of the result of clustering search engine

merge the information from different sources and mining algorithm results will be useful to let the user make the right decision.

Summary

Since the problems of handling and analyzing large-scale and complex input data always exist in data analytics, several efficient analysis methods were presented to accelerate the computation time or to reduce the memory cost for the KDD process, as shown in Table 2.2. The study of [42] shows that the basic mathematical concepts (i.e., triangle inequality) can be used to reduce the computation cost of a clustering algorithm. Another study [43] shows that the new technologies (i.e., distributed computing by GPU) can also be used to reduce the computation time of data analysis method. In addition to the well-known improved methods for these analysis methods (e.g., triangle inequality or distributed computing), a large proportion of studies designed their efficient methods based on the characteristics of mining algorithms or problem itself, which can be found in [32, 44, 45], and so forth. This kind of improved methods typically was designed for solving the drawback of the mining algorithms or using different ways to solve the mining problem. These situations can be found in most association rules and sequential

Table 2.2 Efficient data analytics methods for data mining

Problem	Method	References
Clustering	BIRCH	[44]
	DBSCAN	[45]
	Incremental DBSCAN	[46]
	RKM	[47]
	TKM	[42]
Classification	SLIQ	[50]
	TLAESA	[51]
	FastNN	[52]
	SFFS	[53]
	CPU-based SVM	[43]
Association rules	CLOSET	[54]
	FP-tree	[32]
	CHARM	[55]
	MAFIA	[56]
	FAST	[57]
Sequential patterns	SPADE	[35]
	CloSpan	[58]
	PrefixSpan	[59]
	SPAM	[60]
	ISE	[61]

patterns problems because the original assumption of these problems is for the analysis of large-scale dataset. Since the earlier frequent pattern algorithm (e.g., apriori algorithm) needs to scan the whole dataset many times which is computationally very expensive. How to reduce the number of times the whole dataset is scanned so as to save the computation cost is one of the most important things in all the frequent pattern studies. The similar situation also exists in data clustering and classification studies because the design concept of earlier algorithms, such as mining the patterns on-the-fly [46], mining partial patterns at different stages [47], and reducing the number of times the whole dataset is scanned [32], are therefore presented to enhance the performance of these mining algorithms. Since some of the data mining problems are NP-hard [48] or the solution space is very large, several recent studies [23, 49] have attempted to use metaheuristic algorithm as the mining algorithm to get the approximate solution within a reasonable time.

Abundant research results of data analysis [20, 27, 62] show possible solutions for dealing with the dilemmas of data mining algorithms. It means that the open issues of data analysis from the literature [2, 63] usually can help us easily find the possible solutions. For instance, the clustering result is extremely sensitive to the initial means, which can be mitigated by using multiple sets of initial means [64]. According to our observation, most data analysis methods have limitations for big data, that can be described as follows:

- *Unscalability and centralization* Most data analysis methods are not for large-scale and complex dataset. The traditional data analysis methods cannot be scaled up because their design does not take into account large or complex datasets. The design of traditional data analysis methods typically assumed they will be performed in a single machine, with all the data in memory for the data analysis process. For this reason, the performance of traditional data analytics will be limited in solving the volume problem of big data.
- *Non-dynamic* Most traditional data analysis methods cannot be dynamically adjusted for different situations, meaning that they do not analyze the input data on-the-fly. For example, the classifiers are usually fixed which cannot be automatically changed. The incremental learning [65] is a promising research trend because it can dynamically adjust the the classifiers on the training process with limited resources. As a result, the performance of traditional data analytics may not be useful to the problem of velocity problem of big data.
- *Uniform data structure* Most of the data mining problems assume that the format of the input data will be the same. Therefore, the traditional data mining algorithms may not be able to deal with the problem that the formats of different input data may be different and some of the data may be incomplete. How to make the input data from different sources the same format will be a possible solution to the variety problem of big data.

Because the traditional data analysis methods are not designed for large-scale and complex data, they are almost impossible to be capable of analyzing the big data. Redesigning and changing the way the data analysis methods are designed are two critical trends for big data analysis. Several important concepts in the design of the big data analysis method will be given in the following sections.

Big Data Analytics

Nowadays, the data that need to be analyzed are not just large, but they are composed of various data types, and even including streaming data [66]. Since big data has the unique features of "massive, high dimensional, heterogeneous, complex, unstructured, incomplete, noisy, and erroneous," which may change the statistical and data analysis approaches [67]. Although it seems that big data makes it possible for us to collect more data to find more useful information, the truth is that more data do not necessarily mean more useful information. It may contain more ambiguous or abnormal data. For instance, a user may have multiple accounts, or an account may be used by multiple users, which may degrade the accuracy of the mining results [68]. Therefore, several new issues for data analytics come up, such as privacy, security, storage, fault tolerance, and quality of data [69].

The big data may be created by handheld device, social network, internet of things, multimedia, and many other new applications that all have the characteristics of volume, velocity, and variety. As a result, the whole data analytics has to be re-examined from the following perspectives:

- From the volume perspective, the deluge of input data is the very first thing that we need to face because it may paralyze the data analytics. Different from traditional data analytics, for the wireless sensor network data analysis, Baraniuk [70] pointed out that the bottleneck of big data analytics will be shifted from sensor to processing, communications, storage of sensing data, as shown in Fig. 2.6. This is because sensors can gather much more data, but when uploading such large data to upper layer system, it may create bottlenecks everywhere.
- In addition, from the velocity perspective, real-time or streaming data bring up the problem of large quantity of data coming into the data analytics within a short duration but the device and system may not be able to handle these input data. This situation is similar to that of the network flow analysis for which we typically cannot mirror and analyze everything we can gather.
- From the variety perspective, because the incoming data may use different types or have incomplete data, how to handle them also bring up another issue for the input operators of data analytics.

In this section, we will turn the discussion to the big data analytics process.

Fig. 2.6 The comparison between traditional data analysis and big data analysis on wireless sensor network

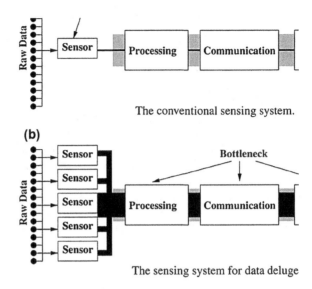

The conventional sensing system.

The sensing system for data deluge

Big Data Input

The problem of handling a vast quantity of data that the system is unable to process is not a brand-new research issue; in fact, it appeared in several early approaches [2, 21, 71], e.g., marketing analysis, network flow monitor, gene expression analysis, weather forecast, and even astronomy analysis. This problem still exists in big data analytics today; thus, preprocessing is an important task to make the computer, platform, and analysis algorithm be able to handle the input data. The traditional data preprocessing methods [72] (e.g., compression, sampling, feature selection, and so on) are expected to be able to operate effectively in the big data age. However, a portion of the studies still focus on how to reduce the complexity of the input data because even the most advanced computer technology cannot efficiently process the whole input data by using a single machine in most cases. By using domain knowledge to design the preprocessing operator is a possible solution for the big data. In [73], Ham and Lee used the domain knowledge, B-tree, divide-and-conquer to filter the unrelated log information for the mobile web log analysis. A later study [74] considered that the computation cost of preprocessing will be quite high for massive logs, sensor, or marketing data analysis. Thus, Dawelbeit and McCrindle employed the bin packing partitioning method to divide the input data between the computing processors to handle this high computations of preprocessing on cloud system. The cloud system is employed to preprocess the raw data and then output the refined data (e.g., data with uniform format) to make it easier for the data analysis method or system to preform the further analysis work.

Sampling and compression are two representative data reduction methods for big data analytics because reducing the size of data makes the data analytics computationally less expensive, thus faster, especially for the data coming to the system rapidly. In addition to making the sampling data represent the original data effectively [75], how many instances need to be selected for data mining method is another research issue [76] because it will affect the performance of the sampling method in most cases.

To avoid the application-level slow-down caused by the compression process, in [77], Jun et al. attempted to use the FPGA to accelerate the compression process. The I/O performance optimization is another issue for the compression method. For this reason, Zou et al. [78] employed the tentative selection and predictive dynamic selection and switched the appropriate compression method from two different strategies to improve the performance of the compression process. To make it possible for the compression method to efficiently compress the data, a promising solution is to apply the clustering method to the input data to divide them into several different groups and then compress these input data according to the clustering information. The compression method described in [79] is one of this kind of solutions, it first clusters the input data and then compresses these input data via the clustering results while the study [80] also used clustering method to improve the performance of the compression process.

In summary, in addition to handling the large and fast data input, the research issues of heterogeneous data sources, incomplete data, and noisy data may also affect the performance of the data analysis. The input operators will have a stronger impact on the data analytics at the big data age than it has in the past. As a result, the design of big data analytics needs to consider how to make these tasks (e.g., data clean, data sampling, data compression) work well.

Big Data Analysis Frameworks and Platforms

Various solutions have been presented for the big data analytics which can be divided [81] into (1) Processing/Compute: Hadoop [82], Nvidia CUDA [83], or Twitter Storm [84], (2) Storage: Titan or HDFS, and (3) Analytics: MLPACK [85] or Mahout [86]. Although there exist commercial products for data analysis [82–85], most of the studies on the traditional data analysis are focused on the design and development of efficient and/or effective "ways" to find the useful things from the data. But when we enter the age of big data, most of the current computer systems will not be able to handle the whole dataset all at once; thus, how to design a good data analytics framework or platform[3] and how to design analysis methods are both important things for the data analysis process. In this section, we will start with a brief introduction to data analysis frameworks and platforms, followed by a comparison of them.

---

[3]In this paper, the analysis framework refers to the whole system, from raw data gathering, data reformat, data analysis, all the way to knowledge representation.

Researches in Frameworks and Platforms

To date, we can easily find tools and platforms presented by well-known organizations. The cloud computing technologies are widely used on these platforms and frameworks to satisfy the large demands of computing power and storage. As shown in Fig. 2.7, most of the works on KDD for big data can be moved to cloud system to speed up the response time or to increase the memory space. With the advance of these works, handling and analyzing big data within a reasonable time has become not so far away. Since the foundation functions to handle and manage the big data were developed gradually; thus, the data scientists nowadays do not have to take care of everything, from the raw data gathering to data analysis, by themselves if they use the existing platforms or technologies to handle and manage the data. The data scientists nowadays can pay more attention to finding out the useful information from the data even thought this task is typically like looking for a needle in a haystack. That is why several recent studies tried to present efficient and effective framework to analyze the big data, especially on find out the useful things.

Performance-oriented From the perspective of platform performance, Huai [87] pointed out that most of the traditional parallel processing models improve the performance of the system by using a new larger computer system to replace the old computer system, which is usually referred to as "scale up", as shown in Fig. 2.8a. But for the big data analytics, most researches improve the performance of the system by adding more milar computer systems to make it possible for a system to handle all the tasks that cannot be loaded or computed in a single computer system (called "scale out"), as shown in Fig. 2.8b where M1, M2, and M3 represent computer systems that have different computing power, respectively. For the scale up based solution, the computing power of the three systems is in the order of

Fig. 2.7 The basic idea of big data analytics on cloud system

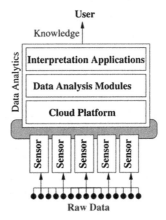

Fig. 2.8 The comparisons
between scale up and scale
out

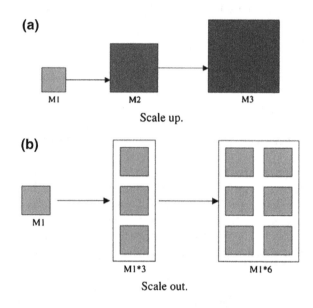

M3 > M1; but for the scale out based system, all we have to do is to keep adding more similar computer systems to to a system to increase its ability. To build a scalable and fault-tolerant manager for big data analysis, Huai et al. [87] presented a matrix model which consists of three matrices for data set (D), concurrent data processing operations (O), and data transformations (T), called DOT. The big data is divided into n subsets each of which is processed by a computer node (worker) in such a way that all the subsets are processed concurrently, and then the results from these n computer nodes are collected and transformed to a computer node. By using this framework, the whole data analysis framework is composed of several DOT blocks. The system performance can be easily enhanced by adding more DOT blocks to the system.

Another efficient big data analytics was presented in [88], called generalized linear aggregates distributed engine (GLADE). The GLADE is a multi-level tree-based data analytics system which consists of two types of computer nodes that are a coordinator workers. The simulation results [89] show that the GLADE can provide a better performance than Hadoop in terms of the execution time. Because Hadoop requires large memory and storage for data replication and it is a single master,[4] Essa et al. [90] presented a mobile agent based framework to solve these two problems, called the map reduce agent mobility (MRAM). The main reason is that each mobile agent can send its code and data to any other machine; therefore, the whole system will not be down if the master failed. Compared to Hadoop, the

---

[4]The whole system may be down when the master machine crashed for a system that has only one master.

architecture of MRAM was changed from client/server to a distributed agent. The load time for MRAM is less than Hadoop even though both of them use the map-reduce solution and Java language. In [91], Herodotou et al. considered issues of the user needs and system workloads. They presented a selftuning analytics system built on Hadoop for big data analysis. Since one of the major goals of their system is to adjust the system based on the user needs and system workloads to provide good performance automatically, the user usually does not need to understand and manipulate the Hadoop system. The study [92] was from the perspectives of data centric architecture and operational models to presented a big data architecture framework (BDAF) which includes: big data infrastructure, big data analytics, data structures and models, big data lifecycle management, and big data security. According to the observations of Demchenko et al. [92], cluster services, Hadoop related services, data analytics tools, databases, servers, and massively parallel processing databases are typically the required applications and services in big data analytics infrastructure.

Result-oriented Fisher et al. [5] presented a big data pipeline to show the workflow of big data analytics to extract the valuable knowledge from big data, which consists of the acquired data, choosing architecture, shaping data into architecture, coding/debugging, and reflecting works. From the perspectives of statistical computation and data mining, Ye et al. [93] presented an architecture of the services platform which integrates R to provide better data analysis services, called cloud-based big data mining and analyzing services platform (CBDMASP). The design of this platform is composed of four layers: the infrastructure services layer, the virtualization layer, the dataset processing layer, and the services layer. Several large-scale clustering problems (the datasets are of size from 0.1 G up to 25.6 G) were also used to evaluate the performance of the CBDMASP. The simulation results show that using map-reduce is much faster than using a single machine when the input data become too large. Although the size of the test dataset cannot be regarded as a big dataset, the performance of the big data analytics using mapreduce can be sped up via this kind of testings. In this study, map-reduce is a better solution when the dataset is of size more than 0.2 G, and a single machine is unable to handle a dataset that is of size more than 1.6 G.

Another study [94] presented a theorem to explain the big data characteristics, called HACE: the characteristics of big data usually are large-volume, Heterogeneous, Autonomous sources with distributed and decentralized control, and we usually try to find out some useful and interesting things from complex and evolving relationships of data. Based on these concerns and data mining issues, Wu and his colleagues [94] also presented a big data processing framework which includes data accessing and computing tier, data privacy and domain knowledge tier, and big data mining algorithm tier. This work explains that the data mining algorithm will become much more important and much more difficult; thus, challenges will also occur on the design and implementation of big data analytics platform. In addition to the platform performance and data mining issues, the privacy issue for big data analytics was a promising research in recent years. In [95], Laurila et al. explained that the privacy is an essential problem when we try to find something from the data that are gathered from mobile devices; thus, data security and data anonymization should also be considered in analyzing this kind of data. Demirkan and Delen [96] presented a service-oriented decision support system (SODSS) for big data analytics which includes information source, data management, information management, and operations management.

Comparison Between the Frameworks/Platforms of Big Data

In [97], Talia pointed out that cloud-based data analytics services can be divided into data analytics software as a service, data analytics platform as a service, and data analytics infrastructure as a service. A later study [98] presented a general architecture of big data analytics which contains multi-source big data collecting, distributed big data storing, and intra/inter big data processing. Since many kinds of data analytics frameworks and platforms have been presented, some of the studies attempted to compare them to give a guidance to choose the applicable frameworks or platforms for relevant works. To give a brief introduction to big data analytics, especially the platforms and frameworks, in [99], Cuzzocrea et al. first discuss how recent studies responded the "computational emergency" issue of big data analytics. Some open issues, such as data source heterogeneity and uncorrelated data filtering, and possible research directions are also given in the same study. In [100], Zhang and Huang used the 5Ws model to explain what kind of framework and method we need for different big data approaches. Zhang and Huang further explained that the 5Ws model represents what kind of data, why we have these data, where the data come from, when the data occur, who receive the data, and how the data are transferred. A later study [101] used the features (i.e., owner, workload, source code, low latency, and complexity) to compare the frameworks of Hadoop [82], Storm [84] and Drill [102]. Thus, it can be easily seen that the framework of Apache Hadoop has high latency compared with the other two frameworks. To better understand the strong and weak points of solutions of big data, Chalmers et al. [81] then employed the volume, variety, variability, velocity, user skill/experience, and infrastructure to evaluate eight solutions of big data analytics.

In [103], in addition to defining that a big data system should include data generation, data acquisition, data storage, and data analytics modules, Hu et al. also mentioned that a big data system can be decomposed into infrastructure, computing, and application layers. Moreover, a promising research for NoSQL storage systems was also discussed in this study which can be divided into key-value, column, document, and row databases. Since big data analysis is generally regarded as a high computation cost work, the high performance computing cluster system (HPCC) is also a possible solution in early stage of big data analytics. Sagiroglu and Sinanc [104] therefore compare the characteristics between HPCC and Hadoop. They then emphasized that HPCC system uses the multikey and multivariate indexes on distributed file system while Hadoop uses the column-oriented database. In [17], Chen et al. give a brief introduction to the big data analytics of business intelligence (BI) from the perspective of evolution, applications, and emerging research topics. In their survey, Chen et al. explained that the revolution of business intelligence and analytics (BI&I) was from BI&I 1.0, BI&I 2.0, to BI&I 3.0 which are DBMS-based and structured content, web-based and unstructured content, and mobile and sensor based content, respectively.

Big Data Analysis Algorithms

Mining Algorithms for Specific Problem

Because the big data issues have appeared for nearly 10 years, in [105], Fan and Bifet pointed out that the terms "big data" [106] and "big data mining" [107] were first presented in 1998, respectively. The big data and big data mining almost appearing at the same time explained that finding something from big data will be one of the major tasks in this research domain. Data mining algorithms for data analysis also play the vital role in the big data analysis, in terms of the computation cost, memory requirement, and accuracy of the end results. In this section, we will give a brief discussion from the perspective of analysis and search algorithms to explain its importance for big data analytics.

Clustering algorithms In the big data age, traditional clustering algorithms will become even more limited than before because they typically require that all the data be in the same format and be loaded into the same machine so as to find some useful things from the whole data. Although the problem [63] of analyzing large-scale and high-dimensional dataset has attracted many researchers from various disciplines in the last century, and several solutions [2, 108] have been presented presented in recent years, the characteristics of big data still brought up several new challenges for the data clustering issues. Among them, how to reduce the data complexity is one of the important issues for big data clustering. In [109], Shirkhorshidi et al. divided the big data clustering into two categories: single-machine clustering (i.e., sampling and dimension reduction solutions), and multiple-machine clustering (parallel and MapReduce solutions). This means that traditional reduction solutions can also be used in the big data age because the complexity and memory space needed for the process of data analysis will be decreased by using sampling and dimension reduction methods. More precisely, sampling can be regarded as reducing the "amount of data" entered into a data analyzing process while dimension reduction can be regarded as "downsizing the whole dataset" because irrelevant dimensions will be discarded before the data analyzing process is carried out.

CloudVista [110] is a representative solution for clustering big data which used cloud computing to perform the clustering process in parallel. BIRCH [44] and sampling method were used in CloudVista to show that it is able to handle large-scale data, e.g., 25 million census records. Using GPU to enhance the performance of a clustering algorithm is another promising solution for big data mining. The multiple species flocking (MSF) [111] was applied to the CUDA platform from NVIDIA to reduce the computation time of clustering algorithm in [112]. The simulation results show that the speedup factor can be increased from 30 up to 60 by using GPU for data clustering. Since most traditional clustering algorithms (e.g., k-means) require a computation that is centralized, how to make them capable of handling big data clustering problems is the major concern of Feldman et al. [113] who use a tree construction for generating the coresets in parallel which is called the "merge-and-reduce" approach. Moreover, Feldman et al. pointed out that by using this solution for clustering, the update time per datum and memory of the traditional clustering algorithms can be significantly reduced.

Classification algorithms Similar to the clustering algorithm for big data mining, several studies also attempted to modify the traditional classification algorithms to make them work on a parallel computing environment or to develop new classification algorithms which work naturally on a parallel computing environment. In [114], the design of classification algorithm took into account the input data that are gathered by distributed data sources and they will be processed by a heterogeneous set of learners.[5] In this study, Tekin et al. presented a novel classification algorithm called "classify or send for classification" (CoS). They assumed that each learner can be used to process the input data in two different ways in a distributed data classification system. One is to perform a classification function by itself while the other is to forward the input data to another learner to have them labeled. The information will be exchanged between different learners. In brief, this kind of solutions can be regarded as a cooperative learning to improve the accuracy in solving the big data classification problem. An interesting solution uses the quantum computing to reduce the memory space and computing cost of a classification algorithm. For example, in [115], Rebentrost et al. presented a quantumbased support vector machine for big data classification and argued that the classification algorithm they proposed can be implemented with a time complexity O(log NM) where N is the number of dimensions and M is the number of training data. There are bright prospects for big data mining by using quantum-based search algorithm when the hardware of quantum computing has become mature.

Frequent pattern mining algorithms Most of the researches on frequent pattern mining (i.e., association rules and sequential pattern mining) were focused on handling large-scale dataset at the very beginning because some early approaches of them were attempted to analyze the data from the transaction data of large shopping mall. Because the number of transactions usually is more than "tens of thousands", the issues about how to handle the large scale data were studied for several years,

---

[5]The learner typically represented the classification function which will create the classifier to help us classify the unknown input data.

such as FP-tree [32] using the tree structure to include the frequent patterns to further reduce the computation time of association rule mining. In addition to the traditional frequent pattern mining algorithms, of course, parallel computing and cloud computing technologies have also attracted researchers in this research domain. Among them, the map-reduce solution was used for the studies [116–118] to enhance the performance of the frequent pattern mining algorithm. By using the map-reduce model for frequent pattern mining algorithm, it can be easily expected that its application to "cloud platform" [119, 120] will definitely become a popular trend in the forthcoming future. The study of [118] no only used the map-reduce model, it also allowed users to express their specific interest constraints in the process of frequent pattern mining. The performance of these methods by using map-reduce model for big data analysis is, no doubt, better than the traditional frequent pattern mining algorithms running on a single machine.

Machine Learning for Big Data Mining

The potential of machine learning for data analytics can be easily found in the early literature [22, 49]. Different from the data mining algorithm design for specific problems, machine learning algorithms can be used for different mining and analysis problems because they are typically employed as the "search" algorithm of the required solution. Since most machine learning algorithms can be used to find an approximate solution for the optimization problem, they can be employed for most data analysis problems if the data analysis problems can be formulated as an optimization problem. For example, genetic algorithm, one of the machine learning algorithms, can not only be used to solve the clustering problem [25], it can also be used to solve the frequent pattern mining problem [33]. The potential of machine learning is not merely for solving different mining problems in data analysis operator of KDD; it also has the potential of enhancing the performance of the other parts of KDD, such as feature reduction for the input operators [71].

A recent study [67] shows that some traditional mining algorithms, statistical methods, preprocessing solutions, and even the GUI's have been applied to several representative tools and platforms for big data analytics. The results show clearly that machine learning algorithms will be one of the essential parts of big data analytics. One of the problems in using current machine learning methods for big data analytics is similar to those of most traditional data mining algorithms which are designed for sequential or centralized computing. However, one of the most possible solutions is to make them work for parallel computing. Fortunately, some of the machine learning algorithms (e.g., population-based algorithms) can essentially be used for parallel computing, which have been demonstrated for several years, such as parallel computing version of genetic algorithm [121]. Different from the traditional GA, as shown in Fig. 2.9a, the population of island model genetic algorithm, one of the parallel GA's, can be divided into several subpopulations, as

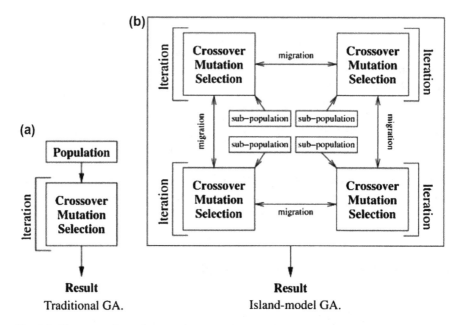

Fig. 2.9 The comparisons between basic idea of traditional GA (TGA) and parallel genetic algorithm (PGA)

shown in Fig. 2.9b. This means that the sub-populations can be assigned to different threads or computer nodes for parallel computing, by a simple modification of the GA.

For this reason, in [122], Kiran and Babu explained that the framework for distributed data mining algorithm still needs to aggregate the information from different computer nodes. As shown in Fig. 2.10, the common design of distributed data mining algorithm is as follows: each mining algorithm will be performed on a computer node (worker) which has its locally coherent data, but not the whole data. To construct a globally meaningful knowledge after each mining algorithm finds its local model, the local model from each computer node has to be aggregated and integrated into a final model to represent the complete knowledge. Kiran and Babu [122] also pointed out that the communication will be the bottleneck when using this kind of distributed computing framework.

Bu et al. [123] found some research issues when trying to apply machine learning algorithms to parallel computing platforms. For instance, the early version of map-reduce framework does not support "iteration" (i.e., recursion). But the good news is that some recent works [86, 124] have paid close attention to this problem and tried to fix it. Similar to the solutions for enhancing the performance of the traditional data mining algorithms, one of the possible solutions to enhancing the performance of a machine learning algorithm is to use CUDA, i.e., a GPU, to

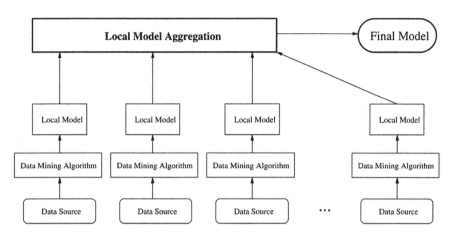

Fig. 2.10 A simple example of distributed data mining framework [85]

reduce the computing time of data analysis. Hasan et al. [125] used CUDA to implement the self-organizing map (SOM) and multiple back-propagation (MBP) for the classification problem. The simulation results show that using GPU is faster than using CPU. More precisely, SOM running on a GPU is three times faster than SOM running on a CPU, and MPB running on a GPU is twenty-seven times faster than MPB running on a. Another study [126] attempted to apply the ant-based algorithm to grid computing platform. Since the proposed mining algorithm is extended by the ant clustering algorithm of Deneubourg et al. [127],[6] Ku-Mahamud modified the ant behavior of this ant clustering algorithm for big data clustering. That is, each ant will be randomly placed on the grid. This means that the ant clustering algorithm then can be used on a parallel computing environment.

The trends of machine learning studies for big data analytics can be divided into twofold: one attempts to make machine learning algorithms run on parallel platforms, such as Radoop [128], Mahout [86], and PIMRU [123]; the other is to redesign the machine learning algorithms to make them suitable for parallel computing or to parallel computing environment, such as neural network algorithms for GPU [125] and ant-based algorithm for grid [126]. In summary, both of them make it possible to apply the machine learning algorithms to big data analytics although still many research issues need to be solved, such as the communication cost for different computer nodes [85] and the large computation cost most machine learning algorithms require [125].

---

[6]The basic idea of [128] is that each ant will pick up and drop data items in terms of the similarity of its local neighbors.

Output the Result of Big Data Analysis

The benchmarks of PigMix [129], GridMix [130], TeraSort and GraySort [131], TPC-C, TPC-H, TPC-DS [132], and yahoo cloud serving benchmark (YCSB) [133] have been presented for evaluating the performance of the cloud computing and big data analytics systems. Ghazal et al. [134] presented another benchmark (called BigBench) to be used as an end-to-end big data benchmark which covers the characteristics of 3V of big data and uses the loading time, time for queries, time for procedural processing queries, and time for the remaining queries as the metrics. By using these benchmarks, the computation time is one of the intuitive metrics for evaluating the performance of different big data analytics platforms or algorithms. That is why Cheptsov [135] compered the high performance computing (HPC) and cloud system by using the measurement of computation time to understand their scalability for text file analysis. In addition to the computation time, the throughput (e.g., the number of operations per second) and read/write latency of operations are the other measurements of big data analytics [136]. In the study of [137], Zhao et al. believe that the maximum size of data and the maximum number of jobs are the two important metrics to understand the performance of the big data analytics platform. Another study described in [138] presented a systematic evaluation method which contains the data throughput, concurrency during map and reduce phases, response times, and the execution time of map and reduce. Moreover, most benchmarks for evaluating the performance of big data analytics typically can only provide the response time or the computation cost; however, the fact is that several factors need to be taken into account at the same time when building a big data analytics system. The hardware, bandwidth for data transmission, fault tolerance, cost, power consumption of these systems are all issues [69, 103] to be taken into account at the same time when building a big data analytics system. Several solutions available today are to install the big data analytics on a cloud computing system or a cluster system. Therefore, the measurements of fault tolerance, task execution, and cost of cloud computing systems can then be used to evaluate the performance of the corresponding factors of big data analytics.

How to present the analysis results to a user is another important work in the output part of big data analytics because if the user cannot easily understand the meaning of the results, the results will be entirely useless. Business intelligent and network monitoring are the two common approaches because their user interface plays the vital role of making them workable. Zhang et al. [139] pointed out that the tasks of the visual analytics for commercial systems can be divided into four categories which are exploration, dashboards, reporting, and alerting. The study [140] showed that the interface for electroencephalography (EEG) interpretation is another noticeable research issue in big data analytics. The user interface for cloud system [141, 142] is the recent trend for big data analytics. This usually plays vital roles in big data analytics system, one of which is to simplify the explanation of the needed knowledge to the users while the other is to make it easier for the users to handle the data analytics system to work with their opinions. According to our observations, a flexible user interface is needed because although the big data analytics can help us to find some hidden information, the information found usually is not knowledge. This situation is just like the example we mentioned in "Output the result". The mining or statistical techniques can be employed to know the flu situation of each region, but data scientists sometimes need additional ways to display the information to find out the knowledge they need or to prove their assumption. Thus, the user interface can be adjusted by the user to display the knowledge that is needed urgently for big data analytics.

Summary of Process of Big Data Analytics

This discussion of big data analytics in this section was divided into input, analysis, and output for mapping the data analysis process of KDD. For the input (see also in "Big data input") and output (see also "Output the result of big data analysis") of big data, several methods and solutions proposed before the big data age (see also "Data input") can also be employed for big data analytics in most cases.

However, there still exist some new issues of the input and output that the data scientists need to confront. A representative example we mentioned in "Big data input" is that the bottleneck will not only on the sensor or input devices, it may also appear in other places of data analytics [70]. Although we can employ traditional compression and sampling technologies to deal with this problem, they can only mitigate the problems instead of solving the problems completely. Similar situations also exist in the output part. Although several measurements can be used to evaluate the performance of the frameworks, platforms, and even data mining algorithms, there still exist several new issues in the big data age, such as information fusion from different information sources or information accumulation from different times.

Several studies attempted to present an efficient or effective solution from the perspective of system (e.g., framework and platform) or algorithm level. A simple comparison of these big data analysis technologies from different perspectives is described in Table 2.3, to give a brief introduction to the current studies and trends of data analysis technologies for the big data. The "Perspective" column of this table explains that the study is focused on the framework or algorithm level; the "Description" column gives the further goal of the study; and the "Name" column is an abbreviated names of the methods or platform/framework. From the analysis framework perspective, this table shows that big data *framework*, *platform*, and *machine learning* are the current research trends in big data analytics system. For the mining algorithm perspective, the *clustering*, *classification*, and *frequent pattern mining* issues play the vital role of these researches because several data analysis problems can be mapped to these essential issues.

Table 2.3 The big data analysis frameworks and methods

$\mathcal{P}$	Name	References	Year	Description	$\mathcal{T}$
Analysis framework	DOT	[87]	2011	Add more computation resources via scale out solution	Framework
	GLADE	[88]	2011	Multi-level tree-based system architecture	
	Starfish	[91]	2012	Self-turning analytics system	
	ODT-MDC	[95]	2012	Privacy issues	
	MRAM	[90]	2013	Mobile agent technologies	
	CBDMASP	[93]	2013	Statistical computation and data mining approaches	
	SODSS	[96]	2013	Decision support system issues	
	BDAF	[92]	2014	Data centric architecture	
	HACE	[94]	2014	Data mining approaches	
	Hadoop	[82]	2011	Parallel computing platform	Platform
	CUDA	[83]	2007	Parallel computing platform	
	Storm	[84]	2014	Parallel computing platform	
	Pregel	[124]	2010	Large-scale graph data analysis	
	MLPACK	[85]	2013	Scalable machine learning library	ML
	Mahout	[86]	2011	Machine-learning algorithms	
	MLAS	[123]	2012	Machine-learning algorithms	
	PIMRU	[123]	2012	Machine Learning algorithms	
	Radoop	[128]	2011	Data analytics, machine learning algorithms, and R statistical tool	
Mining algorithm	DBDC	[143]	2004	Parallel clustering	CLU
	PKM	[144]	2009	Map-reduce-based k means clustering	
	CloudVista	[110]	2012	Cloud computing for clustering	
	MSFUDA	[112]	2013	GPU for clustering	
	BDCAC	[126]	2013	Ant on grid computing environment for clustering	
	Corest	[113]	2013	Use a tree construction for generating the coresets in parallel for clustering	
	SOM-MBP	[125]	2013	Neural network with CGP for clas sification	CLA
	CoS	[114]	2013	Parallel computing for classification	
	SVMGA	[71]	2014	Using GA for reduce the number of dimensions	
	Quantum SVM	[115]	2014	Quantum computing for classification	

(continued)

Table 2.3 (continued)

$\mathcal{P}$	Name	References	Year	Description	$\mathcal{T}$
	DPSP	[120]		Applied frequent pattern algorithm to cloud platform	FP
	DHTRIE	[119]	2011	Applied frequent pattern algorithm to cloud platform	
	SPC, FPC, and DPC	[116]	2012	Map-reduce model for frequent pat-tern mining	
	MFPSAM	[118]	2014	Concerned the specific interest con- straints and applied map-reduce model	

$\mathcal{P}$ perspective, $\mathcal{T}$ taxonomy, *ML* machine learning *CLU* clustering, *CLA* classification, *FP* frequent pattern

A promising trend that can be easily found from these successful examples is to use machine learning as the search algorithm (i.e., mining algorithm) for the data mining problems of big data analytics system. The machine learning-based methods are able to make the mining algorithms and relevant platforms smarter or reduce the redundant computation costs. That parallel computing and cloud computing technologies have a strong impact on the big data analytics can also be recognized as follows: (1) most of the big data analytics frameworks and platforms are using Hadoop and Hadoop relevant technologies to design their solutions; and (2) most of the mining algorithms for big data analysis have been designed for parallel computing via software or hardware or designed for Map-Reduce-based platform.

From the results of recent studies of big data analytics, it is still at the early stage of Nolan's stages of growth model [145] which is similar to the situations for the research topics of cloud computing, internet of things, and smart grid. This is because several studies just attempted to apply the traditional solutions to the new problems/platforms/environments. For example, several studies [113, 144] used k-means as an example to analyze the big data, but not many studies applied the state-of-the-art data mining algorithms and machine learning algorithms to the analysis the big data. This explains that the performance of the big data analytics can be improved by data mining algorithms and metaheuristic algorithms presented in recent years [146]. The relevant technologies for compression, sampling, or even the platform presented in recent years may also be used to enhance the performance of the big data analytics system. As a result, although these research topics still have several open issues that need to be solved, these situations, on the contrary, also illustrate that everything is possible in these studies.

The Open Issues

Although the data analytics today may be inefficient for big data caused by the environment, devices, systems, and even problems that are quite different from traditional mining problems, because several characteristics of big data also exist in the traditional data analytics. Several open issues caused by the big data will be addressed as the platform/framework and data mining perspectives in this section to explain what dilemmas we may confront because of big data. Here are some of the open issues:

Platform and Framework Perspective

Input and Output Ratio of Platform

A large number of reports and researches mentioned that we will enter the big data age in the near future. Some of them insinuated to us that these fruitful results of big data will lead us to a whole new world where "everything" is possible; therefore, the big data analytics will be an omniscient and omnipotent system. From the pragmatic perspective, the big data analytics is indeed useful and has many possibilities which can help us more accurately understand the so-called "things." However, the situation in most studies of big data analytics is that they argued that the results of big data are valuable, but the business models of most big data analytics are not clear. The fact is that assuming we have infinite computing resources for big data analytics is a thoroughly impracticable plan, the input and output ratio (e.g., return on investment) will need to be taken into account before an organization constructs the big data analytics center.

Communication Between Systems

Since most big data analytics systems will be designed for parallel computing, and they typically will work on other systems (e.g., cloud platform) or work with other systems (e.g., search engine or knowledge base), the communication between the big data analytics and other systems will strongly impact the performance of the whole process of KDD. The first research issue for the communication is that the communication cost will incur between systems of data analytics. How to reduce the communication cost will be the very first thing that the data scientists need to care. Another research issue for the communication is how the big data analytics communicates with other systems. The consistency of data between different systems, modules, and operators is also an important open issue on the communication between systems. Because the communication will appear more frequently between systems of big data analytics, how to reduce the cost of communication and how to make the communication between these systems as reliable as possible will be the two important open issues for big data analytics.

Bottlenecks on Data Analytics System

The bottlenecks will be appeared in different places of the data analytics for big data because the environments, systems, and input data have changed which are different from the traditional data analytics. The data deluge of big data will fill up the "input" system of data analytics, and it will also increase the computation load of the data "analysis" system. This situation is just like the torrent of water (i.e., data deluge) rushed down the mountain (i.e., data analytics), how to split it and how to avoid it flowing into a narrow place (e.g., the operator is not able to handle the input data) will be the most important things to avoid the bottlenecks in data analytics system. One of the current solutions to the avoidance of bottlenecks on a data analytics system is to add more computation resources while the other is to split the analysis works to different computation nodes. A complete consideration for the whole data analytics to avoid the bottlenecks of that kind of analytics system is still needed for big data.

Security Issues

Since much more environment data and human behavior will be gathered to the big data analytics, how to protect them will also be an open issue because without a security way to handle the collected data, the big data analytics cannot be a reliable system. In spite of the security that we have to tighten for big data analytics before it can gather more data from everywhere, the fact is that until now, there are still not many studies focusing on the security issues of the big data analytics. According to our observation, the security issues of big data analytics can be divided into fourfold: input, data analysis, output, and communication with other systems. For the input, it can be regarded as the data gathering which is relevant to the sensor, the handheld devices, and even the devices of internet of things. One of the important security issues on the input part of big data analytics is to make sure that the sensors will not be compromised by the attacks. For the analysis and input, it can be regarded as the security problem of such a system. For communication with other system, the security problem is on the communications between big data analytics and other external systems. Because of these latent problems, security has become one of the open issues of big data analytics.

Data Mining Perspective

Data Mining Algorithm for Map-Reduce Solution

As we mentioned in the previous sections, most of the traditional data mining algorithms are not designed for parallel computing; therefore, they are not particularly useful for the big data mining. Several recent studies have attempted to modify the traditional data mining algorithms to make them applicable to Hadoop-based platforms. As long as porting the data mining algorithms to Hadoop is inevitable, making the data mining algorithms work on a map-reduce architecture is the first very thing to do to apply traditional data mining methods to big data analytics. Unfortunately, not many studies attempted to make the data mining and soft computing algorithms work on Hadoop because several different backgrounds are needed to develop and design such algorithms. For instance, the researcher and his or her research group need to have the background in data mining and Hadoop so as to develop and design such algorithms. Another open issue is that most data mining algorithms are designed for centralized computing; that is, they can only work on all the data at the same time. Thus, how to make them work on a parallel computing system is also a difficult work. The good news is that some studies [144] have successfully applied the traditional data mining algorithms to the map-reduce architecture. These results imply that it is possible to do so. According to our observation, although the traditional mining or soft computing algorithms can be used to help us analyze the data in big data analytics, unfortunately, until now, not many studies are focused on it. As a consequence, it is an important open issue in big data analytics.

Noise, Outliers, Incomplete and Inconsistent Data

Although big data analytics is a new age for data analysis, because several solutions adopt classical ways to analyze the data on big data analytics, the open issues of traditional data mining algorithms also exist in these new systems. The open issues of noise, outliers, incomplete, and inconsistent data in traditional data mining algorithms will also appear in big data mining algorithms. More incomplete and inconsistent data will easily appear because the data are captured by or generated from different sensors and systems. The impact of noise, outliers, incomplete and inconsistent data will be enlarged for big data analytics. Therefore, how to mitigate the impact will be the open issues for big data analytics.

Bottlenecks on Data Mining Algorithm

Most of the data mining algorithms in big data analytics will be designed for parallel computing. However, once data mining algorithms are designed or modified for parallel computing, it is the information exchange between different data mining procedures that may incur bottlenecks. One of them is the synchronization issue because different mining procedures will finish their jobs at different times even though they use the same mining algorithm to work on the same amount of data. Thus, some of the mining procedures will have to wait until the others finished their jobs. This situation may occur because the loading of different computer nodes may be different during the data mining process, or it may occur because the convergence speeds are different for the same data mining algorithm. The bottlenecks of data mining algorithms will become an open issue for the big data analytics which explains that we need to take into account this issue when we develop and design a new data mining algorithm for big data analytics.

Privacy Issues

The privacy concern typically will make most people uncomfortable, especially if systems cannot guarantee that their personal information will not be accessed by the other people and organizations. Different from the concern of the security, the privacy issue is about if it is possible for the system to restore or infer personal information from the results of big data analytics, even though the input data are anonymous. The privacy issue has become a very important issue because the data mining and other analysis technologies will be widely used in big data analytics, the private information may be exposed to the other people after the analysis process. For example, although all the gathered data for shop behavior are anonymous (e.g., buying a pistol), because the data can be easily collected by different devices and systems (e.g., location of the shop and age of the buyer), a data mining algorithm can easily infer who bought this pistol. More precisely, the data analytics is able to reduce the scope of the database because location of the shop and age of the buyer provide the information to help the system find out possible persons. For this reason, any sensitive information needs to be carefully protected and used. The anonymous, temporary identification, and encryption are the representative technologies for privacy of data analytics, but the critical factor is how to use, what to use, and why to use the collected data on big data analytics.

Conclusions

In this paper, we reviewed studies on the data analytics from the traditional data analysis to the recent big data analysis. From the system perspective, the KDD process is used as the framework for these studies and is summarized into three parts: input, analysis, and output. From the perspective of big data analytics framework and platform, the discussions are focused on the performance-oriented and results-oriented issues. From the perspective of data mining problem, this paper gives a brief introduction to the data and big data mining algorithms which consist of clustering, classification, and frequent patterns mining technologies. To better understand the changes brought about by the big data, this paper is focused on the data analysis of KDD from the platform/framework to data mining. The open issues on computation, quality of end result, security, and privacy are then discussed to explain which open issues we may face. Last but not least, to help the audience of the paper find *solutions* to welcome the new age of big data, the possible high impact research trends are given below:

- For the computation time, there is no doubt at all that parallel computing is one of the important future trends to make the data analytics work for big data, and consequently the technologies of cloud computing, Hadoop, and map-reduce will play the important roles for the big data analytics. To handle the computation resources of the cloudbased platform and to finish the task of data analysis as fast as possible, the scheduling method is another future trend.
- Using efficient methods to reduce the computation time of input, comparison, sampling, and a variety of reduction methods will play an important role in big data analytics. Because these methods typically do not consider parallel computing environment, how to make them work on parallel computing environment will be a future research trend. Similar to the input, the data mining algorithms also face the same situation that we mentioned in the previous section, how to make them work on parallel computing environment will be a very important research trend because there are abundant research results on traditional data mining algorithms.
- How to model the mining problem to find something from big data and how to display the knowledge we got from big data analytics will also be another two vital future trends because the results of these two researches will decide if the data analytics can practically work for real world approaches, not just a theoretical stuff.
- The methods of extracting information from external and relative knowledge resources to further reinforce the big data analytics, until now, are not very popular in big data analytics. But combining information from different resources to add the value of output knowledge is a common solution in the area of information retrieval, such as clustering search engine or document summarization. For this reason, information fusion will also be a future trend for improving the end results of big data analytics.

- Because the metaheuristic algorithms are capable of finding an approximate solution within a reasonable time, they have been widely used in solving the data mining problem in recent years. Until now, many state-of-the-art metaheuristic algorithms still have not been applied to big data analytics. In addition, compared to some early data mining algorithms, the performance of metaheuristic is no doubt superior in terms of the computation time and the quality of end result. From these observations, the application of metaheuristic algorithms to big data analytics will also be an important research topic.

- Because social network is part of the daily life of most people and because its data is also a kind of big data, how to analyze the data of a social network has become a promising research issue. Obviously, it can be used to predict the behavior of a user. After that, we can make applicable strategies for the user. For instance, a business intelligence system can use the analysis results to encourage particular customers to buy the goods they are interested.

- The security and privacy issues that accompany the work of data analysis are intuitive research topics which contain how to safely store the data, how to make sure the data communication is protected, and how to prevent someone from finding out the information about us. Many problems of data security and privacy are essentially the same as those of the traditional data analysis even if we are entering the big data age. Thus, how to protect the data will also appear in the research of big data analytics.

Authors' Contributions CWT contributed to the paper review and drafted the first version of the manuscript. CFL contributed to the paper collection and manuscript organization. HCC and AVV double checked the manuscript and provided several advanced ideas for this manuscript. All authors read and approved the final manuscript.

Acknowledgments The authors would like to thank the anonymous reviewers for their valuable comments and suggestions on the paper. This work was supported in part by the Ministry of Science and Technology of Taiwan, R.O.C., under Contracts MOST1032221-E-197-034, MOST104-2221-E-197-005, and MOST104-2221-E-197-014.

Compliance with Ethical Guidelines—Competing Interests The authors declare that they have no competing interests.

References

1. Lyman P, Varian H. How much information 2003? Tech. Rep, 2004. [Online]. http://www2.sims.berkeley.edu/research/projects/how-much-info-2003/printable_report.pdf.
2. Xu R, Wunsch D. Clustering. Hoboken: Wiley-IEEE Press; 2009.
3. Ding C, He X. K-means clustering via principal component analysis. In: Proceedings of the twenty-first international conference on machine learning; 2004. pp. 1–9.

4. Kollios G, Gunopulos D, Koudas N, Berchtold S. Efficient biased sampling for approximate clustering and outlier detection in large data sets. IEEE Trans Knowl Data Eng. 2003;15 (5):1170–87.

5. Fisher D, DeLine R, Czerwinski M, Drucker S. Interactions with big data analytics. Interactions. 2012;19(3):50–9.

6. Laney D. 3D data management: controlling data volume, velocity, and variety. META Group, Tech. Rep. 2001. [Online]. http://blogs.gartner.com/doug-laney/files/2012/01/ad949-3D-Data-Management-Controlling-Data-Volume-Velocity-and-Variety.pdf.

7. van Rijmenam M. Why the 3v's are not sufficient to describe big data. BigData Startups, Tech. Rep. 2013. [Online]. http://www.bigdata-startups.com/3vs-sufficient-describe-big-data/ .

8. Borne K. Top 10 big data challenges a serious look at 10 big data v's. Tech. Rep. 2014. [Online]. https://www.mapr.com/blog/top-10-big-data-challenges-look-10-big-data-v.

9. Press G. $16.1 billion big data market: 2014 predictions from IDC and IIA, Forbes, Tech. Rep. 2013. [Online]. http://www.forbes.com/sites/gilpress/2013/12/12/16-1-billion-big-data-market-2014-predictions-from-idc-and-iia/.

10. Big data and analytics—an IDC four pillar research area. IDC, Tech. Rep. 2013. [Online]. http://www.idc.com/prodserv/FourPillars/bigData/index.jsp.

11. Taft DK. Big data market to reach $46.34 billion by 2018. EWEEK, Tech. Rep. 2013. [Online]. http://www.eweek.com/database/big-data-market-to-reach-46.34-billion-by-2018. html.

12. Research A. Big data spending to reach $114 billion in 2018; look for machine learning to drive analytics, ABI Research. Tech. Rep. 2013. [Online]. https://www.abiresearch.com/press/ big-data-spending-to-reach-114-billion-in-2018-loo.

13. Furrier J. Big data market $50 billion by 2017—HP vertica comes out #1—according to wikibon research, SiliconANGLE. Tech. Rep. 2012. [Online]. http://siliconangle.com/blog/2012/02/15/ big-data-market-15-billion-by-2017-hp-vertica-comes-out-1-according-to-wikibon-research/.

14. Kelly J, Vellante D, Floyer D. Big data market size and vendor revenues. Wikibon, Tech. Rep. 2014. [Online]. http://wikibon.org/wiki/v/Big_Data_Market_Size_and_Vendor_Revenues.

15. Kelly J, Floyer D, Vellante D, Miniman S. Big data vendor revenue and market forecast 2012–2017, Wikibon. Tech. Rep. 2014. [Online]. http://wikibon.org/wiki/v/Big_Data_Vendor_Revenue_and_Market_Forecast_2012-2017.

16. Mayer-Schonberger V, Cukier K. Big data: a revolution that will transform how we live, work, and think. Boston: Houghton Mifflin Harcourt; 2013.

17. Chen H, Chiang RHL, Storey VC. Business intelligence and analytics: from big data to big impact. MIS Quart. 2012;36(4):1165–88.

18. Kitchin R. The real-time city? big data and smart urbanism. Geo J. 2014;79(1):1–14.

19. Fayyad UM, Piatetsky-Shapiro G, Smyth P. From data mining to knowledge discovery in databases. AI Mag. 1996;17(3):37–54.

20. Han J. Data mining: concepts and techniques. San Francisco: Morgan Kaufmann Publishers Inc.; 2005.

21. Agrawal R, Imieliński T, Swami A. Mining association rules between sets of items in large databases. Proc ACM SIGMOD Int Conf Manag Data. 1993;22(2):207–16.

22. Witten IH, Frank E. Data mining: practical machine learning tools and techniques. San Francisco: Morgan Kaufmann Publishers Inc.; 2005.

23. Abbass H, Newton C, Sarker R. Data mining: a heuristic approach. Hershey: IGI Global; 2002.

24. Cannataro M, Congiusta A, Pugliese A, Talia D, Trunfio P. Distributed data mining on grids: services, tools, and applications. IEEE Trans Syst Man Cyber Part B Cyber. 2004;34 (6):2451–65.

25. Krishna K, Murty MN. Genetic k-means algorithm. IEEE Trans Syst Man Cyber Part B Cyber. 1999;29(3):433–9.

26. Tsai C-W, Lai C-F, Chiang M-C, Yang L. Data mining for internet of things: a survey. IEEE Commun Surv Tutor. 2014;16(1):77–97.
27. Jain AK, Murty MN, Flynn PJ. Data clustering: a review. ACM Comp Surv. 1999;31 (3):264–323.
28. McQueen JB. Some methods of classification and analysis of multivariate observations. In: Proceedings of the Berkeley symposium on mathematical statistics and probability; 1967. pp. 281–297.
29. Safavian S, Landgrebe D. A survey of decision tree classifier methodology. IEEE Trans Syst Man Cyber. 1991;21(3):660–74.
30. McCallum A, Nigam K. A comparison of event models for naive bayes text classification. In: Proceedings of the national conference on artificial intelligence; 1998. pp. 41–48.
31. Boser BE, Guyon IM, Vapnik VN. A training algorithm for optimal margin classifiers. In: Proceedings of the annual workshop on computational learning theory; 1992. pp. 144–152.
32. Han J, Pei J, Yin Y. Mining frequent patterns without candidate generation. In: Proceedings of the ACM SIGMOD international conference on management of data; 2000. pp. 1–12.
33. Kaya M, Alhajj R. Genetic algorithm based framework for mining fuzzy association rules. Fuzzy Sets Syst. 2005;152(3):587–601.
34. Srikant R, Agrawal R. Mining sequential patterns: generalizations and performance improvements. In: Proceedings of the international conference on extending database technology: advances in database technology; 1996. pp. 3–17.
35. Zaki MJ. Spade: an efficient algorithm for mining frequent sequences. Mach Learn. 2001;42 (1–2):31–60.
36. Baeza-Yates RA, Ribeiro-Neto B. Modern Information Retrieval. Boston: Addison-Wesley Longman Publishing Co., Inc; 1999.
37. Liu B. Web data mining: exploring hyperlinks, contents, and usage data. Berlin: Springer; 2007.
38. d'Aquin M, Jay N. Interpreting data mining results with linked data for learning analytics: motivation, case study and directions. In: Proceedings of the international conference on learning analytics and knowledge. pp. 155–164.
39. Shneiderman B. The eyes have it: a task by data type taxonomy for information visualizations. In: Proceedings of the IEEE symposium on visual languages; 1996. pp. 336–343.
40. Mani I, Bloedorn E. Multi-document summarization by graph search and matching. In: Proceedings of the national conference on artificial intelligence and ninth conference on innovative applications of artificial intelligence; 1997. pp. 622–628.
41. Kopanakis I, Pelekis N, Karanikas H, Mavroudkis T. Visual techniques for the interpretation of data mining outcomes. In: Proceedings of the Panhellenic conference on advances in informatics; 2005. pp. 25–35.
42. Elkan C. Using the triangle inequality to accelerate k-means. In: Proceedings of the international conference on machine learning; 2003. pp. 147–153.
43. Catanzaro B, Sundaram N, Keutzer K. Fast support vector machine training and classification on graphics processors. In: Proceedings of the international conference on machine learning; 2008. pp. 104–111.
44. Zhang T, Ramakrishnan R, Livny M. BIRCH: an efficient data clustering method for very large databases. In: Proceedings of the ACM SIGMOD international conference on management of data; 1996. pp. 103–114.
45. Ester M, Kriegel HP, Sander J, Xu X. A density-based algorithm for discovering clusters in large spatial databases with noise. In: Proceedings of the Second International Conference on Knowledge Discovery and Data Mining; 1996. pp. 226–231.
46. Ester M, Kriegel HP, Sander J, Wimmer M, Xu X. Incremental clustering for mining in a data warehousing environment. In: Proceedings of the International Conference on Very Large Data Bases; 1998. pp. 323–333.
47. Ordonez C, Omiecinski E. Efficient disk-based k-means clustering for relational databases. IEEE Trans Knowl Data Eng. 2004;16(8):909–21.

48. Kogan J. Introduction to clustering large and high-dimensional data. Cambridge: Cambridge Univ Press; 2007.
49. Mitra S, Pal S, Mitra P. Data mining in soft computing framework: a survey. IEEE Trans Neural Netw. 2002;13(1):3–14.
50. Mehta M, Agrawal R, Rissanen J. SLIQ: a fast scalable classifier for data mining. In: Proceedings of the 5th international conference on extending database technology: advances in database technology; 1996. pp. 18–32.
51. Micó L, Oncina J, Carrasco RC. A fast branch and bound nearest neighbour classifier in metric spaces. Pattern Recogn Lett. 1996;17(7):731–9.
52. Djouadi A, Bouktache E. A fast algorithm for the nearest-neighbor classifier. IEEE Trans Pattern Anal Mach Intel. 1997;19(3):277–82.
53. Ververidis D, Kotropoulos C. Fast and accurate sequential floating forward feature selection with the bayes classifier applied to speech emotion recognition. Signal Process. 2008;88 (12):2956–70.
54. Pei J, Han J, Mao R. CLOSET: an efficient algorithm for mining frequent closed itemsets. In: Proceedings of the ACM SIGMOD workshop on research issues in data mining and knowledge discovery; 2000. pp. 21–30.
55. Zaki MJ, Hsiao C-J. Efficient algorithms for mining closed itemsets and their lattice structure. IEEE Trans Knowl Data Eng. 2005;17(4):462–78.
56. Burdick D, Calimlim M, Gehrke J. MAFIA: a maximal frequent itemset algorithm for transactional databases. In: Proceedings of the international conference on data engineering; 2001. pp. 443–452.
57. Chen B, Haas P, Scheuermann P. A new two-phase sampling based algorithm for discovering association rules. In: Proceedings of the ACM SIGKDD international conference on knowledge discovery and data mining; 2002. pp. 462–468.
58. Yan X, Han J, Afshar R. CloSpan: mining closed sequential patterns in large datasets. In: Proceedings of the SIAM international conference on data mining; 2003. pp. 166–177.
59. Pei J, Han J, Asl MB, Pinto H, Chen Q, Dayal U, Hsu MC. PrefixSpan mining sequential patterns efficiently by prefix projected pattern growth. In: Proceedings of the international conference on data engineering; 2001. pp. 215–226.
60. Ayres J, Flannick J, Gehrke J, Yiu T. Sequential pattern Mining using a bitmap representation. In: Proceedings of the ACM SIGKDD international conference on knowledge discovery and data mining; 2002. pp. 429–435.
61. Masseglia F, Poncelet P, Teisseire M. Incremental mining of sequential patterns in large databases. Data Knowl Eng. 2003;46(1):97–121.
62. Xu R, Wunsch-II DC. Survey of clustering algorithms. IEEE Trans Neural Netw. 2005;16 (3):645–78.
63. Chiang M-C, Tsai C-W, Yang C-S. A time-efficient pattern reduction algorithm for k-means clustering. Inform Sci. 2011;181(4):716–31.
64. Bradley PS, Fayyad UM. Refining initial points for k-means clustering. In: Proceedings of the international conference on machine learning; 1998. pp. 91–99.
65. Laskov P, Gehl C, Krüger S, Müller K-R. Incremental support vector learning: analysis, implementation and applications. J Mach Learn Res. 2006;7:1909–36.
66. Russom P. Big data analytics. TDWI: Tech. Rep; 2011.
67. Ma C, Zhang HH, Wang X. Machine learning for big data analytics in plants. Trends Plant Sci. 2014;19(12):798–808.
68. Boyd D, Crawford K. Critical questions for big data. Inform Commun Soc. 2012;15(5):662–79.
69. Katal A, Wazid M, Goudar R. Big data: issues, challenges, tools and good practices. In: Proceedings of the international conference on contemporary computing; 2013. pp. 404–409.
70. Baraniuk RG. More is less: signal processing and the data deluge. Science. 2011;331 (6018):717–9.

71. Lee J, Hong S, Lee JH. An efficient prediction for heavy rain from big weather data using genetic algorithm. In: Proceedings of the international conference on ubiquitous information management and communication; 2014. pp. 25:1–25:7.

72. Famili A, Shen W-M, Weber R, Simoudis E. Data preprocessing and intelligent data analysis. Intel Data Anal. 1997;1(1–4):3–23.

73. Zhang H. A novel data preprocessing solution for large scale digital forensics investigation on big data, Master's thesis, Norway; 2013.

74. Ham YJ, Lee H-W. International journal of advances in soft computing and its applications. Calc Paralleles Reseaux et Syst Repar. 2014;6(1):1–18.

75. Cormode G, Duffield N. Sampling for big data: a tutorial. In: Proceedings of the ACM SIGKDD international conference on knowledge discovery and data mining; 2014. pp. 1975–1975.

76. Satyanarayana A. Intelligent sampling for big data using bootstrap sampling and chebyshev inequality. In: Proceedings of the IEEE Canadian conference on electrical and computer engineering; 2014. pp. 1–6.

77. Jun SW, Fleming K, Adler M, Emer JS. Zip-io: architecture for application-specific compression of big data. In: Proceedings of the international conference on field-programmable technology; 2012. pp. 343–351.

78. Zou H, Yu Y, Tang W, Chen HM. Improving I/O performance with adaptive data compression for big data applications. In: Proceedings of the international parallel and distributed processing symposium workshops; 2014. pp. 1228–1237.

79. Yang C, Zhang X, Zhong C, Liu C, Pei J, Ramamohanarao K, Chen J. A spatiotemporal compression based approach for efficient big data processing on cloud. J Comp Syst Sci. 2014;80(8):1563–83.

80. Xue Z, Shen G, Li J, Xu Q, Zhang Y, Shao J. Compression-aware I/O performance analysis for big data clustering. In: Proceedings of the international workshop on big data, streams and heterogeneous source mining: algorithms, systems, programming models and applications; 2012. pp. 45–52.

81. Pospiech M, Felden C. Big data—a state-of-the-art. In: Proceedings of the Americas conference on information systems; 2012. pp. 1–23. [Online]. http://aisel.aisnet.org/amcis2012/proceedings/DecisionSupport/22.

82. Apache Hadoop, February 2, 2015. [Online]. http://hadoop.apache.org.

83. Cuda, February 2, 2015. [Online]. http://www.nvidia.com/object/cuda_home_new.html.

84. Apache Storm, February 2, 2015. [Online]. http://storm.apache.org/.

85. Curtin RR, Cline JR, Slagle NP, March WB, Ram P, Mehta NA, Gray AG. MLPACK: a scalable C++ machine learning library. J Mach Learn Res. 2013;14:801–5.

86. Apache Mahout, February 2, 2015. [Online]. http://mahout.apache.org/.

87. Huai Y, Lee R, Zhang S, Xia CH, Zhang X. DOT: a matrix model for analyzing, optimizing and deploying software for big data analytics in distributed systems. In: Proceedings of the ACM symposium on cloud computing; 2011. pp. 4:1–4:14.

88. Rusu F, Dobra A. GLADE: a scalable framework for efficient analytics. In: Proceedings of LADIS workshop held in conjunction with VLDB; 2012. pp. 1–6.

89. Cheng Y, Qin C, Rusu F. GLADE: big data analytics made easy. In: Proceedings of the ACM SIGMOD international conference on management of data; 2012. pp. 697–700.

90. Essa YM, Attiya G, El-Sayed A. Mobile agent based new framework for improving big data analysis. In: Proceedings of the international conference on cloud computing and big data; 2013. pp. 381–386.

91. Wonner J, Grosjean J, Capobianco A, Bechmann. D Starfish: a selection technique for dense virtual environments. In: Proceedings of the ACM symposium on virtual reality software and technology; 2012. pp. 101–104.

92. Demchenko Y, de Laat C, Membrey P. Defining architecture components of the big data ecosystem. In: Proceedings of the international conference on collaboration technologies and systems; 2014. pp. 104–112.

93. Ye F, Wang ZJ, Zhou FC, Wang YP, Zhou YC. Cloud-based big data mining and analyzing services platform integrating r. In: Proceedings of the international conference on advanced cloud and big data; 2013. pp. 147–151.
94. Wu X, Zhu X, Wu G-Q, Ding W. Data mining with big data. IEEE Trans Knowl Data Eng. 2014;26(1):97–107.
95. Laurila JK, Gatica-Perez D, Aad I, Blom J, Bornet O, Do T, Dousse O, Eberle J, Miettinen M. The mobile data challenge: big data for mobile computing research. In: Proceedings of the mobile data challenge by Nokia workshop; 2012. pp. 1–8.
96. Demirkan H, Delen D. Leveraging the capabilities of service-oriented decision support systems: putting analytics and big data in cloud. Decis Support Syst. 2013;55(1):412–21.
97. Talia D. Clouds for scalable big data analytics. Computer. 2013;46(5):98–101.
98. Lu R, Zhu H, Liu X, Liu JK, Shao J. Toward efficient and privacy-preserving computing in big data era. IEEE Netw. 2014;28(4):46–50.
99. Cuzzocrea A, Song IY, Davis KC. Analytics over large-scale multidimensional data: the big data revolution!. In: Proceedings of the ACM international workshop on data warehousing and OLAP; 2011. pp. 101–104.
100. Zhang J, Huang ML. 5Ws model for big data analysis and visualization. In: Proceedings of the international conference on computational science and engineering; 2013. pp. 1021–1028.
101. Chandarana P, Vijayalakshmi M. Big data analytics frameworks. In: Proceedings of the international conference on circuits, systems, communication and information technology applications; 2014. pp. 430–434.
102. Apache Drill February 2, 2015. [Online]. http://drill.apache.org/.
103. Hu H, Wen Y, Chua T-S, Li X. Toward scalable systems for big data analytics: a technology tutorial. IEEE Access. 2014;2:652–87.
104. Sagiroglu S, Sinanc D, Big data: a review. In: Proceedings of the international conference on collaboration technologies and systems; 2013. pp. 42–47.
105. Fan W, Bifet A. Mining big data: current status, and forecast to the future. ACM SIGKDD Explor Newslett. 2013;14(2):1–5.
106. Diebold FX. On the origin(s) and development of the term "big data". Penn Institute for Economic Research, Department of Economics, University of Pennsylvania, Tech. Rep. 2012. [Online]. http://economics.sas. upenn.edu/sites/economics.sas.upenn. edu/files/12-037.pdf.
107. Weiss SM, Indurkhya N. Predictive data mining: a practical guide. San Francisco: Morgan Kaufmann Publishers Inc.; 1998.
108. Fahad A, Alshatri N, Tari Z, Alamri A, Khalil I, Zomaya A, Foufou S, Bouras A. A survey of clustering algorithms for big data: taxonomy and empirical analysis. IEEE Trans Emerg Topics Comp. 2014;2(3):267–79.
109. Shirkhorshidi AS, Aghabozorgi SR, Teh YW, Herawan T. Big data clustering: a review. In: Proceedings of the international conference on computational science and its applications; 2014. pp. 707–720.
110. Xu H, Li Z, Guo S, Chen K. Cloudvista: interactive and economical visual cluster analysis for big data in the cloud. Proc VLDB Endow. 2012;5(12):1886–9.
111. Cui X, Gao J, Potok TE. A flocking based algorithm for document clustering analysis. J Syst Archit. 2006;52(89):505–15.
112. Cui X, Charles JS, Potok T. GPU enhanced parallel computing for large scale data clustering. Future Gener Comp Syst. 2013;29(7):1736–41.
113. Feldman D, Schmidt M, Sohler C. Turning big data into tiny data: constant-size coresets for k-means, pca and projective clustering. In: Proceedings of the ACM-SIAM symposium on discrete algorithms; 2013. pp. 1434–1453.
114. Tekin C, van der Schaar M. Distributed online big data classification using context information. In: Proceedings of the Allerton conference on communication, control, and computing; 2013. pp. 1435–1442.

115. Rebentrost P, Mohseni M, Lloyd S. Quantum support vector machine for big feature and big data classification. CoRR, vol. abs/1307.0471; 2014. [Online]. http://dblp.uni-trier.de/db/journals/corr/corr1307.html#RebentrostML13.

116. Lin MY, Lee PY, Hsueh SC. Apriori-based frequent itemset mining algorithms on mapreduce. In: Proceedings of the international conference on ubiquitous information management and communication; 2012. pp. 76:1–76:8.

117. Riondato M, DeBrabant JA, Fonseca R, Upfal E. PARMA: a parallel randomized algorithm for approximate association rules mining in mapreduce. In: Proceedings of the ACM international conference on information and knowledge management; 2012. pp. 85–94.

118. Leung CS, MacKinnon R, Jiang F. Reducing the search space for big data mining for interesting patterns from uncertain data. In: Proceedings of the international congress on big data; 2014. pp. 315–322.

119. Yang L, Shi Z, Xu L, Liang F, Kirsh I. DH-TRIE frequent pattern mining on hadoop using JPA. In: Proceedings of the international conference on granular computing; 2011. pp. 875–878.

120. Huang JW, Lin SC, Chen MS. DPSP: Distributed progressive sequential pattern mining on the cloud. In: Proceedings of the advances in knowledge discovery and data mining, vol. 6119; 2010. pp. 27–34.

121. Paz CE. A survey of parallel genetic algorithms. Calc Paralleles Reseaux et Syst Repar. 1998;10(2):141–71.

122. Kranthi Kiran B, Babu AV. A comparative study of issues in big data clustering algorithm with constraint based genetic algorithm for associative clustering. Int J Innov Res Comp Commun Eng. 2014;2(8):5423–32.

123. Bu Y, Borkar VR, Carey MJ, Rosen J, Polyzotis N, Condie T, Weimer M, Ramakrishnan R. Scaling datalog for machine learning on big data, CoRR, vol. abs/1203.0160, 2012. [Online]. http://dblp.uni-trier.de/db/journals/corr/corr1203.html#abs-1203-0160.

124. Malewicz G, Austern MH, Bik AJ, Dehnert JC, Horn I, Leiser N, Czajkowski G. Pregel: A system for large-scale graph processing. In: Proceedings of the ACM SIGMOD international conference on management of data; 2010. pp. 135–146.

125. Hasan S, Shamsuddin S, Lopes N. Soft computing methods for big data problems. In: Proceedings of the symposium on GPU computing and applications; 2013. pp. 235–247.

126. Ku-Mahamud KR. Big data clustering using grid computing and ant-based algorithm. In: Proceedings of the international conference on computing and informatics; 2013. pp. 6–14.

127. Deneubourg JL, Goss S, Franks N, Sendova-Franks A, Detrain C, Chrétien L. The dynamics of collective sorting robot-like ants and ant-like robots. In: Proceedings of the international conference on simulation of adaptive behavior on from animals to animats; 1990. pp. 356–363.

128. Radoop [Online]. https://rapidminer.com/products/radoop/. Accessed 2 Feb 2015.

129. PigMix [Online]. https://cwiki.apache.org/confluence/display/PIG/PigMix. Accessed 2 Feb 2015.

130. GridMix [Online]. http://hadoop.apache.org/docs/r1.2.1/gridmix.html. Accessed 2 Feb 2015.

131. TeraSoft [Online]. http://sortbenchmark.org/. Accessed 2 Feb 2015.

132. TPC, transaction processing performance council [Online]. http://www.tpc.org/. Accessed 2 Feb 2015.

133. Cooper BF, Silberstein A, Tam E, Ramakrishnan R, Sears R. Benchmarking cloud serving systems with YCSB. In: Proceedings of the ACM symposium on cloud computing; 2010. pp. 143–154.

134. Ghazal A, Rabl T, Hu M, Raab F, Poess M, Crolotte A, Jacobsen HA. BigBench: towards an industry standard benchmark for big data analytics. In: Proceedings of the ACM SIGMOD international conference on management of data; 2013. pp. 1197–1208.

135. Cheptsov A. Hpc in big data age: An evaluation report for java-based data-intensive applications implemented with hadoop and openmpi. In: Proceedings of the European MPI Users' Group Meeting; 2014. pp. 175:175–175:180.

136. Yuan LY, Wu L, You JH, Chi Y. Rubato db: A highly scalable staged grid database system for OLTP and big data applications. In: Proceedings of the ACM international conference on conference on information and knowledge management; 2014. pp. 1–10.
137. Zhao JM, Wang WS, Liu X, Chen YF. Big data benchmark—big DS. In: Proceedings of the advancing big data benchmarks; 2014. pp. 49–57.
138. Saletore V, Krishnan K, Viswanathan V, Tolentino M. HcBench: Methodology, development, and full-system characterization of a customer usage representative big data/hadoop benchmark. In: Advancing big data benchmarks; 2014. pp. 73–93.
139. Zhang L, Stoffel A, Behrisch M, Mittelstadt S, Schreck T, Pompl R, Weber S, Last H, Keim D. Visual analytics for the big data era—a comparative review of state-of-the-art commercial systems. In: Proceedings of the IEEE conference on visual analytics science and technology; 2012. pp. 173–182.
140. Harati A, Lopez S, Obeid I, Picone J, Jacobson M, Tobochnik S. The TUH EEG CORPUS: A big data resource for automated EEG interpretation. In: Proceeding of the IEEE signal processing in medicine and biology symposium; 2014. pp. 1–5.
141. Thusoo A, Sarma JS, Jain N, Shao Z, Chakka P, Anthony S, Liu H, Wyckoff P, Murthy R. Hive: a warehousing solution over a map-reduce framework. Proc VLDB Endow. 2009;2 (2):1626–9.
142. Beckmann M, Ebecken NFF, de Lima BSLP,Costa MA. A user interface for big data with rapidminer. RapidMiner World, Boston, MA, Tech. Rep.; 2014. [Online]. http://www.slideshare.net/RapidMiner/a-user-interface-for-big-data-with-rapidminer-marcelo-beckmann.
143. Januzaj E, Kriegel HP, Pfeifle M. DBDC: Density based distributed clustering. In: Proceedings of the advances in database technology, vol. 2992; 2004. pp. 88–105.
144. Zhao W, Ma H, He Q. Parallel k-means clustering based on mapreduce. Proc Cloud Comp. 2009;5931:674–9.
145. Nolan RL. Managing the crises in data processing. Harvard Bus Rev. 1979;57(1):115–26.
146. Tsai CW, Huang WC, Chiang MC. Recent development of metaheuristics for clustering. In: Proceedings of the mobile, ubiquitous, and intelligent computing, vol. 274. 2014; pp. 629–636.

Chapter 3
Transfer Learning Techniques

Karl Weiss, Taghi M. Khoshgoftaar and DingDing Wang

Introduction

The field of data mining and machine learning has been widely and successfully used in many applications where patterns from past information (training data) can be extracted in order to predict future outcomes [1]. Traditional machine learning is characterized by training data and testing data having the same input feature space and the same data distribution. When there is a difference in data distribution between the training data and test data, the results of a predictive learner can be degraded [2]. In certain scenarios, obtaining training data that matches the feature space and predicted data distribution characteristics of the test data can be difficult and expensive. Therefore, there is a need to create a high-performance learner for a target domain trained from a related source domain. This is the motivation for transfer learning.

Transfer learning is used to improve a learner from one domain by transferring information from a related domain. We can draw from real-world non-technical experiences to understand why transfer learning is possible. Consider an example of two people who want to learn to play the piano. One person has no previous experience playing music, and the other person has extensive music knowledge through playing the guitar. The person with an extensive music background will be able to learn the piano in a more efficient manner by transferring previously learned music knowledge to the task of learning to play the piano [3]. One person is able to take information from a previously learned task and use it in a beneficial way to learn a related task.

Looking at a concrete example from the domain of machine learning, consider the task of predicting text sentiment of product reviews where there exists an abundance of labeled data from digital camera reviews. If the training data and the

---

This chapter has been adopted from the Journal of Big Data, Borko Furht and Taghi Khoshgoftar, Editors-in-Chief. Springer, June 2016.

B. Furht and F. Villanustre, *Big Data Technologies and Applications*,
DOI 10.1007/978-3-319-44550-2_3

target data are both derived from digital camera reviews, then traditional machine learning techniques are used to achieve good prediction results. However, in the case where the training data is from digital camera reviews and the target data is from food reviews, then the prediction results are likely to degrade due to the differences in domain data. Digital camera reviews and food reviews still have a number of characteristics in common, if not exactly the same. They both are written in textual form using the same language, and they both express views about a purchased product. Because these two domains are related, transfer learning can be used to potentially improve the results of a target learner [3]. An alternative way to view the data domains in a transfer learning environment is that the training data and the target data exist in different sub-domains linked by a high-level common domain. For example, a piano player and a guitar player are subdomains of a musician domain. Further, a digital camera review and a food review are subdomains of a review domain. The high-level common domain determines how the subdomains are related.

As previously mentioned, the need for transfer learning occurs when there is a limited supply of target training data. This could be due to the data being rare, the data being expensive to collect and label, or the data being inaccessible. With big data repositories becoming more prevalent, using existing datasets that are related to, but not exactly the same as, a target domain of interest makes transfer learning solutions an attractive approach. There are many machine learning applications that transfer learning has been successfully applied to including text sentiment classification [4], image classification [5–7], human activity classification [8], software defect classification [9], and multi-language text classification [10–12].

This survey paper aims to provide a researcher interested in transfer learning with an overview of related works, examples of applications that are addressed by transfer learning, and issues and solutions that are relevant to the field of transfer learning. This survey paper provides an overview of current methods being used in the field of transfer learning as it pertains to data mining tasks for classification, regression, and clustering problems; however, it does not focus on transfer learning for reinforcement learning (for more information on reinforcement learning see Taylor and Stone [13]. Information pertaining to the history and taxonomy of transfer learning is not provided in this survey paper, but can be found in the paper by Pan and Yang [3]. Since the publication of the transfer learning survey paper by Pan and Yang [3] in 2010, there have been over 700 academic papers written addressing advancements and innovations on the subject of transfer learning. These works broadly cover the areas of new algorithm development, improvements to existing transfer learning algorithms, and algorithm deployment in new application domains. The selected surveyed works in this paper are meant to be diverse and representative of transfer learning solutions in the past five years. Most of the surveyed papers provide a generic transfer learning solution; however, some surveyed papers provide solutions that are specific to individual applications. This paper is written with the assumption the reader has a working knowledge of machine learning. For more information on machine learning see Witten and Frank [1]. The surveyed works in this paper are intended to present a high-level description of proposed solutions with unique and salient points being highlighted. Experiments from the surveyed papers are described with respect to applied applications, other competing solutions tested, and overall relative results of the experiments. This survey paper provides a section on heterogeneous transfer learning which, to the best of our knowledge, is unique. Additionally, a list of software downloads for various surveyed papers is provided, which is unique to this paper.

The remainder of this paper is organized as follows. In `Definitions of Transfer Learning' section provides definitions and notations of transfer learning. In `Homogeneous Transfer Learning' and `Heterogeneous Transfer Learning' sections provide solutions on homogeneous and heterogeneous transfer learning, respectively. In Negative Transfer section provides information on negative transfer as it pertains to transfer learning. In Transfer Learning Applications section provides examples of transfer learning applications. In Conclusion and Discussion section summarizes and discusses potential future research work. The Appendix provides information on software downloads for transfer learning.

Definitions of Transfer Learning

The following section lists the notation and definitions used for the remainder of this paper. The notation and definitions in this section match those from the survey paper by Pan and Yang [3], if present in both papers, to maintain consistency across both surveys. To provide illustrative examples of the definitions listed below, a machine learning application of software module defect classification is used where a learner is trained to predict whether a software module is defect prone or not.

A domain D is defined by two parts, a feature space X and a marginal probability distribution $P(X)$, where $X = \{x_1, ..., x_n\} \in \mathcal{X}$. For example, if the machine learning application is software module defect classification and each software metric is taken as a feature, then x_i is the i-th feature vector (instance) corresponding to the i-th software module, n is the number of feature vectors in X, $\mathcal{X}$ is the space of all possible feature vectors, and X is a particular learning sample. For a given domain D, a task T is defined by two parts, a label space Y, and a predictive function $f(\cdot)$, which is learned from the feature vector and label pairs $\{x_i, y_i\}$ where $x_i \in X$ and $y_i \in Y$. Referring to the software module defect classification application, Y is the set of labels and in this case contains true and false, y_i takes on a value of true or false, and f (x) is the learner that predicts the label value for the software module x. From the definitions above, a domain $D = \{X, P(X)\}$ and a task $T = \{Y, f(\cdot)\}$. Now, D_S is defined as the source domain data where $D_S = \{ (x_{S1}, y_{S1}) ..., (x_{Sn}, y_{Sn})\}$, where $x_{Si} \in X_S$ is the i-th data instance of D_S and $y_{Si} \in Y_S$ is the corresponding class label for x_{Si}. In the same way, D_T is defined as the target domain data where $D_T = \{ (x_{T1}, y_{T1}) ..., (x_{Tn}, y_{Tn})\}$, where $x_{Ti}, \in X_T$ is the i-th data instance of D_T and $y_{Ti}, \in Y_T$ is the corresponding class label for x_{Ti}. Further, the source task is notated as T_S, the target task as T_T, the source predictive function as $f_S(\cdot)$, and the target predictive function as $f_T(\cdot)$.

Transfer Learning is now formally defined. Given a source domain D_S with a corresponding source task T_S and a target domain D_T with a corresponding task T_T, transfer learning is the process of improving the target predictive function $f_T(\cdot)$ by using the related information from D_S and T_S, where $D_S \neq D_T$ or $T_S \neq T_T$. The single source domain defined here can be extended to multiple source domains. Given the definition of transfer learning, since $D_S = \{X_S, P(X_S)\}$ and $D_T = \{X_T, P(X_T)\}$, the condition where $D_S \neq D_T$ means that $X_S \neq X_T$ and/or $P(X_S) \neq P(X_T)$. The case where $X_S \neq X_T$ with respect to transfer learning is defined as heterogeneous transfer learning. The case where $X_S = X_T$ with respect to transfer learning is defined as homogeneous transfer learning. Going back to the example of software module defect classification, heterogeneous transfer learning is the case where the source software project has different metrics (features) than the target software project. Alternatively, homogeneous transfer learning is when the software metrics are the same for both the source and the target software projects. Continuing with the definition of transfer learning, the case where $P(X_S) \neq P(X_T)$ means the marginal distributions in the input spaces are different between the source and the target domains. Shimodaira [2] demonstrated that a learner trained with a given source domain will not perform optimally on a target domain when the marginal distributions of the input domains are different. Referring to the software module defect classification application, an example of marginal distribution differences is when the source software program is written for a user interface system and the target software program is written for DSP signaling decoder algorithm. Another possible condition of transfer learning (from the definition above) is $T_S \neq T_T$, and it was stated that $T = \{Y, f(\cdot)\}$ or to rewrite this, $T = \{Y, P(Y|X)\}$. Therefore, in a transfer learning environment, it is possible that $Y_S \neq Y_T$ and/or $P(Y_S|X_S) \neq P(Y_T|X_T)$. The case where $P(Y_S|X_S) \neq P(Y_T|X_T)$ means the conditional probability distributions between the source and target domains are different. An example of a conditional distribution mismatch is when a particular software module yields different fault prone results in the source and target domains. The case of $Y_S \neq Y_T$ refers to a mismatch in the class space. An example of this case is when the source software project has a binary label space of true for defect prone and false for not defect prone, and the target domain has a label space that defines five levels of fault prone modules. Another case that can cause discriminative classifier degradation is when $P(Y_S) \neq P(Y_T)$, which is caused by an unbalanced labeled data set between the source and target domains. The case of traditional machine learning is $D_S = D_T$ and $T_S = T_T$. The common notation used in this paper is summarized in Table 3.1.

Table 3.1 Summary of commonly used notation

Notation	Description	Notation	Description
X	Input feature space	P(X)	Marginal distribution
Y	Label space	P(Y|X)	Conditional distribution
T	Predictive learning task	P(Y)	Label distribution
Subscript S	Denotes source	D_S	Source domain data
Subscript T	Denotes target	D_T	Target domain data

To elaborate on the distribution issues that can occur between the source and target domains, the application of natural language processing is used to illustrate. In natural language processing, text instances are often modeled as a bag-of-words where a unique word represents a feature. Consider the example of review text where the source covers movie reviews and the target covers book reviews. Words that are generic and domain independent should occur at a similar rate in both domains. However, words that are domain specific are used more frequently in one domain because of the strong relationship with that domain topic. This is referred to as frequency feature bias and will cause the marginal distribution between the source and target domains to be different ($P(X_S) \neq P(X_T)$). Another form of bias is referred to as context feature bias and this will cause the conditional distributions to be different between the source and target domains ($P(Y_S|X_S) \neq P(Y_T|X_T)$). An example of context feature bias is when a word can have different meanings in two domains. A specific example is the word "monitor" where in one domain it is used as a noun and in another domain it is used as a verb. Another example of context feature bias is with sentiment classification when a word has a positive meaning in one domain and a negative meaning in another domain. The word "small" can have a good meaning if describing a cell phone but a bad meaning if describing a hotel room. A further example of context feature bias is demonstrated in the case of document sentiment classification of reviews where the source domain contains reviews of one product written in German and the target domain contains reviews of a different product written in English. The translated words from the source document may not accurately represent the actual words used in the target documents. An example is the case of the German word "betonen", which translates to the English word "emphasize" by Google translator. However, in the target documents the corresponding English word used is "highlight" [12].

Negative transfer, with regards to transfer learning, occurs when the information learned from a source domain has a detrimental effect on a target learner. More formally, given a source domain D_S, a source task T_S, a target domain D_T, a target task T_T, a predictive learner f_{T1} ($\cdot$) trained only with D_T, and a predictive learner f_{T2} ($\cdot$) trained with a transfer learning process combining D_T and D_S, negative transfer occurs when the performance of f_{T1} ($\cdot$) is greater than the performance of f_{T2} ($\cdot$). The topic of negative transfer addresses the need to quantify the amount of relatedness between the source domain and the target domain and whether an attempt to transfer knowledge from the source domain should be made. Extending the definition above, positive transfer occurs when the performance of f_{T2} ($\cdot$) is greater than the performance of f_{T1} ($\cdot$).

Throughout the literature on transfer learning, there are a number of terminology inconsistencies. Phrases such as transfer learning and domain adaptation are used to refer to similar processes. The following definitions will be used in this paper. Domain adaptation, as it pertains to transfer learning, is the process of adapting one or more source domains for the means of transferring information to improve the performance of a target learner. The domain adaptation process attempts to alter a source domain in an attempt to bring the distribution of the source closer to that of the target. Another area of literature inconsistencies is in characterizing the transfer learning process with respect to the availability of labeled and unlabeled data. For example, Daumé [14] and Chattopadhyay et al. [15] define supervised transfer learning as the case of having abundant labeled source data and limited labeled target data, and semi-supervised transfer learning as the case of abundant labeled source data and no labeled target data. In Gong et al. [16] and Blitzer et al. [17], semi-supervised transfer learning is the case of having abundant labeled source data and limited labeled target data, and unsupervised transfer learning is the case of abundant labeled source data and no labeled target data. Cook et al. [18] and Feuz and Cook [19] provide a different variation where the definition of supervised or unsupervised refers to the presence or absence of labeled data in the source domain and informed or uninformed refers to the presence or absence of labeled data in the target domain. With this definition, a labeled source and limited labeled target domain is referred to as informed supervised transfer learning. Pan and Yang [3] refers to inductive transfer learning as the case of having available labeled target domain data, transductive transfer learning as the case of having labeled source and no labeled target domain data, and unsupervised transfer learning as the case of having no labeled source and no labeled target domain data. This paper will explicitly state when labeled and unlabeled data are being used in the source and target domains.

There are different strategies and implementations for solving a transfer learning problem. The majority of the homogeneous transfer learning solutions employ one of three general strategies which include trying to correct for the marginal distribution difference in the source, trying to correct for the conditional distribution difference in the source, or trying to correct both the marginal and conditional distribution differences in the source. The majority of the heterogeneous transfer learning solutions are focused on aligning the input spaces of the source and target domains with the assumption that the domain distributions are the same. If the domain distributions are not equal, then further domain adaptation steps are needed. Another important aspect of a transfer learning solution is the form of information transfer (or what is being transferred). The form of information transfer is categorized into four general Transfer Categories [3]. The first Transfer Category is transfer learning through instances. A common method used in this case is for instances from the source domain to be reweighted in an attempt to correct for marginal distribution differences. These reweighted instances are then directly used in the target domain for training (examples in [20, 21]). These reweighting algorithms work best when the conditional distribution is the same in both domains. The second Transfer Category is transfer learning through features. Feature-based transfer learning approaches are categorized in two ways. The first approach transforms the features of the source through reweighting to more closely match the target domain (e.g. Pan et al. [22]). This is referred to as asymmetric feature transformation and is depicted in Fig. 3.1b. The second approach discovers underlying meaningful structures between the domains to find a common latent feature space that has predictive qualities while reducing the marginal distribution between the domains (e.g. Blitzer et al. [17]). This is referred to as symmetric feature transformation and is depicted in Fig. 3.1a. The third Transfer Category is to

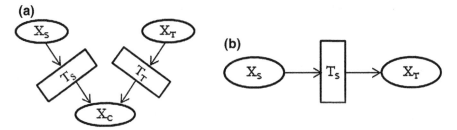

Fig. 3.1 a Shows the symmetric transformation mapping (T_S and T_T) of the source (X_S) and target (X_T) domains into a common latent feature space. **b** Shows the asymmetric transformation (T_T) of the source domain (X_S) to the target domain (X_T)

transfer knowledge through shared parameters of source and target domain learner models or by creating multiple source learner models and optimally combining the reweighted learners (ensemble learners) to form an improved target learner (examples in [23, 24, 25]). The last Transfer Category (and the least used approach) is to transfer knowledge based on some defined relationship between the source and target domains (examples in [26, 27].

Detailed information on specific transfer learning solutions are presented in "Homogeneous Transfer Learning", "Heterogeneous Transfer Learning', and "Negative Transfer" sections. These sections represent the majority of the works surveyed in this paper. In "Homogeneous Transfer Learning', "Heterogeneous Transfer Learning', and "Negative Transfer" sections cover homogeneous transfer learning solutions, heterogeneous transfer learning solutions, and solutions addressing negative transfer, respectively. The section covering transfer learning applications focuses on the general applications that transfer learning is applied to, but does not describe the solution details.

Homogeneous Transfer Learning

This section presents surveyed papers covering homogeneous transfer learning solutions and is divided into subsections that correspond to the Transfer Categories of instance-based, feature-based (both asymmetric and symmetric), parameter-based, and relational-based. Recall that homogeneous transfer learning is the case where $X_S = X_T$. The algorithms surveyed are summarized in Table 3.2 at the end of this section.

The methodology of homogeneous transfer learning is directly applicable to a big data environment. As repositories of big data become more available, there is a desire to use this abundant resource for machine learning tasks, avoiding the timely and potentially costly collection of new data. If there is an available dataset that is drawn from a domain that is related to, but does not an exactly match a target domain of interest, then homogeneous transfer learning can be used to build a predictive model for the target domain as long as the input feature space is the same.

Table 3.2 Homogeneous transfer learning approaches surveyed in Sect. 3 listing different characteristics of each approach

Approach	Transfer category	Source data	Target data	Multiple sources	Generic solution	Negative transfer
CP-MDA [15]	Parameter	Labeled	Limited labels	✓	✓	
2SW-MDA [15]	Instance	Labeled	Unlabeled	✓	✓	
FAM [14]	Asymmetric feature	Labeled	Limited labels	✓	✓	
DTMKL [28]	Asymmetric feature	Labeled	Unlabeled		✓	
JDA [29]	Asymmetric feature	Labeled	Unlabeled		✓	
ARTL [30]	Asymmetric feature	Labeled	Unlabeled		✓	
TCA [31]	Symmetric feature	Labeled	Unlabeled		✓	
SFA [32]	Symmetric feature	Labeled	Limited labels	✓	✓	
SDA [33]	Symmetric feature	Labeled	Unlabeled		✓	
GFK [16]	Symmetric feature	Labeled	Unlabeled		✓	✓
DCP [34]	Symmetric feature	Labeled	Unlabeled		✓	
TCNN [35]	Symmetric feature	Labeled	Limited labels		✓	
MMKT [36]	Parameter	Labeled	Limited labels	✓	✓	✓
DSM [37]	Parameter	Labeled	Unlabeled	✓		✓
MsTrAdaBoost [38]	Instance	Labeled	Limited labels	✓	✓	✓
TaskTrAdaBoost [38]	Parameter	Labeled	Limited labels	✓	✓	✓
RAP [27]	Relational	Labeled	Unlabeled			
SSFE [39]	Hybrid (instance and feature)	Labeled	Limited labels			

Instance-Based Transfer Learning

The paper by Chattopadhyay et al. [15] proposes two separate solutions both using multiple labeled source domains. The first solution is the Conditional Probability based Multi-source Domain Adaptation (CP-MDA) approach, which is a domain adaptation process based on correcting the conditional distribution differences between the source and target domains. The CP-MDA approach assumes a limited amount of labeled target data is available. The main idea is to use a combination of source domain classifiers to label the unlabeled target data. This is accomplished by first building a classifier for each separate source domain. Then a weight value is found for each classifier as a function of the closeness in conditional distribution between each source and the target domain. The weighted source classifiers are summed together to create a learning task that will find the pseudo labels (estimated labels later used for training) for the unlabeled target data. Finally, the target learner is built from the labeled and pseudo labeled target data. The second proposed solution is the Two Stage Weighting framework for Multi-source Domain Adaptation (2SW-MDA) which addresses both marginal and conditional distribution differences between the source and target domains. Labeled target data is not required for the 2SW-MDA approach; however, it can be used if available. In this approach, a weight for each source domain is computed based on the marginal distribution differences between the source and target domains. In the second step, the source domain weights are modified as a function of the difference in the conditional distribution as performed in the CP-MDA approach previously described. Finally, a target classifier is learned based on the reweighted source instances and any labeled target instances that are available. The work presented in Chattopadhyay et al. [15] is an extension of Duan et al. [40] where the novelty is in calculating the source weights as a function of conditional probability. Note, the 2SW-MDA approach is an example of an instance-based Transfer Category, but the CP-MDA approach is more appropriately classified as a parameter-based Transfer Category (see `Heterogeneous Transfer Learning' section). Experiments are performed for muscle fatigue classification using surface electromyography data where classification accuracy is measured as the performance metric. Each source domain represents one person's surface electromyography measurements. A baseline approach is constructed using a Support Vector Machine (SVM) classifier trained on the combination of seven sources used for this test. The transfer learning approaches that are tested against include an approach proposed by Huang et al. [20], Pan et al. [31], Zhong et al. [41], Gao et al. [23], and Duan et al. [40]. The order of performance from best to worst is 2SW-MDA, CP-MDA, Duan et al. [40], Zhong et al. [41], Gao et al. [23], Pan et al. [31], Huang et al. [20], and the baseline approach. All the transfer learning approaches performed better than the baseline approach.

Asymmetric Feature-Based Transfer Learning

In an early and often cited work, Daumé [14] proposes a simple domain adaptation algorithm, referred to as the Feature Augmentation Method (FAM), requiring only 10 lines of Perl script that uses labeled source data and limited labeled target data. In a transfer learning environment, there are scenarios where a feature in the source domain may have a different meaning in the target domain. The issue is referred to as context feature bias, which causes the conditional distributions between the source and target domains to be different. To resolve context feature bias, a method to augment the source and target feature space with three duplicate copies of the original feature set is proposed. More specifically, the three duplicate copies of the original feature set in the augmented source feature space represent a common feature set, a source specific feature set, and a target specific feature set which is always set to zero. In a similar way, the three duplicate copies of the original feature set in the augmented target feature space represent a common feature set, a source specific feature set which is always set to zero, and a target specific feature set. By performing this feature augmentation, the feature space is duplicated three times. From the feature augmentation structure, a classifier learns the individual feature weights for the augmented feature set, which will help correct for any feature bias issues. Using a text document example where features are modeled as a bag-of-words, a common word like "the" would be assigned (through the learning process) a high weight for the common feature set, and a word that is different between the source and target like "monitor" would be assigned a high weight for the corresponding domain feature set. The duplication of features creates feature separation between the source and target domains, and allows the final classifier to learn the optimal feature weights. For the experiments, a number of different natural language processing applications are tested and in each case the classification error rate is measured as the performance metric. An SVM learner is used to implement the Daumé [14] approach. A number of baseline approaches with no transfer learning techniques are measured along with a method by Chelba and Acero [42]. The test results show the Daumé [14] method is able to outperform the other methods tested. However, when the source and target domains are very similar, the Daumé [14] approach tends to underperform. The reason for the underperformance is the duplication of feature sets represents irrelevant and noisy information when the source and target domains are very similar.

Multiple kernel learning is a technique used in traditional machine learning algorithms as demonstrated in the works of Wu et al. [43] and Vedaldi et al. [44]. Multiple kernel learning allows for an optimal kernel function to be learned in a computationally efficient manner. The paper by Duan et al. [28] proposes to implement a multiple kernel learning framework for a transfer learning environment called the Domain Transfer Multiple Kernel Learning (DTMKL). Instead of learning one kernel, multiple kernel learning assumes the kernel is comprised of a linear combination of multiple predefined base kernels. The final classifier and the kernel function are learned simultaneously which has the advantage of using labeled data during the kernel learning process. This is an improvement over Pan et al. [22] and Huang et al. [20] where a two-stage approach is used. The final classifier learning process minimizes the structural risk functional [45] and the marginal distribution between domains using the Maximum Mean Discrepancy measure [46]. Pseudo labels are found for the unlabeled target data to take advantage of this information during the learning process. The pseudo labels are found as a weighted combination of base classifiers (one for each feature) trained from the labeled source data. A regularization term is added to the optimization problem to ensure the predicted values from the final target classifier and the base classifiers are similar for the unlabeled target data. Experiments are performed on the applications of video concept detection, text classification, and email spam detection. The methods tested against include a baseline approach using an SVM classifier trained on the labeled source data, the feature replication method from Daumé [14], an adaptive SVM method from Yang et al. [47], a cross-domain SVM method proposed by Jiang et al. [48], and a kernel mean matching method by Huang et al. [20]. The DTMKL approach uses an SVM learner for the experiments. Average precision and classification accuracy are measured as the performance metrics. The DTMKL method performed the best for all applications, and the baseline approach is consistently the worst performing. The other methods showed better performance over the baseline which demonstrated a positive transfer learning effect.

The work by Long et al. [29] is a Joint Domain Adaptation (JDA) solution that aims to simultaneously correct for the marginal and conditional distribution differences between the labeled source domain and the unlabeled target domain. Principal Component Analysis (PCA) is used for optimization and dimensionality reduction. To address the difference in marginal distribution between the domains, the Maximum Mean Discrepancy distance measure [46] is used to compute the marginal distribution differences and is integrated into the PCA optimization algorithm. The next part of the solution requires a process to correct the conditional distribution differences, which requires labeled target data. Since the target data is unlabeled, pseudo labels (estimated target labels) are found by learning a classifier from the labeled source data. The Maximum Mean Discrepancy distance measure is modified to measure the distance between the conditional distributions and is integrated into the PCA optimization algorithm to minimize the conditional distributions. Finally, the features identified by the modified PCA algorithm are used to train the final target classifier. Experiments are performed for the application of image recognition and classification accuracy is measured as the performance metric. Two baseline approaches of a 1-nearest neighbor classifier and a PCA approach trained on the source data are tested. Transfer learning approaches tested for this experiment include the approach by Pan [31], Gong et al. [16], and Si et al. [49]. These transfer learning approaches only attempt to correct for marginal distribution differences between domains. The Long et al. [29] approach is the best performing, followed by the Pan [31] and Si et al. [49] approaches (a tie), then the Gong et al. [16] approach, and finally the baseline approaches. All transfer learning approaches perform better than the baseline approaches. The possible reason behind the underperformance of the Gong et al. [16] approach is the data smoothness assumption that is made for the Gong et al. [16] solution may not be intact for the data sets tested.

The paper by Long et al. [30] proposes an Adaptation Regularization based Transfer Learning (ARTL) framework for scenarios of labeled source data and unlabeled target data. This transfer learning framework proposes to correct the difference in marginal distribution between the source and target domains, correct the difference in conditional distribution between the domains, and improve classification performance through a manifold regularization [50] process (which

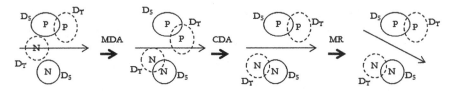

Fig. 3.2 ARTL overview showing marginal distribution adaptation (MDA), conditional distribution adaptation (CDA), and manifold regularization (MR). Diagram adapted from Long [30]

optimally shifts the hyperplane of an SVM learner). This complete framework process is depicted in Fig. 3.2. The proposed ARTL framework will learn a classifier by simultaneously performing structural risk minimization [45], reducing the marginal and conditional distributions between the domains, and optimizing the manifold consistency of the marginal distribution. To resolve the conditional distribution differences, pseudo labels are found for the target data in the same way as proposed by Long et al. [29]. A difference between the ARTL approach and Long et al. [29] is ARTL learns the final classifier simultaneously while minimizing the domain distribution differences, which is claimed by Long et al. [30] to be a more optimal solution. Unfortunately, the solution by Long et al. [29] is not included in the experiments. Experiments are performed on the applications of text classification and image classification where classification accuracy is measured as the performance metric. There are three baseline methods tested where different classifiers are trained with the labeled source data. There are five transfer learning methods tested against, which include methods by Ling et al. [51], Pan et al. [32], Pan et al. [31], Quanz and Huan [52], and Xiao and Guo [53]. The order of performance from best to worst is ARTL, Xiao and Guo [53], Pan et al. [31], Pan et al. [32], Quanz and Huan [52] and Ling et al. [51] (tie), and the baseline approaches. The baseline methods underperformed all other transfer learning approaches tested.

Symmetric Feature-Based Transfer Learning

The paper by Pan et al. [31] proposes a feature transformation approach for domain adaptation called Transfer Component Analysis (TCA), which does not require labeled target data. The goal is to discover common latent features that have the same marginal distribution across the source and target domains while maintaining the intrinsic structure of the original domain data. The latent features are learned between the source and target domains in a Reproducing Kernel Hilbert Space [54] using the Maximum Mean Discrepancy [46] as a marginal distribution measurement criteria. Once the latent features are found, traditional machine learning is used to train the final target classifier. The TCA approach extends the work of Pan et al. [22] by improving computational efficiency. Experiments are conducted for the application of WiFi localization where the location of a particular device is being predicted. The source domain is comprised of data measured from different room and building topologies. The performance metric measured is the average error distance of the position of a device. The transfer learning methods tested against are from Blitzer et al. [17] and Huang et al. [20]. The TCA method performed the best followed by the Huang et al. [20] approach and the Blitzer et al. [17] approach. For the Blitzer et al. [17] approach, the manual definition of the pivot functions (functions that define the correspondence) is important to performance and specific to the end application. There is no mention as to how the pivot functions are defined for WiFi localization.

The work by Pan et al. [32] proposes a Spectral Feature Alignment (SFA) transfer learning algorithm that discovers a new feature representation for the source and target domain to resolve the marginal distribution differences. The SFA method assumes an abundance of labeled source data and a limited amount of labeled target data. The SFA approach identifies domain-specific and domain-independent features and uses the domain-independent features as a bridge to build a bipartite graph modeling the co-occurrence relationship between the domain-independent and domain-specific features. If the graph shows two domain-specific features having connections to common domain-independent feature, then there is a higher chance the domain-specific features are aligned. A spectral clustering algorithm based on graph spectral theory [55] is used on the bipartite graph to align domain-specific features and domain-independent features into a set of clusters representing new features. These clusters are used to reduce the difference between domain-specific features in the source and the target domains. All the data instances are projected into this new feature space and a final target classifier is trained using the new feature representation. The SFA algorithm is a type of correspondence learning where the domain-independent features act as pivot features (see Blitzer et al. [17] and Prettenhofer and Stein [11] for further information on correspondence learning). The SFA method is well-suited for the application of text document classification where a bag-of-words model is used to define features. For this application there are domain-independent words that will appear often in both domains and domain-specific words that will appear often only in a specific domain. This is referred to as frequency feature bias, which causes marginal distribution differences between the domains. An example of domain-specific features being combined is the word "sharp" appearing often in the source domain but not in the target domain, and the word "hooked" appearing often in the target but not in the source domain. These words are both connected to the same domain-independent words (for example "good" and "exciting"). Further, when the words "sharp" or "hooked" appear in text instances, the labels are the same. The idea is to combine (or align) these two features (in this case "sharp" and "hooked") to form a new single invariant feature. The experiments are performed on sentiment classification where classification accuracy is measured as the performance metric. A baseline approach is tested where a classifier is trained only on source data. An upper limit approach is also tested where a classifier is trained on a large amount of labeled target data. The competing transfer learning approach tested against is by Blitzer et al. [17]. The order of performance for the tests from best to worst is the upper limit approach, SFA, Blitzer et al. [17], and baseline approach. Not only does the SFA approach demonstrate better performance than Blitzer et al. [17], the SFA approach does not need to manually define pivot functions as in the Blitzer et al. [17] approach. The SFA approach only addresses the issue of marginal distribution differences and does not address any context feature bias issues, which would represent conditional distribution differences.

The work by Glorot et al. [33] proposes a deep learning algorithm for transfer learning called a Stacked Denoising Autoencoder (SDA) to resolve the marginal distribution differences between a labeled source domain and an unlabeled target domain. Deep learning algorithms learn intermediate invariant concepts between two data sources, which are used to find a common latent feature set. The first step in this process is to train the Stacked Denoising Autoencoders [56] with unlabeled data from the source and target domains. This transforms the input space to discover the common invariant latent feature space. The next step is to train a classifier using the transformed latent features with the labeled source data. Experiments are performed on text review sentiment classification where transfer loss is measured as the performance metric. Transfer loss is defined as the classification error rate using a learner only trained on the source domain and tested on the target minus the classification error rate using a learner only trained on the target domain and tested on the target. There are 12 different source and target domain pairs that are created from four unique review topics. A baseline method is tested where an SVM classifier is trained on the source domain. The transfer learning approaches that are tested include an approach by Blitzer et al. [17], Li and Zong [57], and Pan et al. [32]. The Glorot et al. [33] approach performed the best with the Blitzer et al. [17], Li and Zong [57], and Pan et al. [32] methods all having similar performance and all outperforming the baseline approach.

In the paper by Gong et al. [16], a domain adaptation technique called the Geodesic Flow Kernel (GFK) is proposed that finds a low-dimensional feature space, which reduces the marginal distribution differences between the labeled source and unlabeled target domains. To accomplish this, a geodesic flow kernel is constructed using the source and target input feature data, which projects a large number of subspaces that lie on the geodesic flow curve. The geodesic flow curve represents incremental differences in geometric and statistical properties between the source and target domain spaces. A classifier is then learned from the geodesic flow kernel by selecting the features from the geodesic flow curve that are domain invariant. The work of Gong et al. [16] directly enhances the work of Gopalan et al. [58] by eliminating tuning parameters and improving computational efficiency. In addition, a Rank of Domain (ROD) metric is developed to evaluate which of many source domains is the best match for the target domain. The ROD metric is a function of the geometric alignment between the domains and the Kullback–Leibler divergence in data distributions between the projected source and target subspaces. Experiments are performed for the application of image classification where classification accuracy is measured as the performance metric. The tests use pairs of source and target data sets from four available data sets. A baseline approach is defined that does not use transfer learning, along with the approach defined by Gopalan et al. [58]. Additionally, the Gong et al. [16] approach uses a 1-nearest neighbor classifier. The results in order from best to worst performance are Gong et al. [16], Gopalan et al. [58], and the baseline approach. The ROD measurements between the different source and target domain pairs tested have a high correlation to the actual test results, meaning the domains that are found to be more related with respect to the ROD measurement had higher classification accuracies.

The solution by Shi and Sha [34], referred to as the Discriminative Clustering Process (DCP), proposes to equalize the marginal distribution of the labeled source and unlabeled target domains. A discriminative clustering process is used to discover a common latent feature space that is domain invariant while simultaneously learning the final target classifier. The motivating assumptions for this solution are the data in both domains form well-defined clusters which correspond to unique class labels, and the clusters from the source domain are geometrically close to the target clusters if they share the same label. Through clustering, the source domain labels can be used to estimate the target labels. A one-stage solution is formulated that minimizes the marginal distribution differences while minimizing the predicted classification error in the target domain using a nearest neighbor classifier. Experiments are performed for object recognition and sentiment classification where classification accuracy is measured as the performance metric. The approach described above is tested against a baseline approach taken from Weinberger and Saul [59] with no transfer learning. Other transfer learning approaches tested include an approach from Pan et al. et al. [31], Blitzer et al. [17], and Gopalan et al. [58]. The Blitzer et al. [17] approach is not tested for the object recognition application because the pivot functions are not easily defined for this application. For the object recognition tests, the Shi and Sha [34] method is best in five out of six comparison tests. For the text classification tests, the Shi and Sha [34] approach is the best performing overall, with the Blitzer et al. [17] approach a close second. An important point to note is the baseline method outperformed the Pan et al. [31] and Gopalan et al. [58] methods in both tests. Both the Pan et al. [31] and Gopalan et al. [58] methods are two-stage domain adaptation processes where the first stage reduces the marginal distributions between the domains and the second stage trains a classifier with the adapted domain data. This paper offers a hypothesis that two-stage processes are actually detrimental to transfer learning (causes negative transfer). The one-stage learning process is a novel idea presented by this paper. The hypothesis that the two-stage transfer learning process creates low performing learners does not agree with the results presented in the individual papers by Gopalan et al. [58] and Pan et al. [31] and other previously surveyed works.

Convolutional Neural Networks (CNN) have been successfully used in traditional data mining environments [60]. However, a CNN requires a large amount of labeled training data to be effective, which may not be available. The paper by Oquab et al. [35] proposes a transfer learning method of training a CNN with available labeled source data (a source learner) and then extracting the CNN internal layers (which represent a generic mid-level feature representation) to a target CNN learner. This method is referred to as the Transfer Convolutional Neural

Network (TCNN). To correct for any further distribution differences between the source and the target domains, an adaptation layer is added to the target CNN learner, which is trained from the limited labeled target data. The experiments are run on the application of object image classification where average precision is measured as the performance metric. The Oquab et al. [35] method is tested against a method proposed by Marszalek et al. [61] and a method proposed by Song et al. [62]. Both the Marszalek et al. [61] and Song et al. [62] approaches are not transfer learning approaches and are trained on the limited labeled target data. The first experiment is performed using the Pascal VOC 2007 data set as the target and ImageNet 2012 as the source. The Oquab et al. [35] method outperformed both Song et al. [62] and Marszalek et al. [61] approaches for this test. The second experiment is performed using the Pascal VOC 2012 data set as the target and ImageNet 2012 as the source. In the second test, the Oquab et al. [35] method marginally outperformed the Song et al. [62] method (the Marszalek et al. [61] method was not tested for the second test). The tests successfully demonstrated the ability to transfer information from one CNN learner to another.

Parameter-Based Transfer Learning

The paper by Tommasi et al. [36] addresses the transfer learning environment characterized by limited labeled target data and multiple labeled source domains where each source corresponds to a particular class. In this case, each source is able to build a binary learner to predict that class. The objective is to build a target binary learner for a new class using minimal labeled target data and knowledge transferred from the multiple source learners. An algorithm is proposed to transfer the SVM hyperplane information of each of the source learners to the new target learner. To minimize the effects of negative transfer, the information transferred from each source to the target will be weighted such that the most related source domains receive the highest weighting. The weights are determined through a leave out one process as defined by Cawley [63]. The Tommasi et al. [36] approach, called the Multi-Model Knowledge Transfer (MMKT) method, extends the method proposed by Tommasi and Caputo [64] that only transfers a single source domain. Experiments are performed on the application of image recognition where classification accuracy is measured as the performance metric. Transfer learning methods tested include an average weight approach (same as Tommasi et al. [36] but all source weights are equal), and the Tommasi and Caputo [64] approach. A baseline approach is tested, which is trained on the limited labeled target data. The best performing method is Tommasi et al. [36], followed by the average weight, Tommasi and Caputo [64], and the baseline approach. As the number of labeled target instances goes up, the Tommasi et al. [36] and average weight methods converge to the same performance. This is because the adverse effects of negative transfer are lessened as the labeled target data increases. This result demonstrates the Tommasi et al. [36] approach is able to lessen the effects of negative transfer from unrelated sources.

The transfer learning approach presented in the paper by Duan et al. [37], referred to as the Domain Selection Machine (DSM), is tightly coupled to the application of event recognition in consumer videos. Event recognition in videos is the process of predicting the occurrence of a particular event or topic (e.g. "show" or "performance") in a given video. In this scenario, the target domain is unlabeled and the source information is obtained from annotated images found via web searches. For example, a text query of the event "show" for images on Photosig.com represents one source and the same query on Flickr.com represents another separate source. The Domain Selection Machine proposed in this paper is realized as follows. For each individual source, an SVM classifier is created using SIFT [65] image features. The final target classifier is made up of two parts. The first part is a weighted sum of the source classifier outputs whose input is the SIFT features from key frames of the input video. The second part is a learning function whose inputs are Space-Time features [66] from the input video and is trained from target data where the target labels are estimated (pseudo labels) from the weighted sum of the source classifiers. To combat the effects of negative transfer from unrelated sources, the most relevant source domains are selected by using an alternating optimization algorithm that iteratively solves the target decision function and the domain selection vector. Experiments are performed in the application of event recognition in videos as described above where the mean average precision is measured as the performance metric. A baseline method is created by training a separate SVM classifier on each source domain and then equally combining the classifiers. The other transfer learning approaches tested include the approach by Bruzzone and Marconcini [67], Schweikert et al. [68], Duan et al. [40], and Chattopadhyay et al. [15]. The Duan et al. [40] approach outperforms all the other approaches tested. The other approaches all have similar results, meaning the transfer learning methods did not outperform the baseline approach. The possible reason for this result is the existence of unrelated sources in the experiment. The other transfer learning approaches tested had no mechanism to guard against negative transfer from unrelated sources.

The paper by Yao and Doretto [38] first presents an instance-based transfer learning approach followed by a separate parameter-based transfer learning approach. In the transfer learning process, if the source and target domains are not related enough, negative transfer can occur. Since it is difficult to measure the relatedness between any particular source and target domain, Yao and Doretto [38] proposes to transfer knowledge from multiple source domains using a boosting method in an attempt to minimize the effects of negative transfer from a single unrelated source domain. The boosting process requires some amount of labeled target data. Yao and Doretto [38] effectively extends the work of Dai et al. [69] (TrAdaBoost) by expanding the transfer boosting algorithm to multiple source domains. In the TrAdaBoost algorithm, during every boosting iteration, a so-called weak classifier is built using weighted instance data from the previous iteration. Then, the misclassified source instances are lowered in importance and the misclassified target instances are raised in importance. In the multi-source TrAdaBoost algorithm (called MsTrAdaBoost), each iteration step first finds a weak classifier for each source and target combination, and then the final weak classifier is selected for that iteration by finding the one that minimizes the target classification error. The instance reweighting step remains the same as in the TrAdaBoost. An alternative multi-source boosting method (TaskTrAdaBoost) is proposed that transfers internal learner parameter information from the source to the target. The TaskTrAdaBoost algorithm first finds candidate weak classifiers from each individual source by performing an AdaBoost process on each source domain. Then an AdaBoost process is performed on the labeled target data, and at every boosting iteration, the weak classifier used is selected from the candidate weak source classifiers (found in the previous step) that has the lowest classification error using the labeled target data. Experiments are performed for the application of object category recognition where the area under the curve (AUC) is measured as the performance metric. An AdaBoost baseline approach using only the limited labeled target data is measured along with a TrAdaBoost approach using a single source (the multiple sources are combined to one) and the limited labeled target data. Linear SVM learners are used as the base classifiers in all approaches. Both the MsTrAdaBoost and TaskTrAdaBoost approaches outperform the baseline approach and TrAdaBoost approach. The MsTrAdaBoost and TaskTrAdaBoost demonstrated similar performance.

Relational-Based Transfer Learning

The specific application addressed in the paper by Li et al. [27] is to classify words from a text document into one of three classes (e.g. sentiments, topics, or neither). In this scenario, there exists a labeled text source domain on one particular subject matter and an unlabeled text target domain on a different subject matter. The main idea is that sentiment words remain constant between the source and target domains. By learning the grammatical and sentence structure patterns of the source, a relational pattern is found between the source and target domains, which is used to predict the topic words in the target. The sentiment words act as a common linkage or bridge between the source and target domains. A bipartite word graph is used to represent and score the sentence structure patterns. A bootstrapping algorithm is used to iteratively build a target classifier from the two domains. The bootstrapping process starts with defining seeds which are instances from the source that match frequent patterns in the target. A cross domain classifier is then trained with the seed information and extracted target information (there is no target information in the first iteration). The classifier is used to predict the target labels and the top confidence rated target instances are selected to reconstruct the bipartite word graph. The bipartite word graph is now used to select new target instances that are added to the seed list. This bootstrapping process continues over a selected number of iterations, and the cross domain classifier learned in the bootstrapping process is now available to predict target samples. This method is referred to as the Relational Adaptive bootstraPping (RAP) approach. The experiments tested the Li et al. [27] approach against an upper bound method where a standard classifier is trained with a large amount of target data. Other transfer learning methods tested include an approach by Hu and Liu [70], Qiu et al. [71], Jakob and Gurevych [72], and Dai et al. [69]. The application tested is word classification as described above where the F1 score is measured as the performance metric. The two domains tested are related to movie reviews and product reviews. The Li et al. [27] method performed better than the other transfer learning methods, but fell short of the upper bound method as expected. In its current form, this algorithm is tightly coupled with its underlying text application, which makes it difficult to use for other non-text applications.

Hybrid-Based (Instance and Parameter) Transfer Learning

The paper by Xia et al. [39] proposes a two step approach to address marginal distribution differences and conditional distribution differences between the source and target domains called the Sample Selection and Feature Ensemble (SSFE) method. A sample selection process, using a modified version of Principal Component Analysis, is employed to select labeled source domain samples such that the source and target marginal distributions are equalized. Next, a feature ensemble step attempts to resolve the conditional distribution differences between the source and target domains. Four individual classifiers are defined corresponding to parts of speech of noun, verb, adverb/adjective, and other. The four classifiers are trained using only the features that correspond to that part of speech. The training data is the limited labeled target and the labeled source selected in the previous sample selection step. The four classifiers are weighted as a function of minimizing the classification error using the limited labeled target data. The weighted output of the four classifiers is used as the final target classifier. This work by Xia et al. [39] extends the earlier work of Xia and Zong [73]. The experiments are performed for the application of review sentiment classification using four different review categories, where each category is combined to create 12 different source and target pairs. Classification accuracy is measured as the performance metric. A baseline approach using all the training data from the source is constructed, along with a sample selection approach (only using the first step defined above), a feature ensemble approach (only using the second step defined above) and the complete approach outlined above. The complete approach is the best performing, followed by sample selection and feature ensemble approaches, and the baseline approach. The sample selection and feature ensemble approaches perform equally as well in head-to-head tests. The weighting of the four classifiers (defined by the corresponding parts of speech) in the procedure above gives limited resolution in attempting to adjust for context feature bias issues. A method of having more classifiers in the ensemble step could yield better performance at the expense of higher complexity.

Discussion of Homogeneous Transfer Learning

The previous surveyed homogeneous transfer learning works (summarized in Table 3.2) demonstrate many different characteristics and attributes. Which homogeneous transfer learning solution is best for a particular application? An important characteristic to evaluate in the selection process is what type of differences exist between a given source and target domain. The previous solutions surveyed address domain adaptation by correcting for marginal distribution differences, correcting for conditional distribution differences, or correcting for both marginal and conditional distribution differences. The surveyed works of Duan et al. [28], Gong et al. [16], Pan et al. [31], Li et al. [27], Shi and Sha [34], Oquab et al. [35], Glorot et al. [33], and Pan et al. [32] are focused on solving the differences in marginal distribution between the source and target domains. The surveyed works of Daumé [14], Yao and Doretto [38], Tommasi et al. [36] are focused on solving the differences in conditional distribution between the source and target domains. Lastly, the surveyed works of Long et al. [30], Xia et al. [39], Chattopadhyay et al. [15], Duan et al. [37], and Long et al. [29] correct the differences in both the marginal and conditional distributions. Correcting for the conditional distribution differences between the source and target domain can be problematic as the nature of a transfer learning environment is to have minimal labeled target data. To compensate for the limited labeled target data, many of the recent transfer learning solutions create pseudo labels for the unlabeled target data to facilitate the conditional distribution correction process between the source and target domains. To further help determine which solution is best for a given transfer learning application, the information in Table 3.2 should be used to match the characteristics of the solution to that of the desired application environment. If the application domain contains multiple sources where the sources are not mutually uniformly distributed, a solution that guards against negative transfer may be of greater benefit. A recent trend in the development of transfer learning solutions is for solutions to address both marginal and conditional distribution differences between the source and target domains. Another emerging solution trend is the implementation of a one-stage process as compared to a two-stage process. In the recent works of Long et al. [30], Duan et al. [28], Shi and Sha [34], and Xia et al. [39], a one-stage process is employed that simultaneously performs the domain adaptation process while learning the final classifier. A two-stage solution first performs the domain adaptation process and then independently learns the final classifier. The claim by Long et al. [30] is a one-stage solution achieves enhanced performance because the simultaneous solving of domain adaptation and the classifier establishes mutual reinforcement. The surveyed homogeneous transfer learning works are not specifically applied to big data solutions; however, there is nothing to preclude their use in a big data environment.

Heterogeneous Transfer Learning

Heterogeneous transfer learning is the scenario where the source and target domains are represented in different feature spaces. There are many applications where heterogeneous transfer learning is beneficial. Heterogeneous transfer learning applications that are covered in this section include image recognition [5–7, 74–76], multi-language text classification [5, 10–12, 76], single language text classification [4], drug efficacy classification [74], human activity classification [8], and software defect classification [9]. Heterogeneous transfer learning is also directly applicable to a big data environment. As repositories of big data become more available, there is a desire to use this abundant resource for machine learning tasks, avoiding the timely and potentially costly collection of new data. If there is an available dataset drawn from a target domain of interest that has a different feature space from another target dataset (also drawn from the same target domain), then heterogeneous transfer learning can be used to bridge the difference in the feature spaces and build a predictive model for that target domain. Heterogeneous transfer learning is still a relatively new area of study as the majority of the works covering this topic have been published in the last five years. From a high-level view, there are two main approaches to solving the heterogeneous feature space difference. The first approach, referred to as symmetric transformation shown in Fig. 3.1a, separately transforms the source and target domains into a common latent feature space in an attempt to unify the input spaces of the domains. The second approach, referred to as asymmetric transformation as shown in Fig. 3.1b, transforms the source feature space to the target feature space to align the input feature spaces. The asymmetrical transformation approach is best used when the same class instances in the source and target can be transformed without context feature bias. Many of the heterogeneous transfer learning solutions surveyed make the implicit or explicit assumption that the source and the target domain instances are drawn from the same domain space. With this assumption there should be no significant distribution differences between the domains. Therefore, once the differences in input feature spaces are resolved, no further domain adaptation needs to be performed.

As is the case with homogeneous transfer learning solutions, whether the source and target domains contain labeled data drives the solution formulation for heterogeneous approaches. Data label availability is a function of the underlying application. The solutions surveyed in this paper have different labeled data requirements. For transfer learning to be feasible, the source and the target domains must be related in some way. Some heterogeneous solutions require an explicit mapping of the relationship or correspondence between the source and target domains. For example, the solutions defined for Prettenhofer and Stein [11] and Wei and Pal [77] require manual definitions of source and target correspondence.

Symmetric Feature-Based Transfer Learning

The transfer learning approach proposed by Prettenhofer and Stein [11] addresses the heterogeneous scenario of a source domain containing labeled and unlabeled data, and a target domain containing unlabeled data. The structural correspondence learning technique from Blitzer et al. [17] is applied to this problem. Structural correspondence learning depends on the manual definition of pivot functions that capture correspondence between the source and target domains. Effective pivot functions should use features that occur frequently in both domains and have good predictive qualities. Each pivot function is turned into a linear classifier using data from the source and target domains. From these pivot classifiers, correspondences between features are discovered and a latent feature space is learned. The latent feature space is used to train the final target classifier. The paper by Prettenhofer and Stein [11] uses this solution to solve the problem of text classification where the source is written in one language and the target is written in a different language. In this specific implementation referred to as Cross-Language Structural Correspondence Learning (CLSCL), the pivot functions are defined by pairs of words, one from the target and one from the source, that represent direct word translations from one language to the other. The experiments are performed on the applications of document sentiment classification and document topic classification. English documents are used in the source and other language documents are used in the target. The baseline method used in this test trains a learner on the labeled source documents, then translates the target documents to the source language and tests the translated version. An upper bound method is established by training a learner with the labeled target documents and testing with the target documents. Average classification accuracy is measured as the performance metric. The average results show the upper bound method performing the best and the Prettenhofer and Stein [11] method performing better than the baseline method. An issue with using structural correspondence learning is the difficulty in generalizing the pivot functions. For this solution, the pivot functions need to be manually and uniquely defined for a specific application, which makes it very difficult to port to other applications.

The paper by Shi et al. [74], referred to as Heterogeneous Spectral Mapping (HeMap), addresses the specific transfer learning scenario where the input feature space is different between the source and target ($X_S \neq X_T$), the marginal distribution is different between the source and the target ($P(X_S) \neq P(X_T)$), and the output space is different between the source and the target ($Y_S \neq Y_T$.). This solution uses labeled source data that is related to the target domain and limited labeled target data. The first step is to find a common latent input space between the source and target domains using a spectral mapping technique. The spectral mapping technique is modeled as an optimization objective that maintains the original structure of the data while minimizing the difference between the two domains. The next step is to apply a clustering based sample selection method to select related instances as new training data, which resolves the marginal distribution differences in the latent input space. Finally, a Bayesian based method is used to find the relationship and resolve the differences in the output space. Experiments are performed for the applications of image classification and drug efficacy prediction. Classification error rate is measured as the performance metric. This solution demonstrated better performance as compared to a baseline approach; however, details on the baseline approach are not documented in the paper and no other transfer learning solutions are tested.

The algorithm by Wang and Mahadevan [4], referred to as the Domain Adaptation Manifold Alignment (DAMA) algorithm, proposes using a manifold alignment [78] process to perform a symmetric transformation of the domain input spaces. In this solution, there are multiple labeled source domains and a limited labeled target domain for a total of K domains where all K domains share the same output label space. The approach is to create a separate mapping function for each domain to transform the heterogeneous input space to a common latent input space while preserving the underlying structure of each domain. Each domain is modeled as a manifold. To create the latent input space, a larger matrix model is created that represents and captures the joint manifold union of all input domains. In this manifold model, each domain is represented by a Laplacian matrix that captures the closeness to other instances sharing the same label. The instances with the same labels are forced to be neighbors while separating the instances with different labels. A dimensionality reduction step is performed through a generalized eigenvalue decomposition process to eliminate feature redundancy. The final learner is built in two stages. The first stage is a linear regression model trained on the source data using the latent feature space. The second stage is also a linear regression model that is summed with the first stage. The second stage uses a manifold regularization [50] process to ensure the prediction error is minimized when using the labeled target data. The first stage is trained only using the source data and the second stage compensates for the domain differences caused by the first stage to achieve enhanced target predictions. The experiments are focused on the application of document text classification where classification accuracy is measured as the performance metric. The methods tested against include a Canonical Correlation Analysis approach and a Manifold Regularization approach, which is considered the baseline method. The baseline method uses the limited labeled target domain data and does not use source domain information. The approach presented in this paper substantially outperforms the Canonical Correlation Analysis and baseline approach; however, these approaches are not directly referenced so it is difficult to understand the significance of the test results. A unique aspect of this paper is the modeling of multiple source domains in a heterogeneous solution.

There are scenarios where a large amount of unlabeled heterogeneous source data is readily available that could be used to improve the predictive performance of a particular target learner. The paper by Zhu et al. [7], which presents the method called the Heterogeneous Transfer Learning Image Classification (HTLIC), addresses this scenario with the assumption of having access to a sufficiently large amount of labeled target data. The objective is to use the large supply of available unlabeled source data to create a common latent feature input space that will improve prediction performance in the target classifier. The solution proposed by

Zhu et al. [7] is tightly coupled to the application of image classification and is described as follows. Images with labeled categories (e.g. dog, cake, starfish, etc.) are available in the target domain. To obtain the source data, a web search is performed from Flickr for images that "relate" to the labeled categories. For example, for the category of dog, the words dog, doggy, and greyhound may be used in the Flickr search. As a reference point, the idea of using annotated images from Flickr as unlabeled source data was first proposed by Yang et al. [79]. The retrieved images from Flickr have one or more word tags associated with each image. These tagged image words are then used to search for text documents using Google search. Next, a two-layer bipartite graph is constructed where the first layer represents linkages between the source images and the image tags. The second layer represents linkages between the image tags and the text documents. If an image tag appears in a text document, then a link is created, otherwise there is no link. Images in both the source and the target are initially represented by an input feature set that is derived from the pixel information using SIFT descriptors [65]. Using the initial source image features and the bipartite graph representation derived only from the source image tags and text data, a common latent semantic feature set is learned by employing Latent Semantic Analysis [80]. A learner is now trained with the transformed labeled target instances. Experiments are performed on the proposed approach where 19 different image categories are selected. Binary classification is performed testing different image category pairs. A baseline method is tested using an SVM classifier trained only with the labeled target data. Methods by Raina et al. [81] and by Wang et al. [82] are also tested. The approach proposed by Zhu et al. [7] performed the best overall followed by Raina et al. [81], Wang et al. [82], and baseline approach. The idea of using an abundant source of unlabeled data available through an internet search to improve prediction performance is a very alluring premise. However, this method is very specific to image classification and is enabled by having a web site like Flickr, which essentially provides unlimited labeled image data. This method is difficult to port to other applications.

The transfer learning solution proposed by Qi et al. [75] is another example of an approach that specifically addresses the application of image classification. In the paper by Qi et al. [75], the author claims the application of image classification is inherently more difficult than text classification because image features are not directly related to semantic concepts inherent in class labels. Image features are derived from pixel information, which is not semantically related to class labels, as opposed to word features that have semantic interpretability to class labels. Further, labeled image data is more scarce as compared to labeled text data. Therefore, a transfer learning environment for image classification is desired where an abundance of labeled text data (source) is used to enhance a learner trained on limited labeled image data (target). In this solution, text documents are identified by performing a web search (from Wikipedia for example) on class labels. In order to perform the knowledge transfer from the text documents (source) to the image (target) domain, a bridge in the form of a co-occurrence matrix is used that relates the text and image information. The co-occurrence matrix contains text instances with the corresponding image instances that are found in that particular text document. The co-occurrence matrix can be programmatically built by crawling web pages and extracting the relevant text and image feature information. Using the co-occurrence matrix, a common latent feature space is found between the text and image features, which is used to learn the final target classifier. This approach, called the Text To Image (TTI) method, is similar to Zhu et al. [7]. However, Zhu et al. [7] does not use labeled source data to enhance the knowledge transfer, which will result in degraded performance when there is limited labeled target data. Experiments are performed with the methods proposed by Qi et al. [75], Dai et al. [83], Zhu et al. [7], and a baseline approach using a standard SVM classifier trained on the limited labeled target data. The text documents are collected from Wikipedia, and classification error rate is measured as the performance metric. The results show the Zhu et al. [7] method performing the best in 15 % of the trials, the Dai et al. [83] method being the best in 10 % of the trials, and the Qi et al. [75] method leading in 75 % of the trials. As with the case of Zhu et al. [7], this method is very specific to the application of image classification and is difficult to port to other applications.

The scenario addressed in the paper by Duan et al. [5] is focused on heterogeneous domain adaptation with a single labeled source domain and a target domain with limited labeled samples. The solution proposed is called Heterogeneous Feature Augmentation (HFA). A transformation matrix P is defined for the source and a transformation matrix Q is defined for the target to project the feature spaces to a common latent space. The latent feature space is augmented with the original source and target feature set and zeros where appropriate. This means the source input data projection has the common latent features, the original source features, and zeros for the original target features. The target input data projection has the common latent features, zeros for the original source features, and the original target features. This feature augmentation method was first introduced by Daumé [14] and is used to correct for conditional distribution differences between the domains. For computational simplification, the P and Q matrices are not directly found but combined and represented by an H matrix. An optimization problem is defined by minimizing the structural risk functional [45] of SVM as a function of the H matrix. The final target prediction function is found using an alternating optimization algorithm to simultaneously solve the dual problem of SVM and the optimal transformation H matrix. The experiments are performed for the applications of image classification and text classification. The source contains labeled image data and the target contains limited labeled image data. For the image features, SURF [84] features are extracted from the pixel information and then clustered into different dimension feature spaces creating the heterogeneous source and target environment. For the text classification experiments, the target contains Spanish language documents and the source contains documents in four different languages. The experiments test against a baseline method, which is constructed by training an SVM learner on the limited labeled target data. Other heterogeneous adaptation methods that are tested include the method by Wang and Mahadevan [4], Shi et al. [74], and Kulis et al. [6]. For the image classification test, the HFA method outperforms all the methods tested by an average of one standard deviation with respect to classification accuracy. The Kulis et al. [6] method has comparable results to the baseline method (possibly due to some uniqueness in the data set) and the Wang and Mahadevan [4] method slightly outperforms the baseline method (possibly due to a weak manifold structure in the data set). For the text classification test, the HFA method outperforms all methods tested by an average of 1.5 standard deviation. For this test, the Kulis et al. [6] method is second in performance, followed by Wang and Mahadevan [4], and then the baseline method. The Shi et al. [74] method performed worse than the baseline method in both tests. A possible reason for this result is the Shi et al. [74] method does not specifically use the labeled information from the target when performing the symmetric transformation, which will result in degraded classification performance [76].

The work of Li et al. [76], called the Semi-supervised Heterogeneous Feature Augmentation (SHFA) approach, addresses the heterogeneous scenario of an abundance of labeled source data and limited target data, and directly extends the work of Duan et al. [5]. In this work, the H transformation matrix, which is described above by Duan et al. [5], is decomposed into a linear combination of a set of rank-one positive semi-definite matrices that allow for Multiple Kernel Learning solvers (defined by Kloft et al. [85]) to be used to find a solution. In the process of learning the H transformation matrix, the labels for the unlabeled target data are estimated (pseudo labels created) and used while learning the final target classifier. The pseudo labels for the unlabeled target data are found from an SVM classifier trained on the limited labeled target data. The high-level domain adaptation is shown in Fig. 3.3. Experiments are performed for three applications which include image classification (where 31 unique classes are defined), multi-language text document classification (where six unique classes are defined), and multi-language text sentiment classification. Classification accuracy is measured as the performance metric. The method by Li et al. [76] is tested against a baseline method using an SVM learner and trained on the limited labeled target data. Further, other heterogeneous methods tested include Wang and Mahadevan [4], Duan et al. [5], Kulis et al. [6], Shi et al. [74]. By averaging the three different application test results, the order of performance from best to worst is Li et al. [76], Duan et al. [5], Wang and Mahadevan [4], baseline and Kulis et al. [6] (tie), and Shi et al. [74].

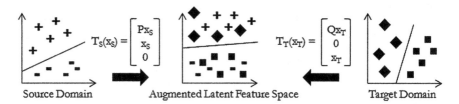

Fig. 3.3 Depicts algorithm approach by Li [76] where the heterogeneous source and target features are transformed to an augmented latent feature space. T_S and T_T are transformation functions. P and Q are projection matrices as described in Duan [5]. Diagram adapted from Li [76]

Asymmetric Feature-Based Transfer Learning

The work of Kulis et al. [6], referred to as the Asymmetric Regularized Cross-domain Transformation (ARC-t), proposes an asymmetric transformation algorithm to resolve the heterogeneous feature space between domains. For this scenario, there is an abundance of labeled source data and limited labeled target data. An objective function is first defined for learning the transformation matrix. The objective function contains a regularizer term and a cost function term that is applied to each pair of cross-domain instances and the learned transformation matrix. The construction of the objective function is responsible for the domain invariant transformation process. The optimization of the objective function aims to minimize the regularizer and the cost function terms. The transformation matrix is learned in a non-linear Gaussian RBF kernel space. The method presented is referred to as the Asymmetric Regularized Cross-domain transformation. Two experiments using this approach are performed for image classification where classification accuracy is measured as the performance metric. There are 31 image classes defined for these experiments. The first experiment (test 1) is where instances of all 31 image classes are included in the source and target training data. In the second experiment (test 2), only 16 image classes are represented in the target training data (all 31 are represented in the source). To test against other baseline approaches, a method is needed to bring the source and target input domains together. A preprocessing step called Kernel Canonical Correlation Analysis (proposed by Shawe-Taylor and Cristianini [86]) is used to project the source and target domains into a common domain space using symmetric transformation. Baseline approaches tested include k-nearest neighbors, SVM, metric learning proposed by Davis et al. [87], feature augmentation proposed by Daumé [14], and a cross domain metric learning method proposed by Saenko et al. [88]. For test 1, the Kulis et al. [6] approach performs marginally better than the other methods tested. For test 2, the Kulis et al. [6] approach performs significantly better compared to the k-nearest neighbors approach (note the other methods cannot be tested against as they require all 31 classes to be represented in the target training data). The Kulis et al. [6] approach is best suited for scenarios where all of the classes are not represented in the target training data as demonstrated in test 2.

The problem domain defined by Harel and Mannor [8] is of limited labeled target data and multiple labeled data sources where an asymmetric transformation is desired for each source to resolve the mismatch in feature space. The first step in the process is to normalize the features in the source and target domains, then group the instances by class in the source and target domains. For each class grouping, the features are mean adjusted to zero. Next, each individual source class group is paired with the corresponding target class group, and a singular value decomposition process is performed to find the specific transformation matrix for that class grouping. Once the transformation is performed, the features are mean shifted back reversing the previous step, and the final target classifier is trained using the transformed data. Finding the transformation matrix using the singular value decomposition process allows for the marginal distributions within the class groupings to be aligned while maintaining the structure of the data. This approach is referred to as the Multiple Outlook MAPping algorithm (MOMAP). The experiments use data taken from wearable sensors for the application of activity classification. There are five different activities defined for the experiment which include walking, running, going upstairs, going downstairs, and lingering. The source domain contains similar (but different) sensor readings as compared to the target. The method proposed by Harel and Mannor [8] is compared against a baseline method that trains a classifier with the limited labeled target data and an upper bound method that uses a significantly larger set of labeled target data to train a classifier. An SVM learner is used as the base classifier and a balanced error rate (due to an imbalance in the test data) is measured as the performance metric. The Harel and Mannor [8] approach outperforms the baseline method in every test and falls short of the upper bound method in every test with respect to the balanced error rate.

The heterogeneous transfer learning scenario addressed by Zhou et al. [10] requires an abundance of labeled source data and limited labeled target data. An asymmetric transformation function is proposed to map the source features to the target features. To learn the transformation matrix, a multi-task learning method based on Ando and Zhang [89] is adopted. The solution, referred to as the Sparse Heterogeneous Feature Representation (SHFR), is implemented by creating a binary classifier for each class in the source and the target domains separately. Each binary classifier is assigned a weight term where the weight terms are learned by combining the weighted classifier outputs, while minimizing the classification error of each domain. The weight terms are now used to find the transformation matrix by minimizing the difference between the target weights and the transformed source weights. The final target classifier is trained using the transformed source data and original target data. Experiments are performed for text document classification where the target domain contains documents written in one language and the source domain contains documents written in different languages. A baseline method using a linear SVM classifier trained on the labeled target is established along with testing against the methods proposed by Wang and Mahadevan [4], Kulis et al. [6], and Duan et al. [5]. The method proposed by Zhou et al. [10] performed the best for all tests with respect to classification accuracy. The results of the other approaches are mixed as a function of the data sets used where the Duan et al. [5] method performed either second or third best.

The application of software module defect prediction is usually addressed by training a classifier with labeled data taken from the software project of interest. The environment described in Nam and Kim [9] for software module defect prediction attempts to use labeled source data from one software project to train a classifier to predict unlabeled target data from another project. The source and target software projects collect different metrics making the source and target feature spaces heterogeneous. The proposed solution, referred to as the Heterogeneous Defect Prediction (HDP) approach, is to first select the important features from the source domain using a feature selection method to eliminate redundant and irrelevant features. Feature selection methods used include gain ratio, chi-square, relief-F, and significance attribute evaluation (see Gao et al. [90] and Shivaji et al. [91]). The next step is to statistically match the selected source domain features to ones in the target using a Kolmogorov-Smirnov test that measures the closeness of the empirical distribution between the two sources. A learner is trained with the source features that exhibit a close statistical match to the corresponding target features. The target data is tested with the trained classifier using the corresponding matched features of the target. Even though the approach by Nam and Kim [9] is applied directly to the application of software module defect prediction, this method can be used for other applications. Experiments are performed using five different software defect data sets with heterogeneous features. The proposed method by Nam and Kim [9] uses Logistic Regression as the base learner. The other approaches tested include a within project defect prediction (WPDP) approach where the learner is trained on labeled target data, a cross project defect prediction (CPDP-CM) approach where the source and target represent different software projects but have homogeneous features, and a cross project defect prediction approach with heterogeneous features (CPDP-IFS) as proposed by He et al. [92]. The results of the experiment show the Nam and Kim [9] method significantly outperformed all other approaches with respect to area under the curve measurement. The WPDP approach is next best followed by the CPDP-CM approach and the CPDP-IFS approach. These results can be misleading as the Nam and Kim [9] approach could only match at least one or more input features between the source and target domains in 37 % of the tests. Therefore, in 63 % of the cases, the Nam and Kim [9] method could not be used and these cases are not counted. The WPDP method represents an upper bound and it is an unexpected result that the Nam and Kim [9] approach would outperform the WPDP method.

The paper by Zhou et al. [12] claims that previous heterogeneous solutions assume the instance correspondence between the source and target domains are statistically representative (distributions are equal), which may not always be the case. An example of this claim is in the application of text sentiment classification where the word bias problem previously discussed causes distribution differences between the source and target domains. The paper by Zhou et al. [12] proposes a solution called the Hybrid Heterogeneous Transfer Learning (HHTL) method for a heterogeneous environment with abundant labeled source data and abundant unlabeled target data. The idea is to first learn an asymmetric transformation from the target to the source domain, which reduces the problem to a homogeneous domain adaptation issue. The next step is to discover a common latent feature space using the transformed data (from the previous step) to reduce the distribution bias between the transformed unlabeled target domain and the labeled source domain. Finally, a classifier is trained using the common latent feature space from the labeled source data. This solution is realized using a deep learning method employing a Marginalized Stacked Denoised Autoencoder as proposed by Chen et al. [93] to learn the asymmetric transformation and the mapping to a common latent feature space. The previous surveyed paper by Glorot et al. [33] demonstrated a deep learning approach finding a common latent feature space for homogeneous source and target feature set. The experiments focused on multiple language text sentiment classification where English is used in the source and three other languages are separately used in the target. Classification accuracy is measured as the performance metric. Other methods tested include a heterogeneous spectral mapping approach proposed by Shi et al. [74], a method proposed by Vinokourov et al. [94], and a multimodal deep learning approach proposed by Ngiam et al. [95]. An SVM learner is used as the base classifier for all methods. The results of the experiment from best to worst performance are Zhou et al. [12], Ngiam et al. [95], Vinokourov et al. [94], and Shi et al. [74].

Improvements to Heterogeneous Solutions

The paper by Yang et al. [96] proposes to quantify the amount of knowledge that can be transferred between domains in a heterogeneous transfer learning environment. In other words, it attempts to measure the "relatedness" of the domains. This is accomplished by first building a co-occurrence matrix for each domain. The co-occurrence matrix contains the set of instances represented in every domain. For example, if one particular text document is an instance in the co-occurrence matrix, that text document is required to be represented in every domain. Next, Principal Component Analysis is used to select the most important features in each domain and assign the principal component coefficient to those features. The principal component coefficients are used to form a directed cyclic network (DCN) where each node represents a domain (either source or target) and each node connection (edge weight) is the conditional dependence from one domain to another. The DCN is built using a Markov Chain Monte Carlo method. The edge weights represent the potential amount of knowledge that can be transferred between domains where a higher value means higher knowledge transfer. These edge weights are then used as tuning parameters in different heterogeneous transfer learning solutions, which include works from Yang et al. [79], Ng et al. [97], and Zhu et al. [7] (the weights are calculated first using Yang et al. [96] and then applied as tuning values in the other solutions). Note, that integrating the edge weight values into a particular approach is specific to the implementation of the solution and cannot be generically applied. The experiments are run on the three different learning solutions comparing the original solution against the solution using the weighted edges of the DCN as the tuned parameters. In all three solutions, the classification accuracy is improved using the DCN tuned parameters. One potential issue with this approach is the construction of the co-occurrence matrix. The co-occurrence matrix contains many instances; however, each instance must be represented in each domain. This may be an unrealistic constraint in many real-world applications.

Experiment Results

In reviewing the experiment results of the previous surveyed papers, there are instances where one solution can show varying results over a range of different experiments. There are many reasons why this can happen which include varying test environments, different test implementations, different applications being tested, and different data sets being used. An interesting area of future work is to evaluate the solutions presented to determine the best performing solutions as a function of specific datasets. To facilitate that goal, a repository of open-source software containing the software implementations for solutions used in each paper would be extremely beneficial. Table 3.3 lists a compilation of head-to-head results for the most commonly tested solutions contained in the Heterogeneous Transfer Learning section. The results listed in Table 3.3 represent a win, loss, and tie performance record of the head-to-head solution comparisons. Note, these results are compiled directly from the surveyed papers. It is difficult to draw exact conclusions from this information because of the reasons just outlined; however, it provides some interesting insight into the comparative performances of the solutions.

Discussion of Heterogeneous Solutions

The previous surveyed heterogeneous transfer learning works demonstrate many different characteristics and attributes. Which heterogeneous transfer learning solution is best for a particular application? The heterogeneous transfer learning solutions use either a symmetric transformation or an asymmetric transformation process in an attempt to resolve the differences between the input feature space (as shown in Fig. 3.1). The asymmetrical transformation approach is best used when the same class instances in the source and target domains can be transformed

Table 3.3 Lists the head-to-head results of experiments performed in the heterogeneous transfer learning works surveyed

Methods	HeMap	ARC-t	DAMA	HFA	SHFR	SHFA
HeMap [74]	–	0-5-0	0-5-0	0-5-0	0-0-0	0-3-0
ARC-t [6]	5-0-0	–	4-2-0	1-7-0	0-3-0	0-3-0
DAMA [4]	5-0-0	2-4-0	–	0-8-0	0-3-0	0-3-0
HFA [5]	5-0-0	7-1-0	8-0-0	–	0-3-0	0-3-0
SHFR [10]	0-0-0	3-0-0	3-0-0	3-0-0	–	0-0-0
SHFA [76]	3-0-0	3-0-0	3-0-0	3-0-0	0-0-0	–

The numbers (x-y-z) in the table indicate the far left column method outperforms the top row method x times, underperforms y times, and has similar performance z times

Table 3.4 Heterogeneous transfer learning approaches surveyed in Sect. 4 listing various characteristics of each approach

Approach	Transfer category	Source data	Target data	Multiple sources	Generic solution	Negative transfer
CLSCL [11]	Symmetric feature	Labeled	Unlabeled			
HeMap [74]	Symmetric feature	Labeled	Limited labels		✓	
DAMA [4]	Symmetric feature	Labeled	Limited labels	✓	✓	
HTLIC [7]	Symmetric feature	Unlabeled	Abundant labels			
TTI [75]	Symmetric feature	Labeled	Limited labels			
HFA [5]	Symmetric feature	Labeled	Limited labels		✓	
SHFA [76]	Symmetric feature	Labeled	Limited labels		✓	
ARC-t [6]	Asymmetric feature	Labeled	Limited labels		✓	
MOMAP [8]	Asymmetric feature	Labeled	Limited labels	✓		
SHFR [10]	Asymmetric feature	Labeled	Limited labels		✓	
HDP [9]	Asymmetric feature	Labeled	Unlabeled		✓	
HHTL [12]	Asymmetric feature	Labeled	Unlabeled		✓	

without context feature bias. Many of the surveyed heterogeneous transfer learning solutions only address the issue of the input feature space being different between the source and target domains and do not address other domain adaptation steps needed for marginal and/or conditional distribution differences. If further domain adaptation needs to be performed after the input feature spaces are aligned, then an appropriate homogeneous solution should be used. To further help determine which solution is best for a given transfer learning application, the information in Table 3.4 should be used to match the characteristics of the solution to that of the desired application environment. None of the surveyed heterogeneous transfer learning solutions have a means to guard against negative transfer effects. However, the paper by Yang et al. [96] demonstrates that negative transfer guards can benefit heterogeneous transfer learning solutions. It seems likely that future heterogeneous transfer learning works will integrate means for negative transfer protection. Many of the same heterogeneous transfer learning solutions are tested in the surveyed solution experiments. These head-to-head comparisons are summarized in Table 3.3 and can be used as a starting point to understand the relative performance between the solutions. As observed as a trend in the previous homogeneous solutions, the recent heterogeneous solution by Duan et al. [5] employs a one-stage solution that simultaneously performs the feature input space alignment process while learning the final classifier. As is the case for the surveyed homogeneous transfer learning works, the surveyed heterogeneous transfer learning works are not specifically applied to big data solutions; however, there is nothing to preclude their use in a big data environment.

Negative Transfer

The high-level concept of transfer learning is to improve a target learner by using data from a related source domain. But what happens if the source domain is not well-related to the target? In this case, the target learner can be negatively impacted by this weak relation, which is referred to as negative transfer. In a big data environment, there may be a large dataset where only a portion of the data is related to a target domain of interest. For this case, there is a need to divide the dataset into multiple sources and employ negative transfer methods when using transfer learning algorithm. In the scenario where multiple datasets are available that initially appear to be related to the target domain of interest, it is desired to select the datasets that provide the best information transfer and avoid the datasets that cause negative transfer. This allows for the best use of the available large datasets. How related do the source and target domains need to be for transfer learning to be advantageous? The area of negative transfer has not been widely researched, but the following papers begin to address this issue.

An early paper by Rosenstein et al. [98] discusses the concept of negative transfer in transfer learning and claims that the source domain needs to be sufficiently related to the target domain; otherwise, the attempt to transfer knowledge from the source can have a negative impact on the target learner. Cases of negative transfer are demonstrated by Rosenstein et al. [98] in experiments using a hierarchical Naive Bayes classifier. The author also demonstrates the chance of negative transfer goes down as the number of labeled target training samples goes up.

The paper by Eaton et al. [99] proposes to build a target learner based on a transferability measure from multiple related source domains. The approach first builds a Logistic Regression learner for each source domain. Next, a model transfer graph is constructed to represent the transferability between each source learner. In this case, transferability from a first learner to a second learner is defined as the performance of the second learner with learning from the first learner minus the performance of the second learner without learning from the first learner. Next, the model transfer graph is modified by adding the transferability measures between the target learner and all the source learners. Using spectral graph theory [55] on the model transfer graph, a transfer function is derived that maintains the geometry of the model transfer graph and is used in the final target learner to determine the level of transfer from each source. Experiments are performed in the applications of document classification and alphabet classification. Source domains are identified that are either related or unrelated to the target domain. The method by Eaton et al. [99] is tested along with a handpicked method where the source domains are manually selected to be related to the target, an average method that uses all sources available, and a baseline method that does not use transfer learning. Classification accuracy is the performance metric measured in the experiments. The source and target domains are represented by a homogeneous feature input space. The results of the experiments are mixed. Overall, the Eaton et al. [99] approach performs the best; however, there are certain instances where Eaton et al. [99] performed worse than the handpicked, average, and baseline methods. In the implementation of the algorithm, the transferability measure between two sources is required to be the same; however, the transferability from source 1 to source 2 is not always equal to the transferability from source 2 to source 1. A suggestion for future improvement is to use directed graphs to specify the bidirectional nature of the transferability measure between two sources.

The paper by Ge et al. [100] claims that knowledge transfer can be inhibited due to the existence of unrelated or irrelevant source domains. Further, current transfer learning solutions are focused on transferring knowledge from source domains to a target domain, but are not concerned about different source domains that could potentially be irrelevant and cause negative transfer. In the model presented by Ge et al. [100], there is a single target domain with limited labeled data and multiple labeled source domains for knowledge transfer. To reduce negative transfer effects from unrelated source domains, each source is assigned a weight (called the Supervised Local Weight) corresponding to how related the source is with the target (the higher the weight the more it is related). The Supervised Local Weight is found by first using a spectral clustering algorithm [55] on the unlabeled target information and propagating labels to the clusters from the labeled target information. Next, each source is separately clustered and labels assigned to the clusters from the labeled source. The Supervised Local Weight of each source cluster is computed by comparing the source and target clusters. This solution further addresses the issue of imbalanced class distribution in source domains by preventing a high-weight class assignment in the case of high-accuracy predictions in a minority target class. The final target learner uses the Supervised Local Weights to attenuate the effects of negative transfer. Experiments are performed in three application areas including Cardiac Arrhythmia Detection, Spam Email Filtering, and Intrusion Detection. Area under the curve is measured as the performance metric. The source and target domains are represented by a homogeneous feature input space. The method presented in this paper is compared against methods by Luo et al. [101], by Gao et al. [102], by Chattopadhyay et al. [15], and by Gao et al. [23]. The Luo et al. [101] and Gao et al. [102] methods are the worst performing, most likely due to the fact that these solutions do not attempt to combat negative transfer effects. The Chattopadhyay et al. [15] and Gao et al. [23] methods are the next best performing, which have means in place to reduce the effects of negative transfer from the source domains. The Chattopadhyay et al. [15] and Gao et al. [23] methods do address the negative transfer problem but do not address the imbalanced distribution issue. The Ge et al. [100] method does exhibit the best overall performance due to the handling of negative transfer and imbalanced class distribution.

The paper by Seah et al. [103] claims the root cause of negative transfer is mainly due to conditional distribution differences between source domains $(P_{S1} (y|x) \neq P_{S2} (y|x))$ and a difference in class distribution (class imbalance) between the source and target $(P_S(y) \neq P_T(y))$. Because the target domain usually contains a small number of labeled instances, it is difficult to find the true class distribution of the target domain. A Predictive Distribution Matching (PDM) framework is proposed to align the conditional distributions of the source domains and target domain in an attempt to minimize negative transfer effects. A positive transferability measure is defined that measures the transferability of instance pairs with the same label from the source and target domains. The first step in the PDM framework is to assign pseudo labels to the unlabeled target data. This is accomplished by an iterative process that forces source and target instances which are similar (as defined by the positive transferability measure) to have the same label. Next, irrelevant source data are removed by identifying data that does not align with the conditional distribution of the pseudo labeled target data for each class. Both Logistic Regression and SVM classifiers are implemented using the PDM framework. Experiments are performed on document classification using the PDM method described in this paper, the approach from Daumé [14], the approach from Huang et al. [20], and the approach from Bruzzone and Marconcini [67]. Classification accuracy is measured as the performance metric. The source and target domains are represented by a homogeneous feature input space. The PDM approach demonstrates better performance as compared to the other approaches tested as these solutions do not attempt to account for negative transfer effects.

A select number of previously surveyed papers contain solutions addressing negative transfer. The paper by Yang et al. [96] addresses the negative transfer issue, which is presented in the Heterogeneous Transfer Learning section. The homogeneous solution by Gong et al. [16] defines an ROD value that measures the relatedness between a source and target domain. The work presented in Chattopadhyay et al. [15] is a multiple source transfer learning approach that calculates the source weights as a function of conditional probability differences between the source and target domains attempting to give the most related sources the highest weights. Duan et al. [37] proposes a transfer learning approach that only uses source domains that are deemed relevant and test data demonstrates better performance compared to methods with no negative transfer protection.

The previous papers attempt to measure how related source data is to the target data in a transfer learning environment and then selectively transfer the information that is highly related. The experiments in the above papers demonstrate that accounting for negative transfer effects from source domain data can improve target learner performance. However, most transfer learning solutions do not attempt to account for negative transfer effects. Robust negative transfer measurements are difficult to define. Since the target domain typically has limited labeled data, it is inherently difficult to find a true measure of the relatedness between the source and target domains. Further, by selectively transferring information that seems related to the limited labeled target domain, a risk of overfitting in the target learner is a concern. The topic of negative transfer is a fertile area for further research.

Transfer Learning Applications

The surveyed works in this paper demonstrate that transfer learning has been applied to many real-world applications. There are a number of application examples pertaining to natural language processing, more specifically in the areas of sentiment classification, text classification, spam email detection, and multiple language text classification. Other well-represented transfer learning applications include image classification and video concept classification. Applications that are more selectively addressed in the previous papers include WiFi localization classification, muscle fatigue classification, drug efficacy classification, human activity classification, software defect classification, and cardiac arrhythmia classification.

The majority of the solutions surveyed are generic, meaning the solution can be easily applied to applications other than the ones implemented and tested in the papers. The application-specific solutions tend to be related to the field of natural language processing and image processing. In the literature, there are a number of transfer learning solutions that are specific to the application of recommendation systems. Recommendation systems provide users with recommendations or ratings for a particular domain (e.g. movies, books, etc.), which are based on historical information. However, when the system does not have sufficient historical information (referred to as the data sparsity issue presented in [104], then the recommendations are not reliable. In the cases where the system does not have sufficient domain data to make reliable predictions (for example when a movie is just released), there is a need to use previously collected information from a different domain (using books for example). The aforementioned problem has been directly addressed using transfer learning methodologies and captured in papers by Moreno et al. [104], Cao et al. [105], Li et al. [106, 107], Pan et al. [108, 110], Zhang et al. [109], Roy et al. [111], Jiang et al. [112], and Zhao et al. [113].

Transfer learning solutions continue to be applied to a diverse number of real-world applications, and in some cases the applications are quite obscure. The application of head pose classification finds a learner trained with previously captured labeled head positions to predict a new head position. Head pose classification is used for determining the attentiveness of drivers, analyzing social behavior, and human interaction with robots. Head positions captured in source training data will have different head tilt ranges and angles than that of the predicted target. The paper by Rajagopal et al. [114] addresses the head pose classification issues using transfer learning solutions.

Other transfer learning applications include the paper by Ma et al. [115] that uses transfer learning for atmospheric dust aerosol particle classification to enhance global climate models. Here the TrAdaBoost algorithm proposed by Dai et al. [69] is used in conjunction with an SVM classifier to improve on classification results. Being able to identify areas of low income in developing countries is important for disaster relief efforts, food security, and achieving sustainable growth. To better predict poverty mapping, Xie et al. [116] proposes an approach similar to Oquab et al. [35] that uses a convolution neural network model. The first prediction model is trained to predict night time light intensity from source image data. The final target prediction model predicts the poverty mapping from source night time light intensity data. In the paper by Ogoe et al. [117], transfer learning in used to enhance disease prediction. In this solution, a rule-based learning approach is formulated to use abstract source domain data to perform modeling of multiple types of gene expression data. Online display web advertising is a growing industry where transfer learning is used to optimally predict targeted ads. In the paper by Perlich et al. [118], a transfer learning approach is employed that uses the weighted outputs of multiple source classifiers to enhance a target classifier trained to predict targeted online display advertising results. The paper by Kan et al. [119] addresses the field of facial recognition and is able to use face image information from one ethnic group to improve the learning of a classifier for a different ethnic group. The paper by Farhadi et al. [120] is focused on the application of sign language recognition where the model is able to learn from different people signing at various angles. Transfer learning is applied to the field of biology in the paper by Widmer and Ratsch [121]. Specifically, a multi-task learning approach is used in the prediction of splice sites in genome biology. Predicting if patients will contract a particular bacteria when admitted to a hospital is addressed in the paper by Wiens et al. [122]. Information taken from different hospitals is used to predict the infection rate for a different hospital. In the paper by Romera-Paredes et al. [123], a multi-task transfer learning approach is used to predict pain levels from an individual's facial expression by using labeled source facial images from other individuals. The paper by Deng et al. [124] applies transfer learning to the application of speech emotion recognition where information is transferred from multiple labeled speech sources. The application of wine quality classification is implemented in Zhang and Yeung [125] using a multi-task transfer learning approach. As a reference, the survey paper by Cook et al. [18] covers transfer learning for the application of activity recognition and the survey papers by Patel et al. [126] and Shao et al. [127] address transfer learning in the domain of image recognition.

Conclusion and Discussion

The subject of transfer learning is a well-researched area as evidenced with more than 700 academic papers addressing the topic in the last five years. This survey paper presents solutions from the literature representing current trends in transfer learning. Homogeneous transfer learning papers are surveyed that demonstrate instance-based, feature-based, parameter-based, and relational-based information transfer techniques. Solutions having various requirements for labeled and unlabeled data are also presented as a key attribute. The relatively new area of heterogeneous transfer learning is surveyed showing the two dominant approaches for domain adaptation being asymmetric and symmetric transformations. Many real-world applications that transfer learning is applied to are listed and discussed in this survey paper. In some cases, the proposed transfer learning solutions are very specific to the underlying application and cannot be generically used for other applications. A list of software downloads implementing a portion of the solutions surveyed is presented in the appendix of this paper. A great benefit to researchers is to have software available from previous solutions so experiments can be performed more efficiently and more reliably. A single open-source software repository for published transfer learning solutions would be a great asset to the research community.

In many transfer learning solutions, the domain adaptation process performed is focused either on correcting the marginal distribution differences or the conditional distribution differences between the source and target domains. Correcting the conditional distribution differences is a challenging problem due to the lack of labeled target data. To address the lack of labeled target data, some solutions estimate the labels for the target data (called pseudo labels), which are then used to correct the conditional distribution differences. This method is problematic because the conditional distribution corrections are being made with the aid of pseudo labels. Improved methods for correcting the conditional distribution differences is a potential area of future research. A number of more recent works attempt to correct both the marginal distribution differences and the conditional distribution differences during the domain adaptation process. An area of future work is to quantify the advantage of correcting both distributions and in what scenarios it is most effective. Further, Long et al. [30] states that the simultaneous solving of marginal and conditional distribution differences is preferred over serial alignment as it reduces the risk of overfitting. Another area of future work is to quantify any performance gains for simultaneously solving both distribution differences. In addition to solving for distribution differences in the domain adaptation process, exploring possible data preprocessing steps using heuristic knowledge of the domain features can be used as a method to improve the target learner performance. The heuristic knowledge would represent a set of complex rules or relations that standard transfer learning techniques cannot account for. In most cases, this heuristic knowledge would be specific to each domain, which would not lead to a generic solution. However, if such a preprocessing step leads to improved target learner performance, it is likely worth the effort.

A trend observed in the formulation of transfer learning solutions is in the implementation of a one-stage process as opposed to a two-stage process. A two-stage solution first performs the domain adaptation process and then independently learns the final classifier. A one-stage process simultaneously performs the domain adaptation process while learning the final classifier. Recent solutions employing a one-stage solution include Long et al. [30], Duan et al. [28], Shi and Sha [34], Xia et al. [39], and Duan et al. [5]. With respect to the one-stage solution, Long et al. [30] claims the simultaneous solving of domain adaptation and the classifier establishes mutual reinforcement for enhanced performance. An area of future work is to better quantify the effects of a one-stage approach over a two-stage approach.

This paper surveys a number of works addressing the topic of negative transfer. The subject of negative transfer is still a lightly researched area. The expanded integration of negative transfer techniques into transfer learning solutions is a natural extension for future research. Solutions supporting multiple source domains enabling the splitting of larger source domains into smaller domains to more easily discriminate against unrelated source data are a logical area for continued research. Additionally, optimal transfer is another fertile area for future research. Negative transfer is defined as a source domain having a negative impact on a target learner. The concept of optimal transfer is when select information from a source domain is transferred to achieve the highest possible performance in a target learner. There is overlap between the concepts of negative transfer and optimal transfer; however, optimal transfer attempts to find the best performing target learner, which goes well beyond the negative transfer concept.

With the recent proliferation of sensors being deployed in cell phones, vehicles, buildings, roadways, and computers, larger and more diverse information is being collected. The diversity in data collection makes heterogeneous transfer learning solutions more important moving forward. Larger data collection sizes highlight the potential for big data solutions being deployed concurrent with current transfer learning solutions. How the diversity and large size of sensor data integrates into transfer learning solutions is an interesting topic of future research. Another area of future work pertains to the scenario where the output label space is different between domains. With new data sets being captured and being made available, this topic could be a needed area of focus for the future. Lastly, the literature has very few transfer learning solutions addressing the scenario of unlabeled source and unlabeled target data, which is certainly an area for expanded research.

Appendix

The majority of transfer learning solutions surveyed are complex and implemented with non-trivial software. It is a great advantage for a researcher to have access to software implementations of transfer learning solutions so comparisons with competing solutions are facilitated more quickly and fairly. Table 3.5 provides a list of available software downloads for a number of the solutions surveyed in this paper. Table 3.6 provides a resource for useful links that point to transfer learning tutorials and other interesting articles on the topic of transfer learning.

Table 3.5 Software downloads for various transfer learning solutions

Approach	Location
Prettenhofer and Stein [11]	https://github.com/pprett/bolt [128]
Zhu et al. [7]	http://www.cse.ust.hk/~yinz/ [129]
Dai et al. [69]	https://github.com/BoChen90/machine-learning-matlab/blob/master/TrAdaBoost.m [130]
Daumé [14]	http://hal3.name/easyadapt.pl.gz [131]
Duan et al. [5]	https://sites.google.com/site/xyzliwen/publications/HFA_release_0315.rar [132]
Kulis et al. [6]	http://vision.cs.uml.edu/adaptation.html [133]
Qi et al. [75]	http://www.eecs.ucf.edu/~gqi/publications.html [134]
Li et al. [76]	http://www.lxduan.info/#sourcecode_hfa [135]
Gong [16]	http://www-scf.usc.edu/~boqinggo/ [136]
Long et al. [30]	http://ise.thss.tsinghua.edu.cn/~mlong/ [137]
Oquab et al. [35]	http://leon.bottou.org/papers/oquab-2014 [138]
Long et al. [29]	http://ise.thss.tsinghua.edu.cn/~mlong/ [137]
Other transfer learning code	http://www.cse.ust.hk/TL/ [139]

Table 3.6 Useful links for transfer learning information

Item	Location
Slides for Nam and Kim [9]	http://www.slideshare.net/hunkim/heterogeneous-defect-prediction-esecfse-2015 [140]
Code for SVMLIB	http://www.csie.ntu.edu.tw/~cjlin/libsvm [141]
Slide for Kulis et al. [6]	https://www.eecs.berkeley.edu/~jhoffman/domainadapt/ [142]
Tutorial on transfer learning	http://tommasit.wix.com/datl14tutorial [143]
Tutorial on transfer learning	http://sifaka.cs.uiuc.edu/jiang4/domain_adaptation/survey/da_survey.html [144]
Overview of Duan et al. [37]	http://lxduan.info/papers/DuanCVPR2012_poster.pdf [145]

References

1. Witten IH, Frank E. Data mining, practical machine learning tools and techniques. 3rd ed. San Francisco, CA: Morgan Kaufmann Publishers; 2011.
2. Shimodaira H. Improving predictive inference under covariate shift by weighting the log-likelihood function. J Stat Plan Inference. 2000;90(2):227–44.
3. Pan SJ, Yang Q. A survey on transfer learning. IEEE Trans Knowl Data Eng. 2010;22 (10):1345–59.
4. Wang C, Mahadevan S. Heterogeneous domain adaptation using manifold alignment. In: Proceedings of the twenty-second international joint conference on artificial intelligence, vol. 2; 2011. p. 1541–6.
5. Duan L, Xu D, Tsang IW. Learning with augmented features for heterogeneous domain adaptation. IEEE Trans Pattern Anal Mach Intell. 2012;36(6):1134–48.
6. Kulis B, Saenko K, Darrell T. What you saw is not what you get: domain adaptation using asymmetric kernel transforms. In: IEEE 2011 conference on computer vision and pattern recognition; 2011. p. 1785–92.
7. Zhu Y, Chen Y, Lu Z, Pan S, Xue G, Yu Y, Yang Q. Heterogeneous transfer learning for image classification. Proc Nat Conf Artif Intell. 2011;2:1304–9.
8. Harel M, Mannor S. Learning from multiple outlooks. In: Proceedings of the 28th international conference on machine learning; 2011. p. 401–408.
9. Nam J, Kim S. Heterogeneous defect prediction. In: Proceedings of the 2015 10th joint meeting on foundations of software engineering; 2015. p. 508–19.
10. Zhou JT, Tsang IW, Pan SJ Tan M. Heterogeneous domain adaptation for multiple classes. In: International conference on artificial intelligence and statistics; 2014. p. 1095–103.
11. Prettenhofer P, Stein B. Cross-language text classification using structural correspondence learning. In: Proceedings of the 48th annual meeting of the association for computational linguistics; 2010. p. 1118–27.
12. Zhou JT, Pan S, Tsang IW, Yan Y. Hybrid heterogeneous transfer learning through deep learning. Proc Nat Conf Artif Intell. 2014;3:2213–20.
13. Taylor ME, Stone P. Transfer learning for reinforcement learning domains: a survey. JMLR. 2009;10:1633–85.
14. Daumé H III. Frustratingly easy domain adaptation. In: Proceedings of 2007 ACL; 2007. p. 256–63.
15. Chattopadhyay R, Ye J, Panchanathan S, Fan W, Davidson I. Multi-source domain adaptation and its application to early detection of fatigue. ACM Trans Knowl Discov Data. 2011;6(4):18.
16. Gong B, Shi Y, Sha F, Grauman K. Geodesic flow kernel for unsupervised domain adaptation. In: Proceedings of the 2012 IEEE conference on computer vision and pattern recognition; 2012. p. 2066–73.
17. Blitzer J, McDonald R, Pereira F. Domain adaptation with structural correspondence learning. In: Proceedings of the 2006 conference on empirical methods in natural language processing; 2006. p. 120–28.
18. Cook DJ, Feuz KD, Krishnan NC. Transfer learning for activity recognition: a survey. Knowl Inf Syst. 2012;36(3):537–56.
19. Feuz KD, Cook DJ. Transfer learning across feature-rich heterogeneous feature spaces via feature-space remapping (FSR). J ACM Trans Intell Syst Technol. 2014;6(1):1–27.
20. Huang J, Smola A, Gretton A, Borgwardt KM, Schölkopf B. Correcting sample selection bias by unlabeled data. In: Proceedings of the 2006 conference in advances in neural information processing systems; 2006. p. 601–8.
21. Jiang J, Zhai C. Instance weighting for domain adaptation in NLP. In: Proceedings of the 45th annual meeting of the association of computational linguistics; 2007. p. 264–271.
22. Pan SJ, Kwok JT, Yang Q. Transfer learning via dimensionality reduction. In: Proceedings of the 23rd national conference on artificial intelligence, vol. 2; 2008. p. 677–82.

23. Gao J, Fan W, Jiang J, Han J. Knowledge transfer via multiple model local structure mapping. In: Proceedings of the 14th ACM SIGKDD international conference on knowledge discovery and data mining; 2008. p. 283–91.
24. Bonilla E, Chai KM, Williams C. Multi-task Gaussian process prediction. In: Proceedings of the 20th annual conference of neural information processing systems; 2008. p. 153–60.
25. Evgeniou T, Pontil M. Regularized multi-task learning. In: Proceedings of the 10th ACM SIGKDD international conference on knowledge discovery and data mining; 2004. p. 109–17.
26. Mihalkova L, Mooney RJ. Transfer learning by mapping with minimal target data. In: Proceedings of the association for the advancement of artificial intelligence workshop transfer learning for complex tasks; 2008. p. 31–6.
27. Li F, Pan SJ, Jin O, Yang Q, Zhu X. Cross-domain co-extraction of sentiment and topic lexicons. In: Proceedings of the 50th annual meeting of the association for computational linguistics long papers, vol. 1; 2012. p. 410–9.
28. Duan L, Tsang IW, Xu D. Domain transfer multiple kernel learning. IEEE Trans Pattern Anal Mach Intell. 2012;34(3):465–79.
29. Long M, Wang J, Ding G, Sun J, Yu PS. Transfer feature learning with joint distribution adaptation. In: Proceedings of the 2013 IEEE international conference on computer vision; 2013. p. 2200–7.
30. Long M, Wang J, Ding G, Pan SJ, Yu PS. Adaptation regularization: a general framework for transfer learning. IEEE Trans Knowl Data Eng. 2014;26(5):1076–89.
31. Pan SJ, Tsang IW, Kwok JT, Yang Q. Domain adaptation via transfer component analysis. IEEE Trans Neural Netw. 2009;22(2):199–210.
32. Pan SJ, Ni X, Sun JT, Yang Q, Chen Z. Cross-domain sentiment classification via spectral feature alignment. In: Proceedings of the 19th international conference on world wide web; 2010. p. 751–60.
33. Glorot X, Bordes A, Bengio Y. Domain adaptation for large-scale sentiment classification: a deep learning approach. In: Proceedings of the twenty-eight international conference on machine learning, vol. 27; 2011. p. 97–110.
34. Shi Y, Sha F. Information-theoretical learning of discriminative clusters for unsupervised domain adaptation. In: Proceedings of the 29th international conference on machine learning; 2012. p. 1–8.
35. Oquab M, Bottou L, Laptev I, Sivic J. Learning and transferring mid-level image representations using convolutional neural networks. In: Proceedings of the 2014 IEEE conference on computer vision and pattern recognition; 2013. p. 1717–24.
36. Tommasi T, Orabona F, Caputo B. Safety in numbers: learning categories from few examples with multi model knowledge transfer. In: 2010 IEEE conference on computer vision and pattern recognition; 2010. p. 3081–8.
37. Duan L, Xu D, Chang SF. Exploiting web images for event recognition in consumer videos: a multiple source domain adaptation approach. In: IEEE 2012 conference on computer vision and pattern recognition; 2012. p. 1338–45.
38. Yao Y, Doretto G. Boosting for transfer learning with multiple sources. In: Proceedings of the IEEE computer society conference on computer vision and pattern recognition; 2010. p. 1855–62.
39. Xia R, Zong C, Hu X, Cambria E. Feature ensemble plus sample selection: Domain adaptation for sentiment classification. IEEE Intell Syst. 2013;28(3):10–8.
40. Duan L, Xu D, Tsang IW. Domain adaptation from multiple sources: a domain-dependent regularization approach. IEEE Trans Neural Netw Learn Syst. 2012;23(3):504–18.
41. Zhong E, Fan W, Peng J, Zhang K, Ren J, Turaga D, Verscheure O. Cross domain distribution adaptation via kernel mapping. In: Proceedings of the 15th ACM SIGKDD; 2009. p 1027–36.
42. Chelba C, Acero A. Adaptation of maximum entropy classifier: little data can help a lot. Comput Speech Lang. 2004;20(4):382–99.

43. Wu X, Xu D, Duan L, Luo J. Action recognition using context and appearance distribution features. In: IEEE 2011 conference on computer vision and pattern recognition; 2011. p. 489–96.
44. Vedaldi A, Gulshan V, Varma M, Zisserman A. Multiple kernels for object detection. In: 2009 IEEE 12th international conference on computer vision; 2009. p. 606–13.
45. Vapnik V. Principles of risk minimization for learning theory. Adv Neural Inf Process Syst. 1992;4:831–8.
46. Borgwardt KM, Gretton A, Rasch MJ, Kriegel HP, Schölkopf B, Smola AJ. Integrating structured biological data by kernel maximum mean discrepancy. Bioinformatics. 2006;22 (4):49–57.
47. Yang J, Yan R, Hauptmann AG. Cross-domain video concept detection using adaptive SVMs. In: Proceedings of the 15th ACM international conference on multimedia; 2007. p. 188–97.
48. Jiang W, Zavesky E, Chang SF, Loui A. Cross-domain learning methods for high-level visual concept classification. In: IEEE 2008 15th international conference on image processing; 2008. p. 161–4.
49. Si S, Tao D, Geng B. Bregman divergence-based regularization for transfer subspace learning. IEEE Trans Knowl Data Eng. 2010;22(7):929–42.
50. Belkin M, Niyogi P, Sindhwani V. Manifold regularization: a geometric framework for learning from examples. J Mach Learn Res Arch. 2006;7:2399–434.
51. Ling X, Dai W, Xue GR, Yang Q, Yu Y. Spectral domain-transfer learning. In: Proceedings of the 14th ACM SIGKDD international conference on Knowledge discovery and data mining; 2008. p. 488–96.
52. Quanz B, Huan J. Large margin transductive transfer learning. In: Proceedings of the 18th ACM conference on information and knowledge management; 2009. p. 1327–36.
53. Xiao M, Guo Y. Semi-supervised kernel matching for domain adaptation. In: Proceedings of the twenty-sixth AAAI conference on artificial intelligence; 2012. p. 1183–89.
54. Steinwart I. On the influence of the kernel on the consistency of support vector machines. JMLR. 2001;2:67–93.
55. Chung FRK. Spectral graph theory. In: Number 92 in CBMS regional conference series in mathematics. American Mathematical Society, Published by AMS; 1994.
56. Vincent P, Larochelle H, Bengio Y, Manzagol PA. Extracting and composing robust features with denoising autoencoders. In: Proceedings of the 25th international conference on machine learning; 2008. p. 1096–103.
57. Li S, Zong C. Multi-domain adaptation for sentiment classification: using multiple classifier combining methods. In: Proceedings of the conference on natural language processing and knowledge engineering; 2008. p. 1–8.
58. Gopalan R, Li R, Chellappa R. Domain adaptation for object recognition: an unsupervised approach. Int Conf Comput Vis. 2011;2011:999–1006.
59. Weinberger KQ, Saul LK. Distance metric learning for large margin nearest neighbor classification. JMLR. 2009;10:207–44.
60. LeCun Y, Bottou L, HuangFu J. Learning methods for generic object recognition with invariance to pose and lighting. In: Proceedings of the 2004 IEEE computer society conference on computer vision and pattern recognition; 2004, vol. 2, p. 97–104.
61. Marszalek M, Schmid C, Harzallah H, Van de Weijer J. Learning object representations for visual object class recognition. In: Visual recognition challenge workshop ICCV; 2007. p. 1–10.
62. Song Z, Chen Q, Huang Z, Hua Y, Yan S. Contextualizing object detection and classification. IEEE Trans Pattern Anal Mach Intell. 2011;37(1):13–27.
63. Cawley G. Leave-one-out cross-validation based model selection criteria for weighted LS-SVMs. In: IEEE 2006 international joint conference on neural network proceedings; 2006. p. 1661–68.
64. Tommasi T, Caputo B. The more you know, the less you learn: from knowledge transfer to one-shot learning of object categories. In: BMVC; 2009. p. 1–11.

65. Lowe DG. Distinctive image features from scale-invariant keypoints. Int J Comput Vision. 2004;60(2):91–110.
66. Wang H, Klaser A, Schmid C, Liu CL. Action recognition by dense trajectories. In: IEEE 2011 conference on computer vision and pattern recognition; 2011. p. 3169–76.
67. Bruzzone L, Marconcini M. Domain adaptation problems: a DASVM classification technique and a circular validation strategy. IEEE Trans Pattern Anal Mach Intell. 2010;32(5):770–87.
68. Schweikert G, Widmer C, Schölkopf B, Rätsch G. An empirical analysis of domain adaptation algorithms for genomic sequence analysis. Adv Neural Inf Process Syst. 2009;21:1433–40.
69. Dai W, Yang Q, Xue GR, Yu Y. Boosting for transfer learning. In: Proceedings of the 24th international conference on Machine learning; 2007. p. 193–200.
70. Hu M, Liu B. Mining and summarizing customer reviews. In: Proceedings of the 10th ACM SIGKDD international conference on knowledge discovery and data mining; 2004. p. 168–77.
71. Qiu G, Liu B, Bu J, Chen C. Expanding domain sentiment lexicon through double propagation. In: Proceedings of the 21st international joint conference on artificial intelligence; 2009. p. 1199–204.
72. Jakob N, Gurevych I. Extracting opinion targets in a single and cross-domain setting with conditional random fields. In: Proceedings of the 2010 conference on empirical methods in NLP; 2010. p. 1035–45.
73. Xia R, Zong C. A POS-based ensemble model for cross-domain sentiment classification. In: Proceedings of the 5th international joint conference on natural language processing; 2011. p. 614–622.
74. Shi X, Liu Q, Fan W, Yu PS, Zhu R. Transfer learning on heterogeneous feature spaces via spectral transformation. IEEE Int Conf Data Mining. 2010;2010:1049–54.
75. Qi GJ, Aggarwal C, Huang T. Towards semantic knowledge propagation from text corpus to web images. In: Proceedings of the 20th international conference on world wide web; 2011. p. 297–306.
76. Li W, Duan L, Xu D, Tsang IW. Learning with augmented features for supervised and semi-supervised heterogeneous domain adaptation. IEEE Trans Pattern Anal Mach Intell. 2014;36(6):1134–48.
77. Wei B, Pal C. Heterogeneous transfer learning with RBMs. In: Proceedings of the twenty-fifth AAAI conference on artificial intelligence; 2011. p. 531–36.
78. Ham JH, Lee DD, Saul LK. Learning high dimensional correspondences from low dimensional manifolds. In: Proceedings of the twentieth international conference on machine learning; 2003. p. 1–8.
79. Yang Q, Chen Y, Xue GR, Dai W, Yu Y. Heterogeneous transfer learning for image clustering via the social web. In: Proceedings of the joint conference of the 47th annual meeting of the ACL; 2009; vol. 1. p. 1–9.
80. Deerwester S, Dumais ST, Furnas GW, Landauer TK, Harshman R. Indexing by latent semantic analysis. J Am Soc Inf Sci. 1990;41:391–407.
81. Raina R, Battle A, Lee H, Packer B, Ng AY. Self-taught learning: transfer learning from unlabeled data. In: Proceedings of the 24th international conference on machine learning; 2007. p. 759–66.
82. Wang G, Hoiem D, Forsyth DA. Building text Features for object image classification. IEEE Conf Comput Vis Pattern Recognit. 2009;2009:1367–74.
83. Dai W, Chen Y, Xue GR, Yang Q, Yu Y. Translated learning: transfer learning across different feature spaces. Adv Neural Inf Process Syst. 2008;21:353–60.
84. Bay H, Tuytelaars T, Gool LV. Surf: speeded up robust features. Comput Vis Image Underst. 2006;110(3):346–59.
85. Kloft M, Brefeld U, Sonnenburg S, Zien A. Lp-norm multiple kernel learning. J Mach Learn Res. 2011;12:953–97.

86. Shawe-Taylor J, Cristianini N. Kernel methods for pattern analysis. Cambridge: Cambridge University Press; 2004.
87. Davis J, Kulis B, Jain P, Sra S, Dhillon I. Information theoretic metric learning. In: Proceedings of the 24th international conference on machine learning; 2007. p. 209–16.
88. Saenko K, Kulis B, Fritz M, Darrell T. Adapting visual category models to new domains. Comput Vis ECCV. 2010;6314:213–26.
89. Ando RK, Zhang T. A framework for learning predictive structures from multiple tasks and unlabeled data. J Mach Learn Res. 2005;6:1817–53.
90. Gao K, Khoshgoftaar TM, Wang H, Seliya N. Choosing software metrics for defect prediction: an investigation on feature selection techniques. J Softw Pract Exp. 2011;41 (5):579–606.
91. Shivaji S, Whitehead EJ, Akella R, Kim S. Reducing features to improve code change-based bug prediction. IEEE Trans Software Eng. 2013;39(4):552–69.
92. He P, Li B, Ma Y. Towards cross-project defect prediction with imbalanced feature sets; 2014. arXiv preprint arXiv:1411.4228.
93. Chen M, Xu ZE, Weinberger KQ, Sha F. Marginalized denoising autoencoders for domain adaptation. ICML; 2012. arXiv preprint arXiv:1206.4683.
94. Vinokourov A, Shawe-Taylor J, Cristianini N. Inferring a semantic representation of text via crosslanguage correlation analysis. Adv Neural Inf Process Syst. 2002;15:1473–80.
95. Ngiam J, Khosla A, Kim M, Nam J, Lee H, Ng AY. Multimodal deep learning. In: The 28th International conference on machine learning; 2011. p. 689–96.
96. Yang L, Jing L, Yu J, Ng MK. Learning transferred weights from co-occurrence data for heterogeneous transfer learning. In: IEEE transaction on neural networks and learning systems; 2015. p. 1–14.
97. Ng MK, Wu Q, Ye Y. Co-transfer learning via joint transition probability graph based method. In: Proceedings of the 1st international workshop on cross domain knowledge discovery in web and social network mining; 2012. p. 1–9.
98. Rosenstein MT, Marx Z, Kaelbling LP, Dietterich TG. To transfer or not to transfer. In: Proceedings NIPS'05 workshop, inductive transfer, 10 years later; 2005. p. 1–4.
99. Eaton E, desJardins M, Lane R. Modeling transfer relationships between learning tasks for improved inductive transfer. Proc Mach Learn Knowl Discov Databases. 2008;5211:317–32.
100. Ge L, Gao J, Ngo H, Li K, Zhang A. On handling negative transfer and imbalanced distributions in multiple source transfer learning. In: Proceedings of the 2013 SIAM international conference on data mining; 2013. p. 254–71.
101. Luo P, Zhuang F, Xiong H, Xiong Y, He Q. Transfer learning from multiple source domains via consensus regularization. In: Proceedings of the 17th ACM conference on information and knowledge management; 2008. p. 103–12.
102. Gao J, Liang F, Fan W, Sun Y, Han J. Graph based consensus maximization among multiple supervised and unsupervised models. Adv Neural Inf Process Syst. 2009;22:1–9.
103. Seah CW, Ong YS, Tsang IW. Combating negative transfer from predictive distribution differences. IEEE Trans Cybern. 2013;43(4):1153–65.
104. Moreno O, Shapira B, Rokach L, Shani G. TALMUD—transfer learning for multiple domains. In: Proceedings of the 21st ACM international conference on Information and knowledge management; 2012. p. 425–34.
105. Cao B, Liu N, Yang Q. Transfer learning for collective link prediction in multiple heterogeneous domains. In: Proceedings of the 27th international conference on machine learning; 2010. p. 159–66.
106. Li B, Yang Q, Xue X. Can movies and books collaborate? Cross-domain collaborative filtering for sparsity reduction. In: Proceedings of the 21st international joint conference on artificial intelligence; 2009. p. 2052–57.
107. Li B, Yang Q, Xue X. Transfer learning for collaborative filtering via a rating-matrix generative model. In: Proceedings of the 26th annual international conference on machine learning; 2009. p. 617–24.

108. Pan W. Xiang EW, Liu NN, Yang Q. Transfer learning in collaborative filtering for sparsity reduction. In: Twenty-fourth AAAI conference on artificial intelligence, vol. 1; 2010. p. 230–5.
109. Zhang Y, Cao B, Yeung D. Multi-domain collaborative filtering. In: Proceedings of the 26th conference on uncertainty in artificial intelligence; 2010. p. 725–32.
110. Pan W, Liu NN, Xiang EW, Yang Q. Transfer learning to predict missing ratings via heterogeneous user feedbacks. In: Proceedings of the 22nd international joint conference on artificial intelligence; 2011. p. 2318–23.
111. Roy SD, Mei T, Zeng W, Li S. Social transfer: cross-domain transfer learning from social streams for media applications. In: Proceedings of the 20th ACM international conference on multimedia; 2012. p. 649–58.
112. Jiang M, Cui P, Wang F, Yang Q, Zhu W, Yang S. Social recommendation across multiple relational domains. In: Proceedings of the 21st ACM international conference on information and knowledge management; 2012. p. 1422–31.
113. Zhao L, Pan SJ, Xiang EW, Zhong E, Lu Z, Yang Q. Active transfer learning for cross-system recommendation. In: Proceedings of the 27th AAAI conference on artificial intelligence; 2013. p. 1205–11.
114. Rajagopal AN, Subramanian R, Ricci E, Vieriu RL, Lanz O, Ramakrishnan KR, Sebe N. Exploring transfer learning approaches for head pose classification from multi-view surveillance images. Int J Comput Vision. 2014;109(1–2):146–67.
115. Ma Y, Gong W, Mao F. Transfer learning used to analyze the dynamic evolution of the dust aerosol. J Quant Spectrosc Radiat Transfer. 2015;153:119–30.
116. Xie M, Jean N, Burke M, Lobell D, Ermon S. Transfer learning from deep features for remote sensing and poverty mapping. In: Proceedings 30th AAAI conference on artificial intelligence; 2015. p. 1–10.
117. Ogoe HA, Visweswaran S, Lu X, Gopalakrishnan V. Knowledge transfer via classification rules using functional mapping for integrative modeling of gene expression data. BMC Bioinformatics. 2015;16:1–15.
118. Perlich C, Dalessandro B, Raeder T, Stitelman O, Provost F. Machine learning for targeted display advertising: transfer learning in action. Mach Learn. 2014;95:103–27.
119. Kan M, Wu J, Shan S, Chen X. Domain adaptation for face recognition: targetize source domain bridged by common subspace. Int J Comput Vis. 2014;109(1–2):94–109.
120. Farhadi A, Forsyth D, White R. Transfer learning in sign language. In: IEEE 2007 conference on computer vision and pattern recognition; 2007. p. 1–8.
121. Widmer C, Ratsch G. Multitask learning in computational biology. JMLR. 2012;27:207–16.
122. Wiens J, Guttag J, Horvitz EJ. A study in transfer learning: leveraging data from multiple hospitals to enhance hospital-specific predictions. J Am Med Inform Assoc. 2013;21(4):699–706.
123. Romera-Paredes B, Aung MSH, Pontil M, Bianchi-Berthouze N, Williams AC de C, Watson P. Transfer learning to account for idiosyncrasy in face and body expressions. In: Proceedings of the 10th international conference on automatic face and gesture recognition (FG); 2013. p. 1–6.
124. Deng J, Zhang Z, Marchi E, Schuller B. Sparse autoencoder based feature transfer learning for speech emotion recognition. In: Humaine association conference on affective computing and intelligent interaction; 2013. p. 511–6.
125. Zhang Y, Yeung DY. Transfer metric learning by learning task relationships. In: Proceedings of the 16th ACM SIGKDD international conference on knowledge discovery and data mining; 2010. p. 1199–208.
126. Patel VM, Gopalan R, Li R, Chellappa R. Visual domain adaptation: a survey of recent advances. IEEE Signal Process Mag. 2014;32(3):53–69.
127. Shao L, Zhu F, Li X. Transfer learning for visual categorization: a survey. IEEE Trans Neural Netw Learn Syst. 2014;26(5):1019–34.
128. Bolt Online Learning Toolbox. http://pprett.github.com/bolt/. Accessed 4 Mar 2016.
129. Zhu Y. http://www.cse.ust.hk/~yinz/. Accessed 4 Mar 2016.

130. BoChen90 Update TrAdaBoost.m. https://github.com/BoChen90/machine-learning-matlab/blob/master/TrAdaBoost.m. Accessed 4 Mar 2016.
131. EasyAdapt.pl.gz (Download). http://hal3.name/easyadapt.pl.gz Accessed 4 Mar 2016.
132. HFA_release_0315.rar (Download). https://sites.google.com/site/xyzliwen/publications/HFA_release_0315.rar. Accessed 4 Mar 2016.
133. Computer Vision and Learning Group. http://vision.cs.uml.edu/adaptation.html. Accessed 4 Mar 2016.
134. Guo-Jun Qi's Publication List. http://www.eecs.ucf.edu/~gqi/publications.html. Accessed 4 Mar 2016.
135. Duan L. http://www.lxduan.info/#sourcecode_hfa. Accessed 4 Mar 2016.
136. Gong B. http://www-scf.usc.edu/~boqinggo/. Accessed 4 Mar 2016.
137. Long MS—Tsinghua University. http://ise.thss.tsinghua.edu.cn/~mlong/. Accessed 4 Mar 2016.
138. Papers:oquab-2014. http://leon.bottou.org/papers/oquab-2014. Accessed 4 Mar 2016.
139. Transfer Learning Resources. http://www.cse.ust.hk/TL/. Accessed 4 Mar 2016.
140. Heterogeneous Defect Prediction. http://www.slideshare.net/hunkim/heterogeneous-defect-prediction-esecfse-2015. Accessed 4 Mar 2016.
141. LIBSVM—A library for support vector machines. http://www.csie.ntu.edu.tw/~cjlin/libsvm. Accessed 4 Mar 2016.
142. Domain Adaptation Project. https://www.eecs.berkeley.edu/~jhoffman/domainadapt/. Accessed 4 Mar 2016.
143. Tutorial on domain adaptation and transfer learning. http://tommasit.wix.com/datl14tutorial. Accessed 4 Mar 2016.
144. A literature survey on domain adaptation of statistical classifiers. http://sifaka.cs.uiuc.edu/jiang4/domain_adaptation/survey/da_survey.html. Accessed 4 Mar 2016.
145. Exploiting web images for event recognition in consumer videos: a multiple source domain adaptation approach. http://lxduan.info/papers/DuanCVPR2012_poster.pdf. Accessed 4 Mar 2016.

Chapter 4
Visualizing Big Data

Ekaterina Olshannikova, Aleksandr Ometov, Yevgeni Koucheryavy
and Thomas Olsson

Introduction

The whole history of humanity is an enormous accumulation of data. Information
has been stored for thousands of years. Data has become an integral part of history,
politics, science, economics and business structures, and now even social lives. This
trend is clearly visible in social networks such as Facebook, Twitter and Instagram
where users produce an enormous stream of different types of information daily
(music, pictures, text, etc.) [1]. Now, government, scientific and technical laboratory data as well as space research information are available not only for review, but
also for public use. For instance, there is the 1000 Genomes Project [2, 3], which
provide 260 terabytes of human genome data. More than 20 terabytes of data are
publicly available at Internet Archive [4, 5], ClueWeb09 [6], among others.

Lately, Big Data processing has become more affordable for companies from
resource and cost points of view. Simply put, revenues generated from it are higher
than the costs, so Big Data processing is becoming more and more widely used in
industry and business [7]. According to International Data Corporation (IDC), data
trading is forming a separate market [8]. Indeed, 70 % of large organizations
already purchase external data, and it is expected to reach 100 % by the beginning
of 2019.

Simultaneously, Big Data characteristics such as volume, velocity, variety [9],
value and veracity [10] require quick decisions in implementation, as the information may become less up to date and can lose value fast. According to IDC [11],
data volumes have grown exponentially, and by 2020 the number of digital bits will
be comparable to the number of stars in the universe. As the size of bits geminates
every 2 years, for the period from 2013 to 2020 worldwide data will increase from

This chapter has been adopted from the Journal of Big Data, Borko Furht and Taghi
Khoshgoftaar, Editors-in-Chief.

4.4 to 44 zettabytes. Such fast data expansion may result in challenges related to human ability to manage the data, extract information and gain knowledge from it.

The complexity of Big Data analysis presents an undeniable challenge: visualization techniques and methods need to be improved. Many companies and open-source projects see the future of Big Data Analytics via Visualization, and are establishing new interactive platforms and supporting research in this area. Husain et al. [12] in their paper provide a wide list of contemporary and recently developed visualization platforms. There are commercial Big Data platforms such as International Business Machines (IBM) Software [13], Microsoft [14], Amazon [15] and Google [16]. There exists an open-source project, Socrata [17], which deals with dynamic data from public, government and private organizations. Another platform is a JavaScript library D3 [18] for dynamic data visualizations. This list can be extended with Cytoscape [19], Tableau [20], Data Wrangler [21] and others. Intel [22] and Statistical Analysis System (SAS) [23] are performing research in data visualization as well but more from a business perspective.

Organizations and social media generate enormous amounts of data every day and, traditionally, represent it in a format consistent with the poorly structured databases: web blogs, text documents, or machine code, such as geospatial data that may be collected in various stores even outside of a company/organization [24]. On the other hand, information stored in a multitude repository and the use of cloud storage or data centers is also widely common [25]. Furthermore, companies have the necessary tools to establish the relationship between data segments in addition to the process of making the basis for meaningful conclusions. As data processing rates are growing continuously, a situation may appear when traditional analytical methods would not be able to stay up to date, especially with the growing amount of constantly updated data, which ultimately opens the way for Big Data technologies [26].

This paper provides information about various types of existing data to which certain techniques are useful for the analysis. Recently, many visualization methods have been developed for a quick representation of data that is already preprocessed. There has been a step away from planar images towards multi-dimensional volumetric visualizations. However, Big Data visualization evolution cannot be considered as finished, inasmuch as new techniques generate new research challenges and solutions that will be discussed in the following paper.

Current activity in the field of Big Data visualization is focused on the invention of tools that allow a person to produce quick and effective results working with large amounts of data. Moreover, it would be possible to assess the analysis of the visualized information from all the angles in novel, scalable ways. Based on Big Data related literature, we identify the main visualization challenges and propose a novel technical approach to visualize Big Data based on the understandings of human perception and new Mixed Reality (MR) technologies. From our perspective, one of the more promising methods for improving upon current Big Data visualization techniques is in its correlation with Augmented Reality (AR) and Virtual Reality (VR) that are suitable for the limited perception capabilities of humans. We identify important steps for the research agenda to implement this approach.

This paper covers various issues and topics, but there are three main directions of this survey:

- Human cognitive limitations in terms of Big Data Visualization.
- Applying Augmented and Virtual reality opportunities towards Big Data Visualization.
- Challenges and benefits of the proposed visualization approach.

The rest of paper is organized as follows: The first section provides a definition of Big Data and looks at currently used methods for Big Data processing and their specifications. Also it indicates the main challenges and issues in Big Data analysis. Next, in the section Visualization methods, the historical background of this field is given, modern visualization techniques for massive amounts of information are presented and the evolution of visualization methods is discussed. Further in the last section, Integration with Augmented and Virtual Reality, the history of AR and VR is detailed with respect to its influence on Big Data. These developmental processes are supported by the proposed oncoming Big Data visualization extension for VR and AR, which can solve actual perception and cognition challenges. Finally, important data visualization challenges and future research agenda are discussed.

Big Data: An Overview

Today large data sources are ubiquitous throughout the world. Data used for processing may be obtained from measuring devices, radio frequency identifiers, social network message flows, meteorological data, remote sensing, location data streams of mobile subscribers and devices, and audio and video recordings. So, as Big Data is more and more used all over the world, a new and important research field is being established. The mass distribution of the technology and innovative models that utilize these different kinds of devices and services, appeared to be a starting point for the penetration of Big Data in almost all areas of human activity, including the commercial sector and public administration [27].

Nowadays, Big Data and the continuing dramatic increase in human and machine generated data associated with it are quite evident. However, do we actually know what Big Data is, and how close are the various definitions put forward for this term? For instance, there was an article in Forbes in 2014 which is related to this controversial question [28]. It gave a brief history of the establishment of the term, and provided several existing explanations and descriptions of Big Data to improve the core understanding of the phenomenon. On the other hand, Berkeley School of Information published a list with more than 40 definitions of the term [29].

As Big Data covers various fields and sectors, the meaning of this term should be specifically defined in accordance with the activity of the specific organization/ person. For instance, in contrast to industry-driven Big Data "Vs" definitions,

Dr. Ivo Dinov for his research scope listed another data's multi-dimensional characteristics [30] such as data size, incompleteness, incongruency, complex representation, multiscale nature and heterogeneity of its sources [31, 32].

In this paper the modified Gartner Inc. definition [33, 34] is used: Big Data is a technology to process high-volume, high-velocity, high-variety data or data-sets to extract intended data value and ensure high veracity of original data and obtained information that demand cost-effective, innovative forms of data and information processing (analytics) for enhanced insight, decision making, and processes control [35].

Big Data Processing Methods

Currently, there exist many different techniques for data analysis [36], mainly based on tools used in statistics and computer science. The most advanced techniques to analyze large amounts of data include: artificial neural networks [37–39]; models based on the principle of the organization and functioning of biological neural networks [40, 41]; methods of predictive analysis [42]; statistics [43, 44]; Natural Language Processing [45]; etc. Big Data processing methods embrace different disciplines including applied mathematics, statistics, computer science and economics. Those are the basis for data analysis techniques such as Data Mining [39, 46–49], Neural Networks [41, 50–52], Machine Learning [53–55], Signal Processing [56–58] and Visualization Methods [59–61]. Most of these methods are interconnected and used simultaneously during data processing, which increases system utilization tremendously (see Fig. 4.1).

Fig. 4.1 Big Data processing methods interconnection. Applied mathematics, statistics, economics and computer science are foundation of the Bid Data processing methods. Meanwhile, data mining, signal processing, neural networks, visualization and machine learning are strongly connected to each other

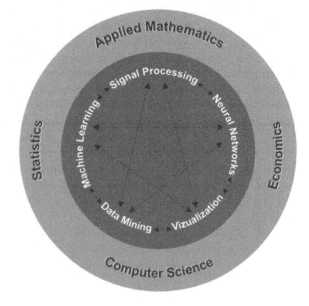

We would like to familiarize reader with the primary methods and techniques in Big Data processing. As this topic is not a focus of the paper, this list is not exhaustive. Nevertheless, the main interconnections between these methods are shown and application examples are given.

Optimization methods are mathematical tools for efficient data analysis. Optimization includes numerical analysis focused on problem solving in various Big Data challenges: volume, velocity, variety and veracity [62] that will be discussed in more detail later. Some widely used analytical techniques are genetic programming [63–65], evolutionary programming [66] and particle swarm optimization [67, 68]. Optimization is focused on the search of the optimal set of actions needed to improve system performance. Notably, genetic algorithms are also a specific part of machine learning direction [69]. Moreover, statistical testing, predictive and simulation models are applied also as for Statistics methods [70].

Statistics methods are used to collect, organize and interpret data, as well as to outline interconnections between realized objectives. Data-driven statistical analysis concentrates on implementation of statistics algorithms [71, 72]. A/B testing [73] technique is an example of a statistics method. In terms of Big Data there is a possibility to perform a variety of tests. The aim of A/B tests is to detect statistically important differences and regularities between groups of variables to reveal improvements. Besides, statistical techniques contain cluster analysis, data mining and predictive modelling methods. Some techniques in spatial analysis [74] originate from the field of statistics as well. It allows analysis of topological, geometric or geographic characteristics of data sets.

Data mining includes cluster analysis, classification, regression and association rule learning techniques. This method is aimed at identifying and extracting beneficial information from extensive data or datasets. Cluster analysis [75, 76] is based on principles of similarities to classify objects. This technique belongs to unsupervised learning [77, 78] where training data [79] is used. Classification [80] is a set of techniques which are aimed at recognizing categories with new data points. In contrast to cluster analysis, a classification technique uses training data sets to discover predictive relationships. Regression [81] is a set of a statistical techniques that are aimed at determining changes between dependent and independent variables. This technique is mostly used for prediction or forecasting. Association rule learning [82, 83] is set of techniques designed to detect valuable relationships or association rules among variables in databases.

Machine Learning is a significant area in computer science which aims to create algorithms and protocols. The main goal of this method is to improve computers' behaviors on the basis of empirical data. Its implementation allows recognition of complicated patterns and automatic application of intelligent decision-making based on. Pattern recognition, natural language processing, ensemble learning and sentiment analysis are examples of machine learning techniques. Pattern recognition [84, 85] is a set of techniques that use a certain algorithm to associate an output value with a given input value. Classification technique is an example of this. Natural language processing [86] takes its origins from computer science within the fields of artificial intelligence and linguistics. This set of techniques performs analysis of human language. Sometimes it uses a sentiment analysis [87] that is able to identify and extract specific information from text materials evaluating words, degree and strength of a sentiment. Ensemble learning [88, 89] in automated decision-making systems is a useful technique for diminishing variance and increase accuracy. It aims to solve diverse machine learning issues such as confidence estimation, missing feature and error correction, etc.

Signal processing consists of various techniques that are part of electrical engineering and applied mathematics. The key aspect of this method is the analysis of discrete and continuous signals. In other words, it enables the analog representation of physical quantities (e.g. radio signals or sounds, etc.). Signal detection theory [90] is applied to evaluate the capacity for distinguishing between signal and noise in some techniques. A time series analysis [91, 92] includes techniques from both statistics and signal processing. Primarily, it is designed to analyze sequences of data points with a demonstration of data values at consistent times. This technique is useful to predict future data values based on knowledge of past ones. Signal processing techniques can be applied to implement some types of data fusion [93]. Data fusion combines multiple sources to obtain improved information that is more relevant or less expensive and has higher quality [94].

Visualization methods concern the design of graphical representation, i.e. to visualize the innumerate amount of the analytical results as diagrams, tables and images. Visualization for Big Data differs from all of the previously mentioned processing methods and also from traditional visualization techniques. To visualize large-scale data, feature extraction and geometric modelling can be implemented. These processes are needed to decrease the data size before actual rendering [95]. Intuitively, visual representation is more likely to be accepted by a human in comparison with unstructured textual information. The era of Big Data has been rapidly promoting the data visualization market. According to Mordor Intelligence [96] the visualization market will increase at a compound annual growth rate (CAGR) of 9.21 % from $4.12 billions in 2014 to $6.40 billions by the end of 2019. SAS Institute provides results of an International Data Group (IDG) research study in the white paper [97]. The research is focused on how companies are performing Big Data analysis. It shows that 98 % of the most effective companies working with Big Data are presenting results of the analysis via visualization. Statistical data from this research provides evidence of the visualization benefits in terms of decision making improvement, better ad hoc data analysis, improved collaboration and information sharing inside/outside an organization.

Nowadays, different groups of people including designers, software developers and scientists are in the process of searching for new visualization tools and opportunities. For example, Amazon, Twitter, Apple, Facebook and Google are companies that utilize data visualization in order to make appropriate business decisions [98]. Visualization solutions can provide insights from different business perspectives. First of all, implementation of advanced visualization tools enables rapid exploration of all customers/users data to improve customer-company relationships. It allows marketers to create more precise customer segments based on data from purchasing history or life stage and other factors. Besides, correlation mapping may assist in the analysis of customer/user behavior to identify and analyze the most profitable of them. Secondly, visualization capabilities allow companies opportunities to reveal correlations between product, sales and customer profiles. Based on gathered metrics, organizations may provide novel special offers to their customers. Moreover, visualization enables tracking of revenue trends and can be useful for risk analysis. Thirdly, visualization as a tool provides better understanding of data. Higher efficiency is reached by obtaining relevant, consistent and accurate information. So, visualized data could assist organizations to find different effective marketing solutions. In this section we familiarized the reader with the main techniques of data analysis and described their strong correlation to each other. Nevertheless, the Big Data era is still in the beginning stage of its evolution. Therefore, Big Data processing methods are evolving to solve the problems of Big Data and new solutions are continuously being developed. By this statement we mean that big world of Big Data requires multiple multidisciplinary methods and techniques that lead to better understanding of the complicated structures and interconnections between them.

Big Data Challenges

Big Data has some inherent challenges and problems that can be primarily divided into three groups according to Akerkar [36]: (1) data, (2) processing, and (3) management challenges (see Fig. 4.2). While dealing with large amounts of information we face such challenges as volume, variety, velocity and veracity that are also known as 5 V of Big Data. As those Big Data characteristics are well examined in scientific literature [99–101] we will only discuss them briefly. Volume refers to the large amount of data, especially, machine-generated. This characteristic defines a size of the data set that makes its storage and analysis problematic utilizing conventional database technology. Variety is related to different types and forms of data sources: structured (e.g. financial data) and unstructured (social media conversations, photos, videos, voice recordings and others). Multiplicity of the various data results in the issue of its handling. Velocity refers to the speed of new data generation and distribution. This characteristic requires the implementation of real-time processing for the streaming data analysis (e.g. on social media, different types of transactions or trading systems, etc.). Veracity refers to the complexity of data which may lead to a lack of quality and accuracy. This characteristic reveals several challenges: uncertainty, imprecision, missing values, misstatement and data availability. There is also a challenge regarding data discovery that is related to the search of high quality data in data sets.

The second branch of Big Data challenges is called processing challenges. It includes data collection, resolving similarities found in different sources, modification data to a type acceptable for the analysis, the analysis itself and output representation, i.e. the results visualization in a form most suitable for human perception.

Fig. 4.2 Big Data challenges. The picture illustrates three main categories of Big Data challenges that are associated with data, its management and processing issues

The last type of challenge offered by this classification is related to data management. Management challenges usually refer to secured data storage, its processing and collection. Here the main focuses of study are: data privacy, its security, governance and ethical issues. Most of them are controlled based on policies and rules provided by information security institutes on state or international levels.

Over past generations, the results of analyzed data were represented as visualized plots and graphs. It is evident that collections of complex figures are sometimes hard to perceive, even by well-trained minds. Nowadays, the main factors causing difficulties in data visualization continue to be the limitations of human perception and new issues related to display sizes and resolutions. This question is studied in detail further in the section "Integration with Augmented and Virtual Reality". Preparatory to the visualization, the main interaction problem is in the extraction of the useful portion of information from massive volumes. Extracted data is not always accurate and mostly overloaded with excrescent information. Visualization technique is useful for simplifying information and transforming it into a more accessible form for human perception.

In the near future, petascale data may cause analysis failures because of traditional approaches in usage, i.e. when the data is stored on a memory disk continuously waiting for further analysis. Hence, the conservative approach of data compressing may become ineffective in visualization methods. To solve this issue, developers should create a flexible tool for the practice of data collection and analysis. Increases in data size make the multilevel hierarchy approach incapable in data scalability. Hierarchy becomes complex and intensive, making navigation difficult for user perception. In this case, a combination of analytics and Data Visualization may enable more accessible data exploration and interaction, which would allow improving insights, outcomes and decision-making.

Contemporary methods, techniques and tools for data analysis are still not flexible enough to discover valuable information in the most efficient way. The question of data perception and presentation remains open. Scientists face the task of uniting the abstract world of data and the physical world through visual representation. Meanwhile, visualization-based tools should fulfill three requirements [102, 103]: expressiveness (demonstrate exactly the information contained in the data), effectiveness (related to cognitive capabilities of human visual system) and appropriateness (cost-value ratio for visualization benefit assessment). Experience of previously used techniques can be repurposed to achieve more beneficial and novel goals in Big Data perception and representation.

Visualization Methods

Historically, the primary areas of visualization were Science Visualization and Information Visualization. However, during recent decades, the field of Visual Analytics was actively developing.

As a separate discipline, visualization emerged in 1980 [104] as a reaction to the increasing amount of data generated by computer calculations. It was named Science Visualization [105–107], as it displays data from scientific experiments related to physical processes. This is primarily a realistic three-dimensional visualization, which has been used in architecture, medicine, biology, meteorology, etc. This visualization is also known as Spatial Data visualization, which focuses on the visualization of volumes and surfaces.

Information Visualization [108–111] emerged as a branch of the Human-Computer Interaction field in the end of 1980s. It utilizes graphics to assist people in comprehending and interpreting data. As it helps to form mental models of the data, for humans it is easier to reveal specific features and patterns of the obtained information.

Visual Analytics [112–114] combines visualization and data analysis. It has absorbed features of Information Visualization as well as Science Visualization. The main difference from other fields is the development and provision of visualization technologies and tools.

Efficient visualization tools should consider cognitive and perceptual properties of the human brain. Visualization aims to improve the clarity and aesthetic appeal of the displayed information and allows a person to understand large amount of data and interact with it. Significant purposes of Big Data visual representation are: to identify hidden patterns or anomalies in data; to increase flexibility while searching of certain values; to compare various units in order to obtain relative difference in quantities; to enable realtime human interaction (touring, scaling, etc.).

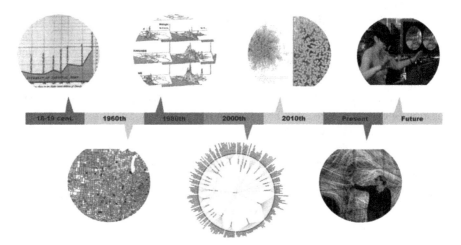

Fig. 4.3 The evolution of visualization methodology. Development of visualization methods originates from eighteenth century and its is rapidly improving today due to technical sophistication

Visualization methods have evolved much over the last decades (see Fig. 4.3), the only limit for novel techniques being human imagination. To anticipate the next steps of data visualization development, it is necessary to take into account the successes of the past. It is considered that quantitative data visualization appeared in the field of statistics and analytics quite recently. However, the main precursors were cartography and statistical graphics, created before the nineteenth century for the expansion of statistical thinking, business planning and other purposes [115]. The evolution in the knowledge of visualization techniques resulted in mathematical and statistical advances as well as in drawing and reproducing images.

By the sixteenth century, tools for accurate observation and measurement were developed. Precisely, in those days the first steps were done in the development of data visualization. The seventeenth century was swept by the problem of space, time and distance measurements. Furthermore, the study of the world's population and economic data had started.

The eighteenth century was marked by the expansion of statistical theory, ideas of data graphical representation and the advent of new graphic forms. At the end of the century thematic maps displaying geological, medical and economic data was used for the first time. For example, Charles de Fourcroy used geometric figures and cartograms to compare areas or demographic quantities [116]. Johann Lambert (1728–1777) was a revolutionary person, who used different types of tables and line graphs to display variable data [117]. The first methods were performed as simple plots followed by onedimensional histograms [118]. Still, those examples are useful only for small amounts of data. By introducing more information, this type of diagram would reach a point of worthlessness.

At the turn of twenty to twenty-first centuries, steps were taken in the development of interactive statistical computing [119] and new paradigms for data analysis [120]. Technological progress was certainly a significant prerequisite for the rapid development of visualization techniques, methods and tools. More precisely, large-scale statistical and graphics software engineering was invented, and computer processing speed and capacity vastly increased [121].

However, the next step presenting a system with the addition of a time dimension appeared as a significant breakthrough. In the beginning of the present century few dimensional visualization methods were in use as a part 2D/3D node-link diagram [122]. Already at this level of abstraction, any user may classify the goal and specify further analytical steps for the research, but unfortunately, data scaling became an essential issue.

Moreover, currently used technologies for data visualization are already causing enormous resource demands which include high memory requirements and extremely high deployment cost. However, the currently existing environment faces a new limitation based on the large amounts of data to be visualized in contrast to past imagination issue. Modern effective methods are focused on representation in specified rooms equipped with widescreen monitors or projectors [123].

Nowadays, there are a fairly large number of data visualization tools offering different possibilities. These tools can be classified based on three factors: by the data type, by visualization technique type and by the interoperability. The first refers to the different types of data to be visualized [124]:

- *Univariate data* One dimensional arrays, time series, etc.
- *Two-dimensional data* Point two-dimensional graphs, geographical coordinates, etc.
- *Multidimensional data* Financial indicators, results of experiments, etc.
- *Texts and hypertexts* Newspaper articles, web documents, etc.
- *Hierarchical and links* The structure subordination in the organization, e-mails, documents and hyperlinks, etc.
- *Algorithms and programs* Information flows, debug operations, etc.

The second factor is based on visualization techniques and samples to represent different types of data. Visualization techniques can be both elementary (line graphs, charts, bar charts) and complex (based on the mathematical apparatus). Furthermore, visualization can be performed as a combination of various methods. However, visualized representation of data is abstract and extremely limited by one's perception capabilities and requests (see Fig. 4.4).

Types of visualization techniques are listed below:

1. 2D/3D standard figure [125]. May be implemented as bars, line graphs, various charts, etc. (see Fig. 4.5). The main drawback of this type is the complexity of the acceptable visualization for complicated data structures;
2. Geometric transformations [126]. This technique represents information as scatter diagram (see Fig. 4.6). This type is geared towards a multi-dimensional

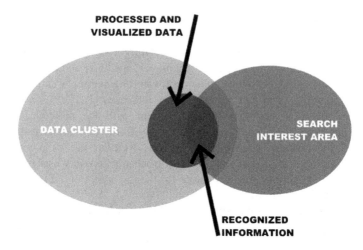

Fig. 4.4 Human perception capability issue. Human perceptional capabilities are not sufficient to embrace large amount of data

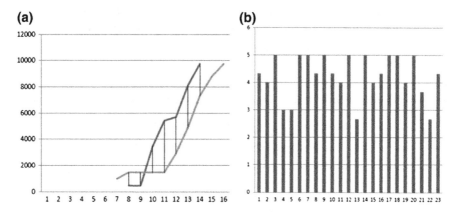

Fig. 4.5 An example of the 2D/3D standard figures visualization techniques. **a** The simple *line graph* and **b** example of a *bar chart*

data set's transformation in order to display it in Cartesian and non-Cartesian geometric spaces. This class includes methods of mathematical statistics;

3. Display icons [127]. Ruled shapes (needle icons) and star icons. Basically, this type displays the values of elements of multidimensional data in properties of images (see Fig. 4.7). Such images may include human faces, arrows, stars, etc. Images can be grouped together for holistic analysis. The result of the visualization is a texture pattern, which varies according to the specific characteristics of the data;

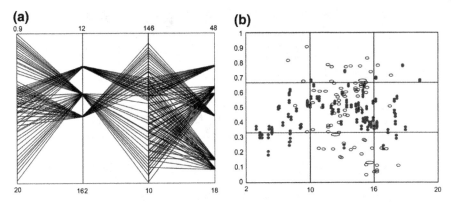

Fig. 4.6 An example of the geometric transformations visualization techniques. **a** Example of a parallel coordinates and **b** the scatter plot

Fig. 4.7 An example of the display icons visualization techniques. Picture demonstrates the visualization of various social connections in Australia

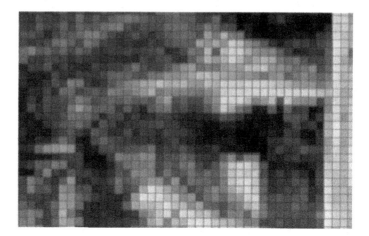

Fig. 4.8 An example of the methods focused on the pixels. Picture demonstrates an amount of data visualized in pixels. Each *color* has its specific meaning

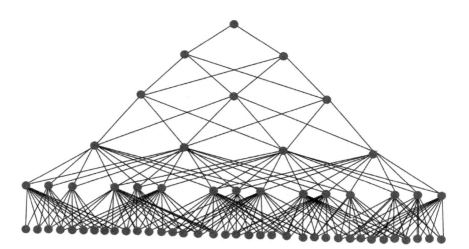

Fig. 4.9 An example of the hierarchical images. Picture illustrates a tree map of data

4. Methods focused on the pixels [128]. Recursive templates and cyclic segments. The main idea is to display the values in each dimension into the colored pixel and to merge some of them according to specific measurements (see Fig. 4.8). Since one pixel is used to display a single value, therefore visualization of large amounts of data can be reachable with this methodology;

5. Hierarchical images [129]. Tree maps and overlay measurements (see Fig. 4.9). These type methods are used with the hierarchical structured data.

The third factor is related to the interoperability with visual imagery and techniques for better data analysis. The application used for the visualization should present visual forms that capture the essence of data itself. However, it is not always enough for a complete analysis. Data representation should be constructed in order to allow a user to have different visual points of view. Thus, the appropriate compatibility should be performed:

1. *Dynamic projection* [130]. Non-static change of projections in multidimensional data sets is used. An example of the dynamic projection in two-dimensional plane of multidimensional data in a scatter plots. It is necessary to note that the number of possible projections increases exponentially with the number of measurements and, thus, perception suffers more.
2. *Interactive filtering* [131]. In the investigation of large amounts of data there is a need to share data sets and highlight significant subsets in order to filter images. Significantly, that there should be an opportunity to have a visual representation in real time. A subset can be chosen either directly from a list or by determining a subset of the properties of interest;
3. *Scaling images* [132]. Scaling is a well-known method of interaction used in many applications. Especially for Big Data processing, this method is very useful due to the ability to represent data in a compressed form. It provides the ability to simultaneously display any part of an image in a more detailed form. Nevertheless, a lower level entity may be represented by a pixel at a higher level, a certain visual image or an accompanying text label;
4. *Interactive distortion* [133] supports the research process data using distortion scale with partial detail. The basic idea of this method is that a part of the fine granularity displayed data is shown in addition to one with a low level of details. The most popular methods are hyperbolic and spherical distortion;
5. *Interactive combination* [134, 135] brings together a combination of different visualization techniques to overcome specific deficiencies by their conjugation. For example, different points of the dynamic projection can be combined with the techniques of coloring.

To summarize, any visualization method can be classified by data type, visualization technique and interoperability. Each method can support different types of data, various images and varied methods for interaction.

A visual representation of Big Data analysis is crucial for its interpretation. As it was already mentioned, it is evident that human perception is limited. The main purpose of modern data representation methods is related to improvement in forms of images, diagrams or animation. Examples of well known techniques for data visualization are presented below [136]:

- *Tag cloud* [137] is used in text analysis, with a weighting value dependent on the frequency of use (citation) of a particular word or phrase (see Fig. 4.10). It consists of an accumulation of lexical items (words, symbols or combination of the two). This technique is commonly integrated with web sources to quickly familiarize visitors with the content via key words.

Fig. 4.10   An example of the tag cloud. This picture illustrates visualization of the paper abstract

Fig. 4.11   An example of the clustergram. This picture illustrates different state of data in several clusters

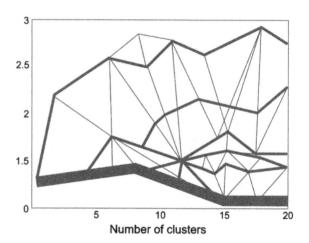

- *Clustergram* [138] is an imaging technique used in cluster analysis by means of representing the relation of individual elements of the data as they change their number (see Fig. 4.11). Choosing the optimal number of clusters is also an important component of cluster analysis.
- *Motion charts* allow effective exploration of large and multivariate data and interact with it utilizing dynamic 2D bubble charts (see Fig. 4.12). The blobs (bubbles—central objects of this technique) can be controlled due to variable mapping for which it is designed. For instance, motion charts graphical data tools are provided by Google [139], amCharts [140] and IBM Many Eyes [141].

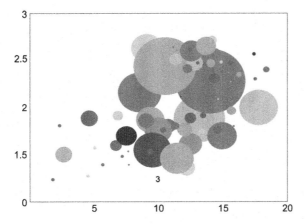

Fig. 4.12 An example of the motion chart. This picture illustrates the data in forms of bubbles that have various meaning based on *color* and *size*

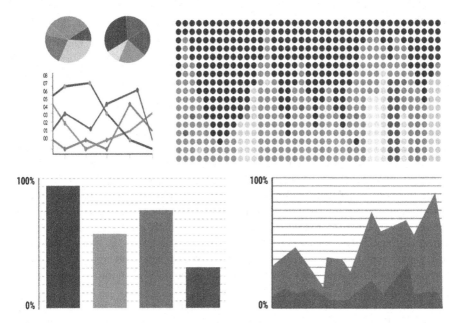

Fig. 4.13 An example of the dashboard. This picuture illustrates pie chart, visualization of data in pixels, *line graph* and *bar chart*

- *Dashboard* [142] enables the display of log files of various formats and filter data based on chosen data ranges (see Fig. 4.13). Traditionally, dashboard consists of three layers [143]: data (raw data), analysis (includes formulas and imported data from data layer to tables) and presentation (graphical representation based on the analysis layer)

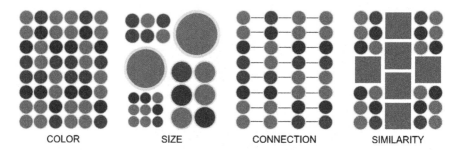

| COLOR | SIZE | CONNECTION | SIMILARITY |

Fig. 4.14 Fundamental cognitive psychology principles. *Color* is used to catch significant differences in the data sets by view; manipulation of visual object sizes may assist persons to identify the most important elements of the information; representation of connections improves patterns identifications and aims to facilitate data analysis; Grouping objects using similarity principle decreases cognitive load

Nowadays, there are many publicly available tools to create meaningful and attractive visualizations. For instance, there is a chart of open visualization tools for data visualization and analysis published by Machils [144]. The author provides a list, which contains more than 30 tools from easiest to most difficult: Zoho Reports, Weave, Infogr. am, Datawrapper and others.

All of these modern methods and tools follow fundamental cognitive psychology principles and use the essential criteria of data successful representation [145] such as manipulation of size, color and connections between visual objects (see Fig. 4.14). In terms of human cognition, the Gestalt Principles [146] are relevant. The basis of Gestalt psychology is a study of visual perception. It suggests that people tend to perceive the world in a form of holistic ordered configuration rather than constituent fragments (e.g. at first, person perceives forest and after that can identify single trees as part of the whole). Moreover, our mind fills in the gaps, seeks to avoid uncertainty and easily recognizes similarities and differences. The main Gestalt principles such as law of proximity (collection of objects forming a group), law of similarity (objects are grouped perceptually if they are similar to each other), symmetry (people tend to perceive object as symmetrical shapes), closure (our mind tends to close up objects that are not complete) and figure-ground law (prominent and recessed roles of visual objects) should be taken into account in Big Data Visualization.

To this end, the most effective visualization method is the one that uses multiple criteria in the optimal manner. Otherwise, too many colors, shapes, and interconnections may cause difficulties in the comprehension of data, or some visual elements may be too complex to recognize.

After observation and discussion about existing visualization methods and tools for Big Data, we can clarify and outline its important disadvantages that are sufficiently discussed by specialists from different fields [147–149]. Various ways of data interpretation make them meaningful. It is easy to distort valuable information in its visualization, because a picture convinces people more effectively than textual content. Existing visualization tools aim to create as simple and abstract images as possible, which can lead to a problem when significant data can be interpreted as disordered information and important connections between data units will be hidden from the user. It is a problem of visibility loss, which also refers to display resolution, where the quality of represented data depends on number of pixels and their density. A solution may be in the use of larger screens [150]. However, this concept brings a problem of human brain cognitive perceptual limitations, as will be discussed in detail in the section Integration with Augmented and Virtual Reality.

Using visual and automated methods in Big Data processing gives a possibility to use human knowledge and intuition. Moreover, it becomes possible to discover novel solutions for complex data visualization [151]. Vast amounts of information motivate researchers and developers to create new tools for quick and accurate analysis. As an example, the rapid development of visualization techniques may be concerned. In the world of interconnected research areas, developers need to combine existing basic, effective visualization methods with new technological opportunities to solve the central problems and challenges of Big Data analysis.

Integration with Augmented and Virtual Reality

It is well known that the vision perception capabilities of the human brain are limited [152]. Furthermore, handling a visualization process on currently used screens requires high costs in both time and health. This leads to the need of its proper usage in the case of image interpretation. Nevertheless, the market is in the process of being flooded with countless numbers of wearable devices [153, 154] as well as various display devices [155, 156].

The term Augmented Reality was invented by Tom Caudell and David Mizel in 1992 and meant to describe data produced by a computer that is superimposed to the real world [157]. Nevertheless, Ivan Sutherland created the first AR/VR system already in 1968. He developed the optical see-through head-mounted display that can reveal simple three-dimensional models in real time [158]. This invention was a predecessor to the modern VR displays and AR helmets [159] that seem to be an established research and industrial area for the coming decade [160]. Applications for use have already been found in military [161], education [162], healthcare [163], industry [164] and gaming fields [165]. At the moment, the Oculus Rift [166] helmet gives many opportunities for AR practice. Concretely, it will make it possible to embed virtual content into the physical world. William Steptoe has already done research in this field. The use of it in the visualization area might solve many issues from narrow visual angle, navigation, scaling, etc. For example, offering a way to have a complete 360° view with a helmet can solve an angle problem. On the other hand, a solution can be obtained with help of specific widescreen rooms, which by definition involves enormous budgets. Focusing on the combination of dynamic projection and interactive filtering visualization methods, AR devices in combination with motion recognition tools might solve a significant scaling problem especially for multidimensional representations that comes to this area from the field of Architecture. Speaking more precisely, designers (specialized in 3D-visualization) work with flat projections in order to produce a visual model [167]. However, the only option to present a final image is in moving around it and thus navigation inside the model seems to be another influential issue [168].

From the Big Data visualization point of view, scaling is a significant issue mainly caused by multidimensional systems where a need to delve into a branch of information in order to obtain some specific value or knowledge takes its place. Unfortunately, it cannot be solved from a static point of view. Likewise, integration with motion detection wearables [169] would highly increase such visualization system usability. For example, the additional use of an MYO armband [170] may be a key to the interaction with visualized data in the most native way. Similar comparison may be given as a pencil-case in which one tries to find a sharpener and spreads stationery with his/her fingers.

However, the use of AR displays and helmets is also limited by specific characteristics of the human eye (visual system), such as field of view and/or diseases like scotoma [171] and blind spots [172]. Central vision [173] is most significant and necessary for human activities such as reading or driving. Additionally, it is responsible for accurate vision in the pointed direction and takes most of the visual cortex in the brain but its retinal size is <1 % [174]. Furthermore, it captures only two degrees of the vision field, which stays the most considerable for text and object recognition. Nevertheless, it is supported with Peripheral vision which is responsible for events outside the center of gaze. Many researchers around the world are currently working with virtual and AR to train young professionals [175–177], develop new areas [178, 179] and analyze the patient's behavior [180].

Despite the well known topics like colorblindness, natural field of view and other physiological abnormalities, recent research by Israel Abramov et al. [181] is over viewing physiological gender and age differences based on the cerebral cortex and its large number of testosterone receptors [182], as a basis for the variety in perception procedures. The study was mainly about the focused image onto the retina at the back of the eyeball and its visual system processing. We overview the main reasons for those differences, starting from prehistoric times, when African habitats in forest regions had limited distance for object detection and identification, thus obtained higher acuity for males may be explained. Also, sex differences might be related to different roles in the survival commune. So that males were mainly hunting (hunter-gatherer hypothesis)—they had to detect enemies and predators much faster [183]. Moreover, there are significant gender differences for far- and near-vision: males have their advantage in a far-space [184]. On the other hand, females are much more susceptible for brightness and color changes in addition to static objects in near-space [185]. However, we can conclude that male/female differences in the sensory capacities are adaptive but should be considered in order to optimize represented and visualized data for end-uses. Additionally, there exists a research area focusing on the human eye movement patterns during the perception of scenes and objects. It can be based on different factors starting from particular culture peculiar properties [186] and up to specific search tasks [187] being in high demand for Big Data visualization purposes.

Further studies shall be focused on the usage of ophthalmology and neurology for the development of the new visualization tools. Basically, such cross-discipline collaboration would support decision making for the image position selection, which is mainly related to the problem of the significant information losses due to the vision angle extension. Moreover, it is highly important to take in account current hardware quality and screens resolution in addition to the software part. Nevertheless, there is a need of the improvement for multicore GPU processors besides the address bus throughput refinement between CPU and GPU or even replacement for wireless transfer computations on cluster systems. Never the less, it is significant to discuss current visualization challenges to support future research.

Future Research Agenda and Data Visualization Challenges

Visualized data can significantly improve the understanding of the preselected information for an average user. In fact, people start to explore the world using visual abilities since birth. Images are often more easier to perceive in comparison to text. In the modern world, we can see clear evolution towards visual data representation and imagery experience. Moreover, visualization software becomes ubiquitous and publicly available for ordinary user. As a result, visual objects are widely distributed—from social media to scientific papers and, thus, the role of visualization while working with large amount of data should be reconsidered. In this section, we overview important challenges and possible solutions related to future agenda for Big Data visualization with AR and VR usage:

1. *Application development integration* In order to operate with visualized objects, it is necessary to create a new interactive system for the user. It should support such actions as: scaling; navigating in visualized 3D space; selecting sub-spaces, objects, groups of visual elements (flow/path elements) and views; manipulating and placing; planning routes of view; generating, extracting and collecting data (based on the reviewed visualized data). A novel system should allow multi-modal control by voice and/or gestures in order to make it more intuitive for users as it is shown in [188, 190, 191]. Nevertheless, one of the main issues regarding this direction of development is the fact that implementing effective gestural and voice interaction is not a trivial matter. There is a need to develop a machine learning system and to define basic intuitive gestures that are currently in research for general [192–194] and more specific (medical) purposes [195].
2. *Equipment and virtual interface* It is necessary to apply certain equipment for the implementation of such an interactive system in practice. Currently, there are optical and video see-trough head-mounted displays (HMD) [196] that merge virtual objects into the real scene view. Both have the following issues: distortion and resolution of the real scene; delay of a system; viewpoint matching;

engineering and cost factors. As for the interaction issue, for an appropriate haptic feedback in an MR environment there is a need to create a framework that would allow an interaction with intuitive gesture. As it is revealed in the section Integration with Augmented and Virtual Reality, glove-based systems [197] are mainly used for virtual object manipulation. The disadvantage of hand-tracking input is so that there is no tactile feedback. In summary, the interface should be redesigned or reinvented in order to simplify user interaction. Software engineers should create new approaches, principles and methods in User Interface Design to make all instruments easily accessible and intuitive to use.

3. *Tracking and recognition system* Objects and tools have to be tracked in virtual space. The position and orientation values of virtual items are dynamic and have to be re-estimated during presentation. Tracking head movement is another significant challenge. It aims to avoid mismatch of the real view scene and computer generated objects. This challenge may be solved by using more flexible software platforms.

4. *Perception and cognition* Actually, the level of computer operation is high but still not sufficiently effective in comparison to human brain performance even in cases of neural networks. As was mentioned earlier in the section Integration with Augmented and Virtual Reality, human perception and cognition have their own characteristics and features, and the consideration of this issue by developers during hardware and interface design for AR is vital. In addition, the user's ability to recognize and understand the data is a central issue. Tasks such as browsing and searching require a certain cognitive activity. Also, there can be issues related to different users' reactions with regard to visualized objects depending on their personal and cultural backgrounds. In this sense, simplicity in information visualization has to be achieved in order to avoid misperceptions and cognitive overload [198]. Psychophysical studies would provide answers to questions regarding perception and would give the opportunity to improve performance by motion prediction.

5. *Virtual and physical objects mismatch* In an Augmented Reality environment, virtual images integrate with real world scenery at the static distance in the display while the distance to real objects varies. Consequently, a mismatch of virtual and physical distances is irreversible and it may result in incorrect focus, contrast and brightness of virtual objects in comparison to real ones. The human eye is capable of recognizing many levels of brightness, saturation and contrast [199], but most contemporary optical technologies cannot display all levels appropriately. Moreover, potential optical illusions arise from conflicts between computer-generated and real environment objects. Using modern equipment would be a solution for this challenge.

6. *Screen limitations* With the current technology development level, visualized information is presented mainly on screens. Even a VR helmet is equipped with two displays. Unfortunately, and because of the close-to-the-eye proximity, users can experience lack of comfort while working with it. It is mainly based on a low display resolution and high graininess and, thus, manufacturers should take it into consideration for further improvement.

7. *Education* As this concept is relatively new, there is a need to specify the value of the data visualization and its contribution to the users' work. The value cannot be so obvious; that is why compelling showcase examples and publicly available tutorials can reveal AR and VR potential in visual analytics. Moreover, users need to be educated and trained for the oncoming interaction with this evolving technology. The visual literacy skill should be improved in order to have high performance while working with visualized objects. A preferable guideline can be chosen as Visual Information-Seeking Mantra: overview first, zoom and filter, then details on demand [200].

Despite all the challenges, the main benefit from the implementation of MR approach is human experience improvement. At the same time, such visualization allows convenient access to huge amounts of data and provides a view from different angles. The navigation is smooth and natural via tangible and verbal interaction. It also minimizes perceptional inaccuracy in data analysis and makes visualization powerful at conveying knowledge to the end user. Furthermore, it ensures actionable insights that improves decision making.

In conclusion, challenges of data visualization for AR and VR are associated not only with current technology development but also with human-centric issues. Interestingly, some researchers are already working on the conjugation of such complex fields as massive data analysis, its visualization and complex control of the visualized environment [201]. It is worthwhile to note that those factors should be taken into account simultaneously in order to achieve the best outcome for the established industrial field.

Conclusion

In practice, there are a lot of challenges for Big Data processing and analysis. As all the data is currently visualized by computers, it leads to difficulties in the extraction of data, followed by its perception and cognition. Those tasks are time-consuming and do not always provide correct or acceptable results.

In this paper we have obtained relevant Big Data Visualization methods classification and have suggested the modern tendency towards visualization-based tools for business support and other significant fields. Past and current states of data visualization were described and supported by analysis of advantages and disadvantages. The approach of utilizing VR, AR and MR for Big Data Visualization is presented and the advantages, disadvantages and possible optimization strategies of those are discussed.

For visualization problems discussed in this work, it is critical to understand the issues related to human perception and limited cognition. Only after that, the field of design can provide more efficient and useful ways to utilize Big Data. It can be concluded that data visualization methodology may be improved by considering fundamental cognitive psychological principles and by implementing most natural interaction with visualized virtual objects. Moreover, extending it with functions to exclude blind spots and decreased vision sectors would highly improve recognition time for people with such a disease. Furthermore, a step towards wireless solutions would extend device battery life in addition to computation and quality improvements.

Authors' Contributions EO performed the primary literature review and analysis for this work as well as designed illustrations. Manuscript was drafted by EO and AO. EK introduced this topic to other authors and coordinate the work process to complete the manuscript. EO, AO and TO worked together to develop the article's framework and focus. All authors read and approved the final manuscript.

Compliance with Ethical Guidelines

Competing Interests The authors EO, AO, EK and TO declare that they have no competing interests.

References

1. Manyika J, Chui M, Brown B, Bughin J, Dobbs R, Roxburgh C, Byers AH. Big Data: the next frontier for innovation, competition, and productivity. June Progress Report. McKinsey Global Institute; 2011.
2. Genomes: a Deep Catalog of Human Genetic Variation. 2015. http://www.1000genomes.org/.
3. Via M, Gignoux C, Burchard EG. The 1000 Genomes Project: new opportunities for research and social challenges. Genome Med. 2010;2(3):1.
4. Internet Archive: Internet Archive Wayback Machine. 2015. http://archive.org/web/web.php.
5. Nielsen J. Comparing content in web archives: differences between the Danish archive Netarkivet and Internet Archive. In: Two-day conference at Aarhus University, Denmark; 2015.
6. The Lemur Project: The ClueWeb09 Dataset. 2015. http://lemurproject.org/clueweb09.php/.
7. Russom P. Managing Big Data. TDWI Best Practices Report, TDWI Research; 2013.
8. Gantz J, Reinsel D. The digital universe in 2020: Big data, bigger digital shadows, and biggest growth in the far east. IDC iView IDC Anal Future. 2012;2007:1–16.
9. Beyer MA, Laney D. The importance of "Big Data": a definition. Stamford: Gartner; 2012.
10. Demchenko Y, Ngo C, Membrey P. Architecture framework and components for the big data ecosystem. J Syst Netw Eng. 2013;1–31 (SNE technical report SNE-UVA-2013-02).
11. Turner V, Reinsel D, Gantz JF, Minton S. The digital universe of opportunities: rich data and the increasing value of the internet of things. IDC Anal Future; 2014.
12. Husain SS, Kalinin A, Truong A, Dinov ID. SOCR data dashboard: an integrated Big Data archive mashing medicare, labor, census and econometric information. J Big Data. 2015;2(1):1–18.
13. Keahey TA. Using visualization to understand Big Data. IBM Bus Anal Adv Vis. 2013.
14. Microsoft Corporation: Power BI—Microsoft. 2015. https://powerbi.microsoft.com/.
15. Amazon.com, Inc. Amazone Web Services. 2015. https://aws.amazon.com/.
16. Google, Inc. Google Cloud Platform. 2015. https://cloud.google.com/.
17. Socrata. Data to the People. 2015. http://www.socrata.com.
18. D3.js: D3 Data-Draven Documents. 2015. http://d3js.org.
19. The Cytoscape Consortium: Network Data Integration, Analysis, and Visualization in Box. 2015. http://www.cytoscape.org.

20. Tableau—Business Intelligence and Analytics. http://tableau.com/.
21. Kandel S, Paepcke A., Hellerstein J, Heer J. Wrangler: interactive visual specification of data transformation scripts. In: Proceedings of the SIGCHI conference on human factors in computing systems, ACM; 2011. p. 3363–72.
22. Schaefer D, Chandramouly A, Carmack B, Kesavamurthy K. Delivering self-service BI, data visualization, and Big Data analytics. Intel IT: Business Intelligence; 2013.
23. Choy J, Chawla V, Whitman L. Data visualization techniques: from basics to Big Data with SAS visual analytics. SAS: White Paper; 2013.
24. Ganore P. Need to know what Big Data is? ESDS—Enabling Futurability; 2012.
25. Agrawal D, Das S, El Abbadi A. Big Data and cloud computing: current state and future opportunities. In: Proceedings of the 14th international conference on extending database technology, ACM; 2011. p. 530–3.
26. Kaur M. Challanges and issues during visualization of Big Data. Int J Technol Res Eng. 2013;1:174–6.
27. Childs H, Geveci B, Schroeder W, Meredith J, Moreland K, Sewell C, Kuhlen T, Bethel EW. Research challenges for visualization software. Computer. 2013;46:34–42.
28. Press G. 12 Big Data definitions: what's yours? Forbes; 2014.
29. Dutcher J. What is Big Data? Berkley School of Information; 2014.
30. Bashour N. The Big Data blog, Part V: interview with Dr. Ivo Dinov. 2014. http://www.aaas.org/news/big-data-blog-part-v-interview-dr-ivo-dinov.
31. Komodakis N, Pesquet JC. Playing with duality: an overview of recent primal-dual approaches for solving largescale optimization problems. 2014. http://arxiv.org/abs/1406.5429.
32. Manicassamy J, Kumar SS, Rangan M, Ananth V, Vengattaraman T, Dhavachelvan P. Gene suppressor: an added phase towards solving large scale optimization problems in genetic algorithm. Appl Soft Comput. 2015;35:214–26.
33. Gartner—IT Glossary. Big Data defintion. http://www.gartner.com/it-glossary/big-data/.
34. Sicular S. Gartner's Big Data definition consists of three parts, not to be confused with three "V"s, Gartner, Inc. Forbes; 2013.
35. Demchenko Y, De Laat C, Membrey P. Defining architecture components of the Big Data Ecosystem. In: Proceedings of international conference on collaboration technologies and systems (CTS), IEEE; 2014. p. 104–12.
36. Akerkar R. Big Data computing. Boca Raton: CRC Press, Taylor & Francis Group; 2013.
37. Sethi IK, Jain AK. Artificial neural networks and statistical pattern recognition: old and new connections, vol 1. New York: Elsevier; 2014.
38. Araghinejad S. Artificial neural networks. Data-driven modeling: using MATLAB in water resources and environmental engineering. Netherlands: Springer; 2014. p. 139–94.
39. Larose DT. Discovering knowledge in data: an introduction to data mining. Hoboken: Wiley; 2014.
40. Maren AJ, Harston CT, Pap RM. Handbook of neural computing applications. Cambridge: Academic Press; 2014.
41. Schmidhuber J. Deep learning in neural networks: an overview. Neural Netw. 2015;61:85–117.
42. McCue C. Data mining and predictive analysis: intelligence gathering and crime analysis. Butterworth-Heinemann; 2014.
43. Rudin C, Dunson D, Irizarry R, Ji H, Laber E, Leek J, McCormick T, Rose S, Schafer C, van der Laan M et al. Discovery with data: leveraging statistics with computer science to transform science and society. 2014.
44. Cressie N. Statistics for spatial data. Hoboken: Wiley; 2015.
45. Lehnert WG, Ringle MH. Strategies for natural language processing. Hove: Psychology Press; 2014.
46. Chu WW, editor. Data mining and knowledge discovery for Big Data. Studies in Big Data, vol 1. Heidelberg: Springer; 2014.

47. Berry MJ, Linoff G. Data mining techniques: for marketing, sales, and customer support. New York: Wiley; 1997.
48. PhridviRaj M, GuruRao C. Data mining-past, present and future-a typical survey on data streams. Procedia Technol. 2014;12:255–63.
49. Zaki MJ, Meira W Jr. Data mining and analysis: fundamental concepts and algorithms. Cambridge: Cambridge University Press; 2014.
50. Sutskever I, Vinyals O, Le QV. Sequence to sequence learning with neural networks. In: Advances in neural information processing systems; 2014. p. 3104–12.
51. Rojas R, Feldman J. Neural networks: a systematic introduction. New York: Springer; 2013.
52. Gurney K. An introduction to neural networks. Milton Park: Taylor & Francis; 2003.
53. Mohri M, Rostamizadeh A, Talwalkar A. Foundations of machine learning. Adaptive computation and machine learning series. Cambridge: MIT Press; 2012.
54. Murphy KP. Machine learning: a probabilistic perspective. Adaptive computation and machine learning series. Cambridge: MIT Press; 2012.
55. Alpaydin E. Introduction to machine learning. Adaptive computation and machine learning series. Cambridge: MIT Press; 2014.
56. Vetterli M, Kovačević J, Goyal VK. Foundations of signal processing. Cambridge: Cambridge University Press; 2014.
57. Xhafa F, Barolli L, Barolli A, Papajorgji P. Modeling and processing for next-generation big-data technologies: with applications and case studies. Modeling and optimization in science and technologies. New York: Springer; 2014.
58. Giannakis GB, Bach F, Cendrillon R, Mahoney M, Neville J. Signal processing for Big Data. Signal Process Mag IEEE. 2014;31(5):15–6.
59. Shneiderman B. The big picture for Big Data: visualization. Science. 2014;343:730.
60. Marr B. Big Data: using SMART Big Data. Analytics and metrics to make better decisions and improve performance. Hoboken: Wiley; 2015.
61. Minelli M, Chambers M, Dhiraj A. Big Data, big analytics: emerging business intelligence and analytic trends for today's businesses. Hoboken: Wiley; 2012.
62. Puget JF. Optimization is ready for Big Data. IBM White Paper. 2015.
63. Poli R, Rowe JE, Stephens CR, Wright AH. Allele diffusion in linear genetic programming and variable-length genetic algorithms with subtree crossover. New York: Springer; 2002.
64. Langdon WB. Genetic programming and data structures: genetic programming + data structures = Automatic Programming!, vol. 1. New York: Springer; 2012.
65. Poli R, Koza J. Genetic programming. New York: Springer; 2014.
66. Kothari DP. Power system optimization. In: Proceedings of 2nd national conference on computational intelligence and signal processing (CISP), IEEE; 2012. p. 18–21.
67. Moradi M, Abedini M. A combination of genetic algorithm and particle swarm optimization for optimal DG location and sizing in distribution systems. Int J Electr Power Energy Syst. 2012;34(1):66–74.
68. Engelbrecht A. Particle swarm optimization. In: Proceedings of the 2014 conference companion on genetic and evolutionary computation companion, ACM; 2014. p. 381–406.
69. Melanie M. An introduction to genetic algorithms. Cambridge, Fifth printing; 1999. p. 3.
70. Kitchin R. The data revolution: big data, open data. Data infrastructures and their consequences. California: SAGE Publications; 2014.
71. Pébay P, Thompson D, Bennett J, Mascarenhas A. Design and performance of a scalable, parallel statistics toolkit. In: Proceedings of international symposium on parallel and distributed processing workshops and Phd forum (IPDPSW), IEEE; 2011. p. 1475–84.
72. Bennett J, Grout R, Pébay P, Roe D, Thompson D. Numerically stable, single-pass, parallel statistics algorithms. In: International conference on cluster computing and workshops, IEEE; 2009. p. 1–8.
73. Lake P, Drake R. Information systems management in the Big Data era. Advanced information and knowledge processing. New York: Springer; 2015.
74. Anselin L, Getis A. Spatial statistical analysis and geographic information systems. Perspectives on spatial data analysis. New York: Springer; 2010. p. 35–47.

75. Kaufman L, Rousseeuw PJ. Finding groups in data: an introduction to cluster analysis, vol. 344. Hoboken: Wiley; 2009.
76. Anderberg MR. Cluster analysis for applications: probability and mathematical statistics—a series of monographs and textbooks, vol. 19. Cambridge: Academic press; 2014.
77. Hastie T, Tibshirani R, Friedman J. Unsupervised learning. New York: Springer; 2009.
78. Fisher DH, Pazzani MJ, Langley P. Concept formation: knowledge and experience in unsupervised learning. Burlington: Morgan Kaufmann; 2014.
79. McKenzie M, Wong S. Subset selection of training data for machine learning: a situational awareness system case study. In: SPIE sensing technology + applications. International society for optics and photonics; 2015.
80. Aggarwal CC. Data classification: algorithms and applications. Boca Raton: CRC Press; 2014.
81. Ryan TP. Modern regression methods. Wiley series in probability and statistics. Hoboken: Wiley; 2008.
82. Zhang C, Zhang S. Association rule mining: models and algorithms. New York: Springer; 2002.
83. Cleophas TJ, Zwinderman AH. Machine learning in medicine: part two. Machine learning in medicine. New York: Springer; 2013.
84. Bishop CM. Pattern recognition and machine learning. New York: Springer; 2006.
85. Devroye L, Györfi L, Lugosi G. A probabilistic theory of pattern recognition, vol. 31. New York: Springer; 2013.
86. Powers DM, Turk CC. Machine learning of natural language. New York: Springer; 2012.
87. Liu B, Zhang L. A survey of opinion mining and sentiment analysis. Mining text data. New York: Springer; 2012. p. 415–63.
88. Polikar R. Ensemble learning. Ensemble machine learning. New York: Springer; 2012. p. 1–34.
89. Zhang C, Ma Y. Ensemble machine learning. New York: Springer; 2012.
90. Helstrom CW. Statistical theory of signal detection: international series of monographs in electronics and instrumentation, vol. 9. Amsterdam: Elsevier; 2013.
91. Shumway RH, Stoffer DS. Time series analysis and its applications. New York: Springer; 2013.
92. Akaike H, Kitagawa G. The practice of time series analysis. New York: Springer; 2012.
93. Viswanathan R. Data fusion. Computer vision. Springer: New York; 2014. p. 166–8.
94. Castanedo F. A review of data fusion techniques. Sci World J. 2013.
95. Thompson D, Levine JA, Bennett JC, Bremer PT, Gyulassy A, Pascucci V, Pébay PP. Analysis of large-scale scalar data using hixels. In: Proceedings of symposium on large data analysis and visualization (LDAV), IEEE. 2011. p. 23–30.
96. Report: Data Visualization Applications Market Future Of Decision Making Trends, Forecasts And The Challengers (2014–2019). Mordor Intelligence; 2014.
97. SAS: Data visualization: making big data approachable and valuable. Market Pulse: White Paper; 2013.
98. Simon P. The visual organization: data visualization, Big Data, and the quest for better decisions. Hoboken: Wiley; 2014.
99. Kaisler S, Armour F, Espinosa JA, Money W. Big Data: issues and challenges moving forward. In: Proceedings of 46th Hawaii international conference on system sciences (HICSS), IEEE. 2013. p. 995–1004.
100. Tole AA, et al. Big Data challenges. Database Syst J. 2013;4(3):31–40.
101. Chen M, Mao S, Zhang Y, Leung VC. Big Data: related technologies. Challenges and future prospects. New York: Springer; 2014.
102. Miksch S, Aigner W. A matter of time: applying a data-users-tasks design triangle to visual analytics of time-oriented data. Comput Graph. 2014;38:286–90.
103. Miiller W, Schumann H. Visualization method for time-dependent data: an overview. In: Proceedings of the 2003 winter simulation conference, vol. 1. IEEE. 2003.
104. Telea AC. Data visualization: principles and practice, 2nd ed. Milton Park: Taylor & Francis; 2014.

105. Wright H. Introduction to scientific visualization. New York: Springer; 2007.
106. Bonneau GP, Ertl T, Nielson G. Scientific visualization: the visual extraction of knowledge from data. Mathematics and visualization. New York: Springer; 2006.
107. Rosenblum L, Rosenblum LJ. Scientific visualization: advances and challenges. Policy Series; 19. Academic; 1994.
108. Ware C. Information visualization: perception for design. Burlington: Morgan Kaufmann; 2013.
109. Kerren A, Stasko J, Fekete JD. Information visualization: human-centered issues and perspectives. LNCS sublibrary: information systems and applications, incl. Internet/Web, and HCI. New York: Springer; 2008.
110. Mazza R. Introduction to information visualization. Computer science. New York: Springer; 2009.
111. Bederson BB, Shneiderman B. The craft of information visualization: readings and reflections. Interactive technologies. Amsterdam: Elsevier Science; 2003.
112. Dill J, Earnshaw R, Kasik D, Vince J, Wong PC. Expanding the frontiers of visual analytics and visualization. SpringerLink: Bücher. New York: Springer; 2012.
113. Simoff S, Böhlen MH, Mazeika A. Visual data mining: theory, techniques and tools for visual analytics. LNCS sublibrary: information systems and applications, incl. Internet/Web, and HCI. New York: Springer; 2008.
114. Zhang Q. Visual analytics and interactive technologies: data, text and web mining applications: data. information science reference: text and web mining applications. Premier reference source; 2010.
115. Few S, EDGE P. Data visualization: past, present, and future. IBM Cognos Innovation Center; 2007.
116. Bertin J. La graphique. Communications. 1970;15:169–85.
117. Gray JJ. Johann Heinrich Lambert, mathematician and scientist, 1728–1777. Historia Math. 1978;5:13–41.
118. Tufte ER. The visual display for quantitative information. Chelshire: Graphics Press; 1983.
119. Kehrer J, Boubela RN, Filzmoser P, Piringer H. A generic model for the integration of interactive visualization and statistical computing using R. In: Conference on visual analytics science and technology (VAST), IEEE. 2012. p. 233–34.
120. Härdle W, Klinke S, Turlach B. XploRe: an interactive statistical computing environment. New York: Springer; 2012.
121. Friendly M. A brief history of data visualization. New York: Springer; 2006.
122. Mering C. Traditional node-link diagram of a network of yeast protein-protein and protein-DNA interactions with over 3,000 nodes and 6,800 links. Nature. 2002;417:399–403.
123. Febretti A, Nishimoto A, Thigpen T, Talandis J, Long L, Pirtle J, Peterka T, Verlo A, Brown M, Plepys D et al. CAVE2: a hybrid reality environment for immersive simulation and information analysis. In: IS&T/SPIE electronic imaging. International Society for Optics and Photonics; 2013.
124. Friendly M. Milestones in the history of data visualization: a case study in statistical historiography. Classification: the ubiquitous challenge. Springer: New York; 2005. p. 34–52.
125. Tory M, Kirkpatrick AE, Atkins MS, Moller T. Visualization task performance with 2D, 3D, and combination displays. IEEE Trans Visual Comput Graph. 2006;12(1):2–13.
126. Stanley R, Oliveria M, Zaiane OR. Geometric data transformation for privacy preserving clustering. Departament of Computing Science; 2003.
127. Healey CG, Enns JT. Large datasets at a glance: combining textures and colors in scientific visualization. IEEE Trans Visual Comput Graph. 1999;5(2):145–67.
128. Keim DA. Designing pixel-oriented visualization techniques: theory and applications. IEEE Trans Visual Comput Graph. 2000;6(1):59–78.
129. Kamel M, Camphilho A. Hierarchic image classification visualization. In: Proceedings of image analysis and recognition 10th international conference, ICIAR. 2013.
130. Buja A, Cook D, Asimov D, Hurley C. Computational methods for high-dimensional rotations in data visualization. Handbook Stat Data Mining Data Visual. 2004;24:391–415.

131. Meijester A, Westenberg MA, Wilkinson MHF. Interactive shape preserving filtering and visualization of volumetric data. In: Proceedings of the fourth IASTED international conference. 2002. p. 640–43.
132. Borg I, Groenen P. Modern multidimensional scaling: theory and applications. J Educ Measure. 2003;40:277–80.
133. Bajaj C, Krishnamurthy B. Data visualization techniques, vol. 6. Hoboken: Wiley; 1999.
134. Plaisant C, Monroe M, Meyer T, Shneiderman B. Interactive visualization. Boca Raton: CRC Press; 2014.
135. Janvrin DJ, Raschke RL, Dilla WN. Making sense of complex data using interactive data visualization. J Acc Educ. 2014;32(4):31–48.
136. Manyika J, Chui M, Brown B, Bughin J, Dobbs R, Roxburgh C, Byers AH. Big data: the next frontier for innovation, competition, and productivity. McKinsey Global Institute. 2011.
137. Ebert A, Dix A, Gershon ND, Pohl M. Human aspects of visualization: second IFIP WG 13.7 workshop on humancomputer interaction and visualization, HCIV (INTERACT), Uppsala, Sweden, August 24, 2009, Revised Selected Papers. LNCS sublibrary: information systems and applications, incl. Internet/Web, and HCI. Springer; 2009. p 2011.
138. Schonlau M. Visualizing non-hierarchical and hierarchical cluster analyses with clustergrams. Comput Stat. 2004;19(1):95–111.
139. Google, Inc.: Google Visualization Guide. 2015. https://developers.google.com.
140. Amcharts.com: amCharts visualization. 2004–2015. http://www.amcharts.com/.
141. Viégas F, Wattenberg M. IBM—Many Eyes Project. 2013. http://www-01.ibm.com/software/analytics/many-eyes/.
142. Körner C. Data Visualization with D3 and AngularJS. Community experience distilled. Birmingham: Packt Publishing; 2015.
143. Azzam T, Evergreen S. J-B PE single issue (program) evaluation, vol. pt. 1. Wiley.
144. Machlis S. Chart and image gallery: 30 + free tools for data visualization and analysis. 2015. http://www.computerworld.com/.
145. Julie Steele NI. Beautiful visualization: looking at data through the eyes of experts. O'Reilly Media; 2010.
146. Guberman S. On Gestalt theory principles. Gestalt Theory. 2015;37(1):25–44.
147. Chen C. Top 10 unsolved information visualization problems. Comput Graph Appl IEEE. 2005;25(4):12–6.
148. Johnson C. Top scientific visualization research problems. Comput Graph Appl IEEE. 2004;24(4):13–7.
149. Tory M, Möller T. Human factors in visualization research. Trans Visual Comput Graph. 2004;10(1):72–84.
150. Andrews C, Endert A, Yost B, North C. Information visualization on large, high-resolution displays: Issues, challenges, and opportunities. Inf Vis. 2011.
151. Suthaharan S. Big Data classification: problems and challenges in network intrusion prediction with machine learning. ACM SIGMETRICS. 2014;41:70–3.
152. Field DJ, Hayes A, Hess RF. Contour integration by the human visual system: evidence for a local "association field". Vision Res. 1993;33:173–93.
153. Picard RW, Healy J. Affective wearables, vol. 1. New York: Springer; 1997. p. 231–40.
154. Mann S et al. Wearable technology. St James Ethics Centre; 2014.
155. Carmigniani J, Furht B, Anisetti M, Ceravolo P, Damiani E, Ivkovic M. Augmented reality technologies, systems and applications. 2010;51:341–77.
156. Papagiannakis G, Singh G, Magnenat-Thalmann N. A survey of mobile and wireless technologies for augmented reality systems. Comput Anim Virtual Worlds. 2008;19(1):3–22.
157. Caudell TP, Mizell DW. Augmented reality: an application of heads-up display technology to manual manufacturing processes. IEEE Syst Sci. 1992;2:659–69.
158. Sutherland I. A head-mounted three dimensional display. In: Proceedings of the fall joint computer conference. 1968. p. 757–64.

159. Chacos B. Shining light on virtual reality: busting the 5 most inaccurate Oculus Rift myths. PCWorld; 2014.
160. Krevelen DWF, Poelman R. A survey of augmented reality technologies, applications and limitations. Int J Virtual Real. 2010;9:1–20.
161. Stevens J, Eifert L. Augmented reality technology in US army training (WIP). In: Proceedings of the 2014 summer simulation multiconference, society for computer simulation international. 2014. p. 62.
162. Bower M, Howe C, McCredie N, Robinson A, Grover D. Augmented reality in education-cases, places and potentials. Educ Media Int. 2014;51(1):1–15.
163. Ma M, Jain LC, Anderson P. Future trends of virtual, augmented reality, and games for health. Virtual, augmented reality and serious games for healthcare, vol. 1. New York: Springer; 2014. p. 1–6.
164. Mousavi M, Abdul Aziz F, Ismail N. Investigation of 3D modelling and virtual reality systems in malaysian automotive industry. In: Proceedings of international conference on computer, communications and information technology. Atlantis Press. 2014.
165. Chung IC, Huang CY, Yeh SC, Chiang WC, Tseng MH. Developing kinect games integrated with virtual reality on activities of daily living for children with developmental delay. Advanced technologies, embedded and multimedia for human-centric computing. New York: Springer; 2014. p. 1091–7.
166. Steptoe W. AR-Rift: stereo camera for the rift and immersive AR showcase. Oculus Developer Forums. 2013.
167. Pina JL, Cerezo E, Seron F. Semantic visualization of 3D urban environments. Multimed Tools Appl. 2012;59:505–21.
168. Fonseca D, Villagrasa S, Marta N, Redondo E, Sanchez A. Visualization methods in architecture education using 3D virtual models and augmented reality in mobile and social networks. Procedia Soc Behav Sci. 2013;93:1337–43.
169. Varkey JP, Pompili D, Walls TA. Human motion recognition using a wireless sensor-based wearable system. Personal Ubiquitous Comput. 2011;16:897–910.
170. Nuwer R. Armband adds a twitch to gesture control. New Sci. 2013;217(2906):21.
171. Timberlake GT, Mainster MA, Peli E, Augliere RA, Essock EA, Arend LE. Reading with a macular scotoma I Retinal location of scotoma and fixation area. Investig Ophthalmol Visual Sci. 1986;27(7):1137–47.
172. Foster PJ, Buhrmann R, Quigley HA, Johnson GJ. The definition and classification of glaucoma in prevalence surveys. Br J Ophthalmol. 2002;86:238–42.
173. Deering MF. The limits of human vision. In: Proceedings the 2nd international immersive projection technology workshop. 1998.
174. Krantz J. Experiencing sensation and perception. Pearson Education (US). 2012.
175. Rajanbabu A, Drudi L, Lau S, Press JZ, Gotlieb WH. Virtual reality surgical simulators-a prerequisite for robotic surgery. Indian J Surg Oncol. 2014;5(2):1–3.
176. Moglia A, Ferrari V, Morelli L, Melfi F, Ferrari M, Mosca F, Cuschieri A. Distribution of innate ability for surgery amongst medical students assessed by an advanced virtual reality surgical simulator. Surg Endosc. 2014;28(6):1830–7.
177. Ahn W, Dargar S, Halic T, Lee J, Li B, Pan J, Sankaranarayanan G, Roberts K, De S. Development of a virtual reality simulator for natural orifice translumenal endoscopic surgery (NOTES) cholecystectomy procedure. Medicine Meets Virtual Reality 21: NextMed/ MMVR21 2014;196, 1.
178. Ma M, Jain LC, Anderson P. Virtual, augmented reality and serious games for healthcare 1. New York: Springer; 2014.
179. Wright WG. Using virtual reality to augment perception, enhance sensorimotor adaptation, and change our minds. Front Syst Neurosci. 2014;8:56.
180. Parsons TD, Trost Z. Virtual reality graded exposure therapy as treatment for pain-related fear and disability in chronic pain. Virtual, augmented reality and serious games for healthcare 1. New York: Springer; 2014. p. 523–46.
181. Abramov I, Gordon J, Feldman O, Chavarga A. Biology of sex differences. p. 1–14.

182. McFadden D. Masculinization effects in the auditory system. Archiv Sexual Behav. 2002;31(1):99–111.
183. Voyer D, Voyer S, Bryden MP. Magnitude of sex differences in spatial abilities: a meta-analysis and consideration of critical variables. Psychol Bull. 1995;117:250–70.
184. Stancey H, Turner M. Close women, distant men: line bisection reveals sex-dimorphic patterns of visuomotor performance in near and far space. Br J Psychol. 2010;101:293–309.
185. Rizzolatti G, Matelli M, Pavesi G. Deficits in attention and movement following the removal of postarcuate (area 6) and prearcuate (area 8) cortex in macaque monkeys. Brain. 1983;106:655–73.
186. Chua HF, Boland JE, Nisbett RE. Cultural variation in eye movements during scene perception. PNA. 2005;102(35):12629–33.
187. Zelinsky GJ, Adeli H, Peng Y, Samaras D. Modelling eye movements in a categorical search task. Philos Trans R Soc. 2013.
188. Piumsomboon T, Clark A, Billinghurst M, Cockburn A. user-defined gestures for augmented reality. In: Human computer interaction–INTERACT 2013, Springer. 2013. p. 282–99.
189. Mistry P, Maes P, Chang L. WUW-wear Ur world: a wearable gestural interface. In: Extended abstracts on human factors in computing systems, ACM. 2009. p. 4111–16.
190. Vanacken D, Beznosyk A, Coninx K. Help systems for gestural interfaces and their effect on collaboration and communication. In: Workshop on gesture-based interaction design: communication and cognition. 2014.
191. Mulling T, Lopes C, Cabreira A. Gestural interfaces touchscreen: thinking interactions beyond the button from interaction design for Gmail Android App. In: Design, sser experience, and usability. User experience design for diverse interaction platforms and environments. Springer. 2014. p. 279–88.
192. Piumsomboon T, Clark A., Billinghurst M. [DEMO] G-SIAR: gesture-speech interface for augmented reality. In: Proceedings of International symposium on mixed and augmented reality (ISMAR), IEEE; 2014. p. 365–66.
193. Vafadar M, Behrad A. A vision based system for communicating in virtual reality environments by recognizing human hand gestures. Multi Tools Appl. 2014;74(18):1–21.
194. Roupé M, Bosch-Sijtsema P, Johansson M. Interactive navigation interface for Virtual Reality using the human body. Comput Environ Urban Syst. 2014;43:42–50.
195. Wen R, Tay WL, Nguyen BP, Chng CB, Chui CK. Hand gesture guided robot-assisted surgery based on a direct augmented reality interface. Comput Method Program Biomed. 2014;116(2):68–80.
196. Rolland JP, Fuchs H. Optical versus video see-through head-mounted displays in medical visualization. Presence Teleoperators Virtual Environ. 2000;9(3):287–309.
197. Silanon K, Suvonvorn N. Real time hand tracking as a user input device. New York: Springer; 2011. p. 178–89.
198. Keim DA, Mansmann F, Schneidewind J, Ziegler H. Challenges in visual data analysis. In: Proceedings of 10th international conference on information visualization, IEEE. 2006. p. 9–16.
199. Sonka M, Hlavac V, Boyle R. Image processing, analysis, and machine vision. Cengage Learning. 2014.
200. Shneiderman B. The eyes have it: a task by data type taxonomy for information visualizations. In: Proceedings of IEEE symposium on visual languages. 1996. p. 336–43.
201. Coffey D, Malbraaten N, Le T, Borazjani I, Sotiropoulos F, Keefe DF. Slice WIM: a multi-surface, multi-touch interface for overview + detail exploration of volume datasets in virtual reality. In: Proceedings of symposium on interactive 3D graphics and games, ACM. 2011. p. 191–98.

Chapter 5
Deep Learning Techniques in Big Data Analytics

Maryam M. Najafabadi, Flavio Villanustre, Taghi M. Khoshgoftaar, Naeem Seliya, Randall Wald and Edin Muharemagc

Introduction

The general focus of machine learning is the representation of the input data and generalization of the learnt patterns for use on future unseen data. The goodness of the data representation has a large impact on the performance of machine learners on the data: a poor data representation is likely to reduce the performance of even an advanced, complex machine learner, while a good data representation can lead to high performance for a relatively simpler machine learner. Thus, feature engineering, which focuses on constructing features and data representations from raw data [1], is an important element of machine learning. Feature engineering consumes a large portion of the effort in a machine learning task, and is typically quite domain specific and involves considerable human input. For example, the Histogram of Oriented Gradients (HOG) [2] and Scale Invariant Feature Transform (SIFT) [3] are popular feature engineering algorithms developed specifically for the computer vision domain. Performing feature engineering in a more automated and general fashion would be a major breakthrough in machine learning as this would allow practitioners to automatically extract such features without direct human input.

Deep Learning algorithms are one promising avenue of research into the automated extraction of complex data representations (features) at high levels of abstraction. Such algorithms develop a layered, hierarchical architecture of learning and representing data, where higher-level (more abstract) features are defined in terms of lower-level (less abstract) features. The hierarchical learning architecture of Deep Learning algorithms is motivated by artificial intelligence emulating the

---

This chapter has been adopted from the Journal of Big Data, Borko Furht and Taghi Khoshgoftar, Editors-in-Chief.

deep, layered learning process of the primary sensorial areas of the neocortex in the human brain, which automatically extracts features and abstractions from the underlying data [4–6]. Deep Learning algorithms are quite beneficial when dealing with learning from large amounts of unsupervised data, and typically learn data representations in a greedy layer-wise fashion [7, 8]. Empirical studies have demonstrated that data representations obtained from stacking up nonlinear feature extractors (as in Deep Learning) often yield better machine learning results, e.g., improved classification modeling [9], better quality of generated samples by generative probabilistic models [10], and the invariant property of data representations [11]. Deep Learning solutions have yielded outstanding results in different machine learning applications, including speech recognition [12–16], computer vision [7, 8, 17], and natural language processing [18–20]. A more detailed overview of Deep Learning is presented in "Deep Learning in Data Mining and Machine Learning" section.

Big Data represents the general realm of problems and techniques used for application domains that collect and maintain massive volumes of raw data for domain-specific data analysis. Modern data-intensive technologies as well as increased computational and data storage resources have contributed heavily to the development of Big Data science [21]. Technology based companies such as Google, Yahoo, Microsoft, and Amazon have collected and maintained data that is measured in exabyte proportions or larger. Moreover, social media organizations such as Facebook, YouTube, and Twitter have billions of users that constantly generate a very large quantity of data. Various organizations have invested in developing products using Big Data Analytics to addressing their monitoring, experimentation, data analysis, simulations, and other knowledge and business needs [22], making it a central topic in data science research.

Mining and extracting meaningful patterns from massive input data for decision making, prediction, and other inferencing is at the core of Big Data Analytics. In addition to analyzing massive volumes of data, Big Data Analytics poses other unique challenges for machine learning and data analysis, including format variation of the raw data, fast moving streaming data, trustworthiness of the data analysis, highly distributed input sources, noisy and poor quality data, high dimensionality, scalability of algorithms, imbalanced input data, unsupervised and un-categorized data, limited supervised/labeled data, etc. Adequate data storage, data indexing/tagging, and fast information retrieval are other key problems in Big Data Analytics. Consequently, innovative data analysis and data management solutions are warranted when working with Big Data. For example, in a recent work we examined the high-dimensionality of bioinformatics domain data and investigated feature selection techniques to address the problem [23]. A more detailed overview of Big Data Analytics is presented in "Big Data Analytics" section.

The knowledge learnt from (and made available by) Deep Learning algorithms has been largely untapped in the context of Big Data Analytics. Certain Big Data domains, such as computer vision [17] and speech recognition [13], have seen the application of Deep Learning largely to improve classification modeling results. The ability of Deep Learning to extract high-level, complex abstractions and data representations from large volumes of data, especially unsupervised data, makes it attractive as a valuable tool for Big Data Analytics. More specifically, Big Data problems such as semantic indexing, data tagging, fast information retrieval, and discriminative modeling can be better addressed with the aid of Deep Learning. More traditional machine learning and feature engineering algorithms are not efficient enough to extract the complex and non-linear patterns generally observed in Big Data. By extracting such features, Deep Learning enables the use of relatively simpler linear models for Big Data analysis tasks, such as classification and prediction, which is important when developing models to deal with the scale of Big Data. The novelty of this study is that it explores the application of Deep Learning algorithms for key problems in Big Data Analytics, motivating further targeted research by experts in these two fields.

The paper focuses on two key topics: (1) how Deep Learning can assist with specific problems in Big Data Analytics, and (2) how specific areas of Deep Learning can be improved to reflect certain challenges associated with Big Data Analytics. With respect to the first topic, we explore the application of Deep Learning for specific Big Data Analytics, including learning from massive volumes of data, semantic indexing, discriminative tasks, and data tagging. Our investigation regarding the second topic focuses on specific challenges Deep Learning faces due to existing problems in Big Data Analytics, including learning from streaming data, dealing with high dimensionality of data, scalability of models, and distributed and parallel computing. We conclude by identifying important future areas needing innovation in Deep Learning for Big Data Analytics, including data sampling for generating useful high-level abstractions, domain (data distribution) adaption, defining criteria for extracting good data representations for discriminative and indexing tasks, semi-supervised learning, and active learning.

The remainder of the paper is structured as follows: "Deep Learning in Data Mining and Machine Learning" section presents an overview of Deep Learning for data analysis in data mining and machine learning; "Big Data Analytics" section presents an overview of Big Data Analytics, including key characteristics of Big Data and identifying specific data analysis problems faced in Big Data Analytics; "Applications of Deep Learning in Big Data Analytics" section presents a targeted survey of works investigating Deep Learning based solutions for data analysis, and discusses how Deep Learning can be applied for Big Data Analytics problems; "Deep Learning Challenges in Big Data Analytics" section discusses some challenges faced by Deep Learning experts due to specific data analysis needs of Big Data; "Future Work on Deep Learning in Big Data Analytics" section presents our insights into further works that are necessary for extending the application of Deep Learning in Big Data, and poses important questions to domain experts; and in "Conclusion" section we reiterate the focus of the paper and summarize the work presented.

Deep Learning in Data Mining and Machine Learning

The main concept in deep leaning algorithms is automating the extraction of representations (abstractions) from the data [5, 24, 25]. Deep learning algorithms use a huge amount of unsupervised data to automatically extract complex representation. These algorithms are largely motivated by the field of artificial intelligence, which has the general goal of emulating the human brain's ability to observe, analyze, learn, and make decisions, especially for extremely complex problems. Work pertaining to these complex challenges has been a key motivation behind Deep Learning algorithms which strive to emulate the hierarchical learning approach of the human brain. Models based on shallow learning architectures such as decision trees, support vector machines, and case-based reasoning may fall short when attempting to extract useful information from complex structures and relationships in the input corpus. In contrast, Deep Learning architectures have the capability to generalize in non-local and global ways, generating learning patterns and relationships beyond immediate neighbors in the data [4]. Deep learning is in fact an important step toward artificial intelligence. It not only provides complex representations of data which are suitable for AI tasks but also makes the machines independent of human knowledge which is the ultimate goal of AI. It extracts representations directly from unsupervised data without human interference.

A key concept underlying Deep Learning methods is distributed representations of the data, in which a large number of possible configurations of the abstract features of the input data are feasible, allowing for a compact representation of each sample and leading to a richer generalization. The number of possible configurations is exponentially related to the number of extracted abstract features. Noting that the observed data was generated through interactions of several known/unknown factors, and thus when a data pattern is obtained through some configurations of learnt factors, additional (unseen) data patterns can likely be described through new configurations of the learnt factors and patterns [5, 24]. Compared to learning based on local generalizations, the number of patterns that can be obtained using a distributed representation scales quickly with the number of learnt factors.

Deep learning algorithms lead to abstract representations because more abstract representations are often constructed based on less abstract ones. An important advantage of more abstract representations is that they can be invariant to the local changes in the input data. Learning such invariant features is an ongoing major goal in pattern recognition (for example learning features that are invariant to the face orientation in a face recognition task). Beyond being invariant such representations can also disentangle the factors of variation in data. The real data used in AI-related tasks mostly arise from complicated interactions of many sources. For example an image is composed of different sources of variations such a light, object shapes, and object materials. The abstract representations provided by deep learning algorithms can separate the different sources of variations in data.

Deep learning algorithms are actually Deep architectures of consecutive layers. Each layer applies a nonlinear transformation on its input and provides a representation in its output. The objective is to learn a complicated and abstract representation of the data in a hierarchical manner by passing the data through multiple transformation layers. The sensory data (for example pixels in an image) is fed to the first layer. Consequently the output of each layer is provided as input to its next layer.

Stacking up the nonlinear transformation layers is the basic idea in deep learning algorithms. The more layers the data goes through in the deep architecture, the more complicated the nonlinear transformations which are constructed. These transformations represent the data, so Deep Learning can be considered as special case of representation learning algorithms which learn representations of the data in a Deep Architecture with multiple levels of representations. The achieved final representation is a highly non-linear function of the input data.

It is important to note that the transformations in the layers of deep architecture are non-linear transformations which try to extract underlying explanatory factors in the data. One cannot use a linear transformation like PCA as the transformation algorithms in the layers of the deep structure because the compositions of linear transformations yield another linear transformation. Therefore, there would be no point in having a deep architecture. For example by providing some face images to the Deep Learning algorithm, at the first layer it can learn the edges in different orientations; in the second layer it composes these edges to learn more complex features like different parts of a face such as lips, noses and eyes. In the third layer it composes these features to learn even more complex feature like face shapes of different persons. These final representations can be used as feature in applications of face recognition. This example is provided to simply explain in an understandable way how a deep learning algorithm finds more abstract and complicated representations of data by composing representations acquired in a hierarchical architecture. However, it must be considered that deep learning algorithms do not necessarily attempt to construct a pre-defined sequence of representations at each layer (such as edges, eyes, faces), but instead more generally perform non-linear transformations in different layers. These transformations tend to disentangle factors of variations in data. Translating this concept to appropriate training criteria is still one of the main open questions in deep learning algorithms [5].

The final representation of data constructed by the deep learning algorithm (output of the final layer) provides useful information from the data which can be used as features in building classifiers, or even can be used for data indexing and other applications which are more efficient when using abstract representations of data rather than high dimensional sensory data.

Learning the parameters in a deep architecture is a difficult optimization task, such as learning the parameters in neural networks with many hidden layers. In 2006 Hinton proposed learning deep architectures in an unsupervised greedy layer-wise learning manner [7]. At the beginning the sensory data is fed as learning data to the first layer. The first layer is then trained based on this data, and the output of the first layer (the first level of learnt representations) is provided as learning data to the second layer. Such iteration is done until the desired number of layers is obtained. At this point the deep network is trained. The representations learnt on the last layer can be used for different tasks. If the task is a classification task usually another supervised layer is put on top of the last layer and its parameters are learnt (either randomly or by using supervised data and keeping the rest of the network fixed). At the end the whole network is fine-tuned by providing supervised data to it.

Here we explain two fundamental building blocks, unsupervised single layer learning algorithms which are used to construct deeper models: Autoencoders and Restricted Boltzmann Machines (RBMs). These are often employed in tandem to construct stacked Autoencoders [8, 26] and Deep belief networks [7], which are constructed by stacking up Autoencoders and Restricted Boltzmann Machines respectively. Autoencoders, also called autoassociators [27], are networks constructed of 3 layers: input, hidden and output. Autoencoders try to learn some representations of the input in the hidden layer in a way that makes it possible to reconstruct the input in the output layer based on these intermediate representations. Thus, the target output is the input itself. A basic Autoencoder learns its parameters by minimizing the reconstruction error. This minimization is usually done by stochastic gradient descent (much like what is done in Multilayer Perceptron). If the hidden layer is linear and the mean squared error is used as the reconstruction criteria, then the Autoencoder will learn the first k principle components of the data. Alternative strategies are proposed to make Autoencoders nonlinear which are appropriate to build deep networks as well as to extract meaningful representations of data rather than performing just as a dimensionality reduction method. Bengio et al. [5] have called these methods "regularized Autoencoders", and we refer an interested reader to that paper for more details on algorithms.

Another unsupervised single layer learning algorithm which is used as a building block in constructing Deep Belief Networks is the Restricted Boltzmann machine (RBM). RBMs are most likely the most popular version of Boltzmann machine [28]. They contains one visible layer and one hidden layer. The restriction is that there is no interaction between the units of the same layer and the connections are solely between units from different layers. The Contrastive Divergence algorithm [29] has mostly been used to train the Boltzmann machine.

Big Data Analytics

Big Data generally refers to data that exceeds the typical storage, processing, and computing capacity of conventional databases and data analysis techniques. As a resource, Big Data requires tools and methods that can be applied to analyze and extract patterns from large-scale data. The rise of Big Data has been caused by increased data storage capabilities, increased computational processing power, and availability of increased volumes of data, which give organization more data than they have computing resources and technologies to process. In addition to the obvious great volumes of data, Big Data is also associated with other specific complexities, often referred to as the four Vs: Volume, Variety, Velocity, and Veracity [22, 30, 31]. We note that the aim of this section is not to extensively cover Big Data, but present a brief overview of its key concepts and challenges while keeping in mind that the use of Deep Learning in Big Data Analytics is the focus of this paper.

The unmanageable large Volume of data poses an immediate challenge to conventional computing environments and requires scalable storage and a distributed strategy to data querying and analysis. However, this large Volume of data is also a major positive feature of Big Data. Many companies, such as Facebook, Yahoo, Google, already have large amounts of data and have recently begun tapping into its benefits [21]. A general theme in Big Data systems is that the raw data is increasingly diverse and complex, consisting of largely un-categorized/unsupervised data along with perhaps a small quantity of categorized/supervised data. Working with the Variety among different data representations in a given repository poses unique challenges with Big Data, which requires Big Data preprocessing of unstructured data in order to extract structured/ordered representations of the data for human and/or downstream consumption. In today's data-intensive technology era, data Velocity—the increasing rate at which data is collected and obtained—is just as important as the Volume and Variety characteristics of Big Data. While the possibility of data loss exists with streaming data if it is generally not immediately processed and analyzed, there is the option to save fast-moving data into bulk storage for batch processing at a later time. However, the practical importance of dealing with Velocity associated with Big Data is the quickness of the feedback loop, that is, process of translating data input into useable information. This is especially important in the case of time-sensitive information processing. Some companies such as Twitter, Yahoo, and IBM have developed products that address the analysis of streaming data [22]. Veracity in Big Data deals with the trustworthiness or usefulness of results obtained from data analysis, and brings to light the old adage "Garbage-In-Garbage-Out" for decision making based on Big Data Analytics. As the number of data sources and types increases, sustaining trust in Big Data Analytics presents a practical challenge.

Big Data Analytics faces a number of challenges beyond those implied by the four Vs. While not meant to be an exhaustive list, some key problem areas include: data quality and validation, data cleansing, feature engineering, high-dimensionality and data reduction, data representations and distributed data sources, data sampling, scalability of algorithms, data visualization, parallel and distributed data processing, real-time analysis and decision making, crowdsourcing and semantic input for improved data analysis, tracing and analyzing data provenance, data discovery and integration, parallel and distributed computing, exploratory data analysis and interpretation, integrating heterogenous data, and developing new models for massive data computation.

Applications of Deep Learning in Big Data Analytics

As stated previously, Deep Learning algorithms extract meaningful abstract representations of the raw data through the use of an hierarchical multi-level learning approach, where in a higher-level more abstract and complex representations are learnt based on the less abstract concepts and representations in the lower level(s) of the learning hierarchy. While Deep Learning can be applied to learn from labeled data if it is available in sufficiently large amounts, it is primarily attractive for learning from large amounts of unlabeled/unsupervised data [4, 5, 25], making it attractive for extracting meaningful representations and patterns from Big Data.

Once the hierarchical data abstractions are learnt from unsupervised data with Deep Learning, more conventional discriminative models can be trained with the aid of relatively fewer supervised/labeled data points, where the labeled data is typically obtained through human/expert input. Deep Learning algorithms are shown to perform better at extracting non-local and global relationships and patterns in the data, compared to relatively shallow learning architectures [4]. Other useful characteristics of the learnt abstract representations by Deep Learning include: (1) relatively simple linear models can work effectively with the knowledge obtained from the more complex and more abstract data representations, (2) increased automation of data representation extraction from unsupervised data enables its broad application to different data types, such as image, textural, audio, etc., and (3) relational and semantic knowledge can be obtained at the higher levels of abstraction and representation of the raw data. While there are other useful aspects of Deep Learning based representations of data, the specific characteristics mentioned above are particularly important for Big Data Analytics.

Considering each of the four Vs of Big Data characteristics, i.e., Volume, Variety, Velocity, and Veracity, Deep Learning algorithms and architectures are more aptly suited to address issues related to Volume and Variety of Big Data Analytics. Deep Learning inherently exploits the availability of massive amounts of data, i.e., Volume in Big Data, where algorithms with shallow learning hierarchies fail to explore and understand the higher complexities of data patterns. Moreover, since Deep Learning deals with data abstraction and representations, it is quite likely suited for analyzing raw data presented in different formats and/or from different sources, i.e., Variety in Big Data, and may minimize need for input from human experts to extract features from every new data type observed in Big Data. While presenting different challenges for more conventional data analysis approaches, Big Data Analytics presents an important opportunity for developing novel algorithms and models to address specific issues related to Big Data. Deep Learning concepts provide one such solution venue for data analytics experts and practitioners. For example, the extracted representations by Deep Learning can be considered as a practical source of knowledge for decision-making, semantic indexing, information retrieval, and for other purposes in Big Data Analytics, and in addition, simple linear modeling techniques can be considered for Big Data Analytics when complex data is represented in higher forms of abstraction.

In the remainder of this section, we summarize some important works that have been performed in the field of Deep Learning algorithms and architectures, including semantic indexing, discriminative tasks, and data tagging. Our focus is that by presenting these works in Deep Learning, experts can observe the novel applicability of Deep Learning techniques in Big Data Analytics, particularly since some of the application domains in the works presented involve large scale data. Deep Learning algorithms are applicable to different kinds of input data; however, in this section we focus on its application on image, textual, and audio data.

Semantic Indexing

A key task associated with Big Data Analytics is information retrieval [21]. Efficient storage and retrieval of information is a growing problem in Big Data, particularly since very large-scale quantities of data such as text, image, video, and audio are being collected and made available across various domains, e.g., social networks, security systems, shopping and marketing systems, defense systems, fraud detection, and cyber traffic monitoring. Previous strategies and solutions for information storage and retrieval are challenged by the massive volumes of data and different data representations, both associated with Big Data. In these systems, massive amounts of data are available that needs semantic indexing rather than being stored as data bit strings. Semantic indexing presents the data in a more efficient manner and makes it useful as a source for knowledge discovery and comprehension, for example by making search engines work more quickly and efficiently.

Instead of using raw input for data indexing, Deep Learning can be used to generate high-level abstract data representations which will be used for semantic indexing. These representations can reveal complex associations and factors (especially when the raw input was Big Data), leading to semantic knowledge and understanding. Data representations play an important role in the indexing of data, for example by allowing data points/instances with relatively similar representations to be stored closer to one another in memory, aiding in efficient information retrieval. It should be noted, however, that the high-level abstract data representations need to be meaningful and demonstrate relational and semantic association in order to actually confer a good semantic understanding and comprehension of the input.

While Deep Learning aids in providing a semantic and relational understanding of the data, a vector representation (corresponding to the extracted representations) of data instances would provide faster searching and information retrieval. More specifically, since the learnt complex data representations contain semantic and relational information instead of just raw bit data, they can directly be used for semantic indexing when each data point (for example a given text document) is presented by a vector representation, allowing for a vector-based comparison which is more efficient than comparing instances based directly on raw data. The data instances that have similar vector representations are likely to have similar semantic meaning. Thus, using vector representations of complex high-level data abstractions for indexing the data makes semantic indexing feasible. In the remainder of this section, we focus on document indexing based on knowledge gained from Deep Learning. However, the general idea of indexing based on data representations obtained from Deep Learning can be extended to other forms of data.

Document (or textual) representation is a key aspect in information retrieval for many domains. The goal of document representation is to create a representation that condenses specific and unique aspects of the document, e.g., document topic. Document retrieval and classification systems are largely based on word counts, representing the number of times each word occurs in the document. Various document retrieval schemas use such a strategy, e.g., TF-IDF [32] and BM25 [33]. Such document representation schemas consider individual words to be dimensions, with different dimensions being independent. In practice, it is often observed that the occurrence of words are highly correlated. Using Deep Learning techniques to extract meaningful data representations makes it possible to obtain semantic features from such high-dimensional textual data, which in turn also leads to the reduction of the dimensions of the document data representations.

Hinton and Salakhutdinov [34] describe a Deep Learning generative model to learn the binary codes for documents. The lowest layer of the Deep Learning network represents the word count vector of the document which accounts as high-dimensional data, while the highest layer represents the learnt binary code of the document. Using 128-bit codes, the authors demonstrate that the binary codes of the documents that are semantically similar lay relatively closer in the Hamming space. The binary code of the documents can then be used for information retrieval. For each query document, its Hamming distance compared to all other documents in the data is computed and the top D similar documents are retrieved. Binary codes require relatively little storage space, and in addition they allow relatively quicker searches by using algorithms such as fast-bit counting to compute the Hamming distance between two binary codes. The authors conclude that using these binary codes for document retrieval is more accurate and faster than semantic-based analysis.

Deep Learning generative models can also be used to produce shorter binary codes by forcing the highest layer in the learning hierarchy to use a relatively small number of variables. These shorter binary codes can then simply be used as memory addresses. One word of memory is used to describe each document in such a way that a small Hammingball around that memory address contains semantically similar documents—such a technique is referred as "semantic hashing" [35]. Using such a strategy, one can perform information retrieval on a very large document set with the retrieval time being independent of the document set size. Techniques such as semantic hashing are quite attractive for information retrieval, because documents that are similar to the query document can be retrieved by finding all the memory addresses that differ from the memory address of the query document by a few bits. The authors demonstrate that "memory hashing" is much faster than locality-sensitive hashing, which is one of the fastest methods among existing algorithms. In addition, it is shown that by providing a document's binary codes to algorithms such as TF-IDF instead of providing the entire document, a higher level of accuracy can be achieved. While Deep Learning generative models can have a relatively slow learning/training time for producing binary codes for document retrieval, the resulting knowledge yields fast inferences which is one major goal of Big Data Analytics. More specifically, producing the binary code for a new document requires just a few vector matrix computations performing a feed-forward pass through the encoder component of the Deep Learning network architecture.

To learn better representations and abstractions, one can use some supervised data in training the Deep Learning model. Ranzato and Szummer [36] present a study in which parameters of the Deep Learning model are learnt based on both supervised and unsupervised data. The advantages of such a strategy are that there is no need to completely label a large collection of data (as some unlabeled data is expected) and that the model has some prior knowledge (via the supervised data) to capture relevant class/label information in the data. In other words, the model is required to learn data representations that produce good reconstructions of the input in addition to providing good predictions of document class labels. The authors show that for learning compact representations, Deep Learning models are better than shallow learning models. The compact representations are efficient because they require fewer computations when used in indexing, and in addition, also need less storage capacity.

Google's "word2vec" tool is another technique for automated extraction of semantic representations from Big Data. This tool takes a large-scale text corpus as input and produces the word vectors as output. It first constructs a vocabulary from the training text data and then learns vector representation of words, upon which the word vector file can be used as features in many Natural Language Processing (NLP) and machine learning applications. Miklov et al. [37] introduce techniques to learn high-quality word vectors from huge datasets with hundreds of millions of words (including some datasets containing 1.6 billion words), and with millions of distinct words in the vocabulary. They focus on artificial neural networks to learn the distributed representation of words. To train the network on such a massive dataset, the models are implemented on top of the large-scale distributed framework "DistBelief" [38]. The authors find that word vectors which are trained on massive amounts of data show subtle semantic relationships between words, such as a city and the country it belongs to—for example, Paris belongs to France and Berlin belongs to Germany. Word vectors with such semantic relationships could be used to improve many existing NLP applications, such as machine translation, information retrieval, and question response systems. For example, in a related work, Miklov et al. [39] demonstrate how word2vec can be applied for natural language translation.

Deep Learning algorithms make it possible to learn complex nonlinear representations between word occurrences, which allow the capture of high-level semantic aspects of the document (which could not normally be learned with linear models). Capturing these complex representations requires massive amounts of data for the input corpus, and producing labeled data from this massive input is a difficult task. With Deep Learning one can leverage unlabeled documents (unsupervised data) to have access to a much larger amount of input data, using a smaller amount of supervised data to improve the data representations and make them more related to the specific learning and inference tasks. The extracted data representations have been shown to be effective for retrieving documents, making them very useful for search engines.

Similar to textual data, Deep Learning can be used on other kinds of data to extract semantic representations from the input corpus, allowing for semantic indexing of that data. Given the relatively recent emergence of Deep Learning, additional work needs to be done on using its hierarchical learning strategy as a method for semantic indexing of Big Data. A remaining open question is what criteria is used to define "similar" when trying to extract data representations for indexing purposes (recall, data points that are semantically similar will have similar data representations in a specific distance space).

Discriminative Tasks and Semantic Tagging

In performing discriminative tasks in Big Data Analytics one can use Deep Learning algorithms to extract complicated nonlinear features from the raw data, and then use simple linear models to perform discriminative tasks using the extracted features as input. This approach has two advantages: (1) extracting features with Deep Learning adds nonlinearity to the data analysis, associating the discriminative tasks closely to Artificial Intelligence, and (2) applying relatively simple linear analytical models on the extracted features is more computationally efficient, which is important for Big Data Analytics. The problem of developing efficient linear models for Big Data Analytics has been extensively investigated in the literature [21]. Hence, developing nonlinear features from massive amounts of input data allows the data analysts to benefit from the knowledge available through the massive amounts of data, by applying the learnt knowledge to simpler linear models for further analysis. This is an important benefit of using Deep Learning in Big Data Analytics, allowing practitioners to accomplish complicated tasks related to Artificial Intelligence, such as image comprehension, object recognition in images, etc., by using simpler models. Thus discriminative tasks are made relatively easier in Big Data Analytics with the aid of Deep Learning algorithms.

Discriminative analysis in Big Data Analytics can be the primary purpose of the data analysis, or it can be performed to conduct tagging (such as semantic tagging) on the data for the purpose of searching. For example, Li et al. [40] explore the Microsoft Research Audio Video Indexing System (MAVIS) that uses Deep Learning (with Artificial Neural Networks) based speech recognition technology to enable searching of audio and video files with speech. To converting digital audio and video signals into words, MAVIS automatically generates closed captions and keywords that can increase accessibility and discovery of audio and video files with speech content.

Considering the development of the Internet and the explosion of online users in recent years, there has been a very rapid increase in the size of digital image collections. These come from sources such as social networks, global positioning satellites, image sharing systems, medical imaging systems, military surveillance, and security systems. Google has explored and developed systems that provide image searches (e.g., the Google Images search service), including search systems that are only based on the image file name and document contents and do not consider/relate to the image content itself [41, 42]. Towards achieving artificial intelligence in providing improved image searches, practitioners should move beyond just the textual relationships of images, especially since textual representations of images are not always available in massive image collection repositories. Experts should strive towards collecting and organizing these massive image data collections, such that they can be browsed, searched, and retrieved more efficiently. To deal with large scale image data collections, one approach to consider is to automate the process of tagging images and extracting semantic information from the images. Deep Learning presents new frontiers towards constructing complicated representations for image and video data as relatively high levels of abstractions, which can then be used for image annotation and tagging that is useful for image indexing and retrieval. In the context of Big Data Analytics, here Deep Learning would aid in the discriminative task of semantic tagging of data.

Data tagging is another way to semantically index the input data corpus. However, it should not be confused with semantic indexing as discussed in the prior section. In semantic indexing, the focus is on using the Deep Learning abstract representations directly for data indexing purposes. Here the abstract data representations are considered as features for performing the discriminative task of data tagging. This tagging on data can also be used for data indexing as well, but the primary idea here is that Deep Leaning makes it possible to tag massive amounts of data by applying simple linear modeling methods on complicated features that were extracted by Deep Learning algorithms. The remainder of this section focuses largely on some results from using Deep Leaning for discriminative tasks that involve data tagging.

At the ImageNet Computer Vision Competition, Krizhevsky et al. [17] demonstrated an approach using Deep Learning and Convolutional Neural Networks which outperformed other existing approaches for image object recognition. Using the ImageNet dataset, one of the largest for image object recognition, Hinton's team showed the importance of Deep Learning for improving image searching. Dean et al. [38] demonstrated further success on ImageNet by using a similar Deep Learning modeling approach with a large-scale software infrastructure for training an artificial neural network.

Some other approaches have been tried for learning and extracting features from unlabeled image data, include Restricted Boltzmann Machines (RBMs) [7], autoencoders [26], and sparse coding [43]. However, these were only able to extract low-level features, such as edge and blob detection. Deep Learning can also be used to build very high-level features for image detection. For example, Google and Stanford formulated a very large deep neural network that was able to learn very high-level features, such as face detection or cat detection from scratch (without any priors) by just using unlabeled data [44]. Their work was a large scale investigation on the feasibility of building high-level features with Deep Learning using only unlabeled (unsupervised) data, and clearly demonstrated the benefits of using Deep Learning with unsupervised data. In Google's experimentation, they trained a 9-layered locally connected sparse autoencoder on 10 million 200×200 images downloaded randomly from the Internet. The model had 1 billion connections and the training time lasted for 3 days. A computational cluster of 1000 machines and 16,000 cores was used to train the network with model parallelism and asynchronous SGD (Stochastic Gradient Descent). In their experiments they obtained neurons that function like face detectors, cat detectors, and human body detectors, and based on these features their approach also outperformed the state-of-the-art and recognized 22,000 object categories from the ImageNet dataset. This demonstrates the generalization ability of abstract representations extracted by Deep Learning algorithms on new/unseen data, i.e., using features extracted from a given dataset to successfully perform a discriminative task on another dataset. While Google's work involved the question of whether it is possible to build a face feature detector by just using unlabeled data, typically in computer vision labeled images are used to learn useful features [45]. For example, a large collection of face images with a bounding box around the faces can be used to learn a face detector feature. However, traditionally it would require a very large amount of labeled data to find the best features. The scarcity of labeled data in image data collections poses a challenging problem.

There are other Deep Learning works that have explored image tagging. Socher et al. [46] introduce recursive neural networks for predicting a tree structure for images in multiple modalities, and is the first Deep Learning method that achieves very good results on segmentation and annotation of complex image scenes. The recursive neural network architecture is able to predict hierarchical tree structures for scene images, and outperforms other methods based on conditional random fields or a combination of other methods, as well as outperforming other existing methods in segmentation, annotation and scene classification. Socher et al. [46] also show that their algorithm is a natural tool for predicting tree structures by using it to parse natural language sentences. This demonstrates the advantage of Deep Learning as an effective approach for extracting data representations from different varieties of data types. Kumar et al. [47] suggest that recurrent neural networks can be used to construct a meaningful search space via Deep Learning, where the search space can then be used for a designed-based search.

Le et al. [48] demonstrate that Deep Learning can be used for action scene recognition as well as video data tagging, by using an independent variant analysis to learn invariant spatio-temporal features from video data. Their approach outperforms other existing methods when combined with Deep Learning techniques such as stacking and convolution to learn hierarchical representations. Previous works used to adapt hand designed feature for images like SIFT and HOG to the video domain. The Le et al. [48] study shows that extracting features directly from video data is a very important research direction, which can be also generalized to many domains.

Deep Learning has achieved remarkable results in extracting useful features (i.e., representations) for performing discriminative tasks on image and video data, as well as extracting representations from other kinds of data. These discriminative results with Deep Learning are useful for data tagging and information retrieval and can be used in search engines. Thus, the high-level complex data representations obtained by Deep Learning are useful for the application of computationally feasible and relatively simpler linear models for Big Data Analytics. However, there is considerable work that remains for further exploration, including determining appropriate objectives in learning good representations for performing discriminative tasks in Big Data Analytics [5, 25].

Deep Learning Challenges in Big Data Analytics

The prior section focused on emphasizing the applicability and benefits of Deep Learning algorithms for Big Data Analytics. However, certain characteristics associated with Big Data pose challenges for modifying and adapting Deep Learning to address those issues. This section presents some areas of Big Data where Deep Learning needs further exploration, specifically, learning with streaming data, dealing with high-dimensional data, scalability of models, and distributed computing.

Incremental Learning for Non-stationary Data

One of the challenging aspects in Big Data Analytics is dealing with streaming and fast-moving input data. Such data analysis is useful in monitoring tasks, such as fraud detection. It is important to adapt Deep Learning to handle streaming data, as there is a need for algorithms that can deal with large amounts of continuous input data. In this section, we discuss some works associated with Deep Learning and streaming data, including incremental feature learning and extraction [49], denoising autoencoders [50], and deep belief networks [51].

Zhou et al. [49] describe how a Deep Learning algorithm can be used for incremental feature learning on very large datasets, employing denoising autoencoders [50]. Denoising autoencoders are a variant of autoencoders which extract features from corrupted input, where the extracted features are robust to noisy data and good for classification purposes. Deep Learning algorithms in general use hidden layers to contribute towards the extraction of features or data representations. In a denoising autoencoder, there is one hidden layer which extracts features, with the number of nodes in this hidden layer initially being the same as the number of features that would be extracted. Incrementally, the samples that do not conform to the given objective function (for example, their classification error is more than a threshold, or their reconstruction error is high) are collected and are used for adding new nodes to the hidden layer, with these new nodes being initialized based on those samples. Subsequently, incoming new data samples are used to jointly retrain all the features. This incremental feature learning and mapping can improve the discriminative or generative objective function; however, monotonically adding features can lead to having a lot of redundant features and overfitting of data. Consequently, similar features are merged to produce a more compact set of features. Zhou et al. [49] demonstrate that the incremental feature learning method quickly converges to the optimal number of features in a large-scale online setting. This kind of incremental feature extraction is useful in applications where the distribution of data changes with respect to time in massive online data streams. Incremental feature learning and extraction can be generalized for other Deep Learning algorithms, such as RBM [7], and makes it possible to adapt to new incoming stream of an online large-scale data. Moreover, it avoids expensive cross-validation analysis in selecting the number of features in large-scale datasets.

Calandra et al. [51] introduce adaptive deep belief networks which demonstrates how Deep Learning can be generalized to learn from online non-stationary and streaming data. Their study exploits the generative property of deep belief networks to mimic the samples from the original data, where these samples and the new observed samples are used to learn the new deep belief network which has adapted to the newly observed data. However, a downside of an adaptive deep belief network is the requirement for constant memory consumption.

The targeted works presented in this section provide empirical support to further explore and develop novel Deep Learning algorithms and architectures for analyzing large-scale, fast moving streaming data, as is encountered in some Big Data application domains such as social media feeds, marketing and financial data feeds, web click stream data, operational logs, and metering data. For example, Amazon Kinesis is a managed service designed to handle real-time streaming of Big Data—though it is not based on the Deep Learning approach.

High-Dimensional Data

Some Deep Learning algorithms can become prohibitively computationally-expensive when dealing with high-dimensional data, such as images, likely due to the often slow learning process associated with a deep layered hierarchy of learning data abstractions and representations from a lower-level layer to a higher-level layer. That is to say, these Deep Learning algorithms can be stymied when working with Big Data that exhibits large Volume, one of the four Vs associated with Big Data Analytics. A high-dimensional data source contributes heavily to the volume of the raw data, in addition to complicating learning from the data.

Chen et al. [52] introduce marginalized stacked denoising autoencoders (mSDAs) which scale effectively for high-dimensional data and is computationally faster than regular stacked denoising autoencoders (SDAs). Their approach marginalizes noise in SDA training and thus does not require stochastic gradient descent or other optimization algorithms to learn parameters. The marginalized denoising autoencoder layers to have hidden nodes, thus allowing a closed-form solution with substantial speed-ups. Moreover, marginalized SDA only has two free meta-parameters, controlling the amount of noise as well as the number of layers to be stacked, which greatly simplifies the model selection process. The fast training time, the capability to scale to large-scale and high dimensional data, and implementation simplicity make mSDA a promising method with appeal to a large audience in data mining and machine learning.

Convolutional neural networks are another method which scales up effectively on high dimensional data. Researchers have taken advantages of convolutional neural networks on ImageNet dataset with 256×256 RGB images to achieve state of the art results [17, 26]. In convolutional neural networks, the neurons in the hidden layers units do not need to be connected to all of the nodes in the previous layer, but just to the neurons that are in the same spatial area. Moreover, the resolution of the image data is also reduced when moving toward higher layers in the network.

The application of Deep Learning algorithms for Big Data Analytics involving high dimensional data remains largely unexplored, and warrants development of Deep Learning based solutions that either adapt approaches similar to the ones presented above or develop novel solutions for addressing the high-dimensionality found in some Big Data domains.

Large-Scale Models

From a computation and analytics point of view, how do we scale the recent successes of Deep Learning to much larger-scale models and massive datasets? Empirical results have demonstrated the effectiveness of large-scale models [53–55], with particular focus on models with a very large number of model parameters which are able to extract more complicated features and representations [38, 56].

Dean et al. [38] consider the problem of training a Deep Learning neural network with billions of parameters using tens of thousands of CPU cores, in the context of speech recognition and computer vision. A software framework, DistBelief, is developed that can utilize computing clusters with thousands of machines to train large-scale models. The framework supports model parallelism both within a machine (via multithreading) and across machines (via message passing), with the details of parallelism, synchronization, and communication managed by DistBelief. In addition, the framework also supports data parallelism, where multiple replicas of a model are used to optimize a single objective. In order to make large-scale distributed training possible an asynchronous SGD as well as a distributed batch optimization procedure is developed that includes a distributed implementation of L-BFGS (Limited-memory Broyden-Fletcher-Goldfarb-Shanno, a quasi-Newton method for unconstrained optimization). The primary idea is to train multiple versions of the model in parallel, each running on a different node in the network and analyzing different subsets of data. The authors report that in addition to accelerating the training of conventional sized models, their framework can also train models that are larger than could be contemplated otherwise. Moreover, while the framework focuses on training large-scale neural networks, the underlying algorithms are applicable to other gradient-based learning techniques. It should be noted, however, that the extensive computational resources utilized by DistBelief are generally unavailable to a larger audience. Coates et al.

Coates et al. [56] leverage the relatively inexpensive computing power of a cluster of GPU servers. More specifically, they develop their own system (using neural networks) based on Commodity Off-The-Shelf High Performance Computing (COTS HPC) technology and introduce a high-speed communication infrastructure to coordinate distributed computations. The system is able to train 1 billion parameter networks on just 3 machines in a couple of days, and it can scale to networks with over 11 billion parameters using just 16 machines and where the scalability is comparable to that of DistBelief. In comparison to the computational resources used by DistBelief, the distributed system network based on COTS HPC is more generally available to a larger audience, making it a reasonable alternative for other Deep Learning experts exploring large-scale models.

Large-scale Deep Learning models are quite suited to handle massive volumes of input associated with Big Data, and as demonstrated in the above works they are also better at learning complex data patterns from large volumes of data. Determining the optimal number of model parameters in such large-scale models and improving their computational practicality pose challenges in Deep Learning for Big Data Analytics. In addition to the problem of handling massive volumes of data, large-scale Deep Learning models for Big Data Analytics also have to contend with other Big Data problems, such as domain adaptation (see next section) and streaming data. This lends to the need for further innovations in large-scale models for Deep Learning algorithms and architectures.

Future Work on Deep Learning in Big Data Analytics

In the prior sections, we discussed some recent applications of Deep Learning algorithms for Big Data Analytics, as well as identified some areas where Deep Learning research needs further exploration to address specific data analysis problems observed in Big Data. Considering the low-maturity of Deep Learning, we note that considerable work remains to be done. In this section, we discuss our insights on some remaining questions in Deep Learning research, especially on work needed for improving machine learning and the formulation of the high-level abstractions and data representations for Big Data.

An important problem is whether to utilize the entire Big Data input corpus available when analyzing data with Deep Learning algorithms. The general focus is to apply Deep Learning algorithms to train the high-level data representation patterns based on a portion of the available input corpus, and then utilize the remaining input corpus with the learnt patterns for extracting the data abstractions and representations. In the context of this problem, a question to explore is what volume of input data is generally necessary to train useful (good) data representations by Deep Learning algorithms which can then be generalized for new data in the specific Big Data application domain.

Upon further exploring the above problem, we recall the Variety characteristic of Big Data Analytics, which focuses on the variation of the input data types and domains in Big Data. Here, by considering the shift between the input data source (for training the representations) and the target data source (for generalizing the representations), the problem becomes one of domain adaptation for Deep Learning in Big Data Analytics. Domain adaptation during learning is an important focus of study in Deep Learning [57, 58], where the distribution of the training data (from which the representations are learnt) is different from the distribution of the test data (on which the learnt representations are deployed).

Glorot et al. [57] demonstrate that Deep Learning is able to discover intermediate data representations in a hierarchical learning manner, and that these representations are meaningful to, and can be shared among, different domains. In their work, a stacked denoising autoencoder is initially used to learn features and patterns from unlabeled data obtained from different source domains. Subsequently, a support vector machine (SVM) algorithm utilizes the learnt features and patterns for application on labeled data from a given source domain, resulting in a linear classification model that outperforms other methods. This domain adaptation study is successfully applied on a large industrial strength dataset consisting of 22 source domains. However, it should be noted that their study does not explicitly encode the distribution shift of the data between the source domain and the target domains. Chopra et al. [58] propose a Deep Learning model (based on neural networks) for domain adaptation which strives to learn a useful (for prediction purposes) representation of the unsupervised data by taking into consideration information available from the distribution shift between the training and test data. The focus is to hierarchically learn multiple intermediate representations along an interpolating path between the training and testing domains. In the context of object recognition, their study demonstrates an improvement over other methods. The two studies presented above raise the question about how to increase the generalization capacity of Deep Learning data representations and patterns, noting that the ability to generalize learnt patterns is an important requirement in Big Data Analytics where often there is a distribution shift between the input domain and the target domain.

Another key area of interest would be to explore the question of what criteria is necessary and should be defined for allowing the extracted data representations to provide useful semantic meaning to the Big Data. Earlier, we discussed some studies that utilize the data representations extracted through Deep Learning for semantic indexing. Bengio et al. [5] present some characteristics of what constitutes good data representations for performing discriminative tasks, and point to the open question regarding the definition of the criteria for learning good data representations in Deep Learning. Compared to more conventional learning algorithms where misclassification error is generally used as an important criterion for model training and learning patterns, defining a corresponding criteria for training Deep Learning algorithms with Big Data is unsuitable since most Big Data Analytics involve learning from largely unsupervised data. While availability of supervised data in some Big Data domains can be helpful, the question of defining the criteria for obtaining good data abstractions and representations still remains largely unexplored in Big Data Analytics. Moreover, the question of defining the criteria required for extracting good data representations leads to the question of what would constitute a good data representation that is effective for semantic indexing and/or data tagging.

In some Big Data domains, the input corpus consists of a mix of both labeled and unlabeled data, e.g., cyber security [59], fraud detection [60], and computer vision [45]. In such cases, Deep Learning algorithms can incorporate semi-supervised training methods towards the goal of defining criteria for good data representation learning. For example, following learning representations and patterns from the unlabeled/unsupervised data, the available labeled/supervised data can be exploited to further tune and improve the learnt representations and patterns for a specific analytics task, including semantic indexing or discriminative modeling. A variation of semi-supervised learning in data mining, active learning methods could also be applicable towards obtaining improved data representations where input from crowd sourcing or human experts can be used to obtain labels for some data samples which can then be used to better tune and improve the learnt data representations.

Conclusion

In contrast to more conventional machine learning and feature engineering algorithms, Deep Learning has an advantage of potentially providing a solution to address the data analysis and learning problems found in massive volumes of input data. More specifically, it aids in automatically extracting complex data representations from large volumes of unsupervised data. This makes it a valuable tool for Big Data Analytics, which involves data analysis from very large collections of raw data that is generally unsupervised and un-categorized. The hierarchical learning and extraction of different levels of complex, data abstractions in Deep Learning provides a certain degree of simplification for Big Data Analytics tasks, especially for analyzing massive volumes of data, semantic indexing, data tagging, information retrieval, and discriminative tasks such a classification and prediction.

In the context of discussing key works in the literature and providing our insights on those specific topics, this study focused on two important areas related to Deep Learning and Big Data: (1) the application of Deep Learning algorithms and architectures for Big Data Analytics, and (2) how certain characteristics and issues of Big Data Analytics pose unique challenges towards adapting Deep Learning algorithms for those problems. A targeted survey of important literature in Deep

Learning research and application to different domains is presented in the paper as a means to identify how Deep Learning can be used for different purposes in Big Data Analytics.

The low-maturity of the Deep Learning field warrants extensive further research. In particular, more work is necessary on how we can adapt Deep Learning algorithms for problems associated with Big Data, including high dimensionality, streaming data analysis, scalability of Deep Learning models, improved formulation of data abstractions, distributed computing, semantic indexing, data tagging, information retrieval, criteria for extracting good data representations, and domain adaptation. Future works should focus on addressing one or more of these problems often seen in Big Data, thus contributing to the Deep Learning and Big Data Analytics research corpus.

Competing Interests The authors declare that they have no competing interests.

Authors' Contributions MMN performed the primary literature review and analysis for this work, and also drafted the manuscript. RW and NS worked with MMN to develop the article's framework and focus. TMK, FV and EM introduced this topic to MMN and TMK coordinated with the other authors to complete and finalize this work. All authors read and approved the final manuscript.

References

1. Domingos P. A few useful things to know about machine learning. Commun ACM. 2012;55 (10):78–87.
2. Dalal N, Triggs B. Histograms of oriented gradients for human detection. In: IEEE computer society conference on computer vision and pattern recognition, 2005. CVPR 2005. IEEE, vol. 1. 2005;886–93.
3. Lowe DG. Object recognition from local scale-invariant features. In: Computer Vision, 1999. The Proceedings of the seventh IEEE international conference on IEEE computer society, vol. 2. 1999. p. 1150–7.
4. Bengio Y, LeCun Y. Scaling learning algorithms towards, AI. In: Bottou L, Chapelle O, DeCoste D, Weston J, editors. Large scale kernel machines, vol. 34. Cambridge: MIT Press; 2007. p. 321–60. http://www.iro.umontreal.ca/ ~ lisa/pointeurs/bengio+lecun_chapter2007.pdf.
5. Bengio Y, Courville A, Vincent P. Representation learning: a review and new perspectives. IEEE Trans Pattern Anal Mach Intell. 2013;35(8):1798–828. doi:10.1109/TPAMI.2013.50.
6. Arel I, Rose DC, Karnowski TP. Deep machine learning-a new frontier in artificial intelligence research [research frontier]. IEEE Comput Intell. 2010;5:13–8.
7. Hinton GE, Osindero S, Teh Y-W. A fast learning algorithm for deep belief nets. Neural Comput. 2006;18(7):1527–54.
8. Bengio Y, Lamblin P, Popovici D, Larochelle H. Greedy layer-wise training of deep networks. 2007;19.
9. Larochelle H, Bengio Y, Louradour J, Lamblin P. Exploring strategies for training deep neural networks. J Mach Learn Res. 2009;10:1–40.
10. Salakhutdinov R, Hinton GE. Deep boltzmann machines. In: International conference on artificial intelligence and statistics. JMLR.org. 2009. p. 448–55.

11. Goodfellow I, Lee H, Le QV, Saxe A, Ng AY. Measuring invariances in deep networks. Advances in neural information processing systems. Red Hook: Curran Associates, Inc.; 2009. p. 646–54.
12. Dahl G, Ranzato M, Mohamed A-R, Hinton GE. Phone recognition with the mean-covariance restricted boltzmann machine. Advances in neural information processing systems. Red Hook: Curran Associates, Inc.; 2010. p. 469–77.
13. Hinton G, Deng L, Yu D, Mohamed A-R, Jaitly N, Senior A, Vanhoucke V, Nguyen P, Sainath T, Dahl G, Kingsbury B. Deep neural networks for acoustic modeling in speech recognition: the shared views of four research groups. Signal Process Mag IEEE. 2012;29 (6):82–97.
14. Seide F, Li G, Yu D. Conversational speech transcription using context-dependent deep neural networks. In: INTERSPEECH. ISCA. 2011. p. 437–40.
15. Mohamed A-R, Dahl GE, Hinton G. Acoustic modeling using deep belief networks. IEEE Trans Audio Speech Lang Process. 2012;20(1):14–22.
16. Dahl GE, Yu D, Deng L, Acero A. Context-dependent pre-trained deep neural networks for large-vocabulary speech recognition. IEEE Trans Audio Speech Lang Process. 2012;20(1):30–42.
17. Krizhevsky A, Sutskever I, Hinton G. Imagenet classification with deep convolutional neural networks. Advances in neural information processing systems, vol. 25. Red Hook: Curran Associates, Inc.; 2012. p. 1106–14.
18. Mikolov T, Deoras A, Kombrink S, Burget L, Cernocky J. Empirical evaluation and combination of advanced language modeling techniques. In: INTERSPEECH. ISCA. 2011. p. 605–8.
19. Socher R, Huang EH, Pennin J, Manning CD, Ng A. Dynamic pooling and unfolding recursive autoencoders for paraphrase detection. Advances in neural information processing systems. Red Hook: Curran Associates, Inc.; 2011. p. 801–9.
20. Bordes A, Glorot X, Weston J, Bengio Y. Joint learning of words and meaning representations for open-text semantic parsing. In: International conference on artificial intelligence and statistics. JMLR.org. 2012. p. 127–35.
21. National Research Council. Frontiers in Massive Data Analysis. Washington, DC: The National Academies Press. 2013. http://www.nap.edu/openbook.php?record_id=18374.
22. Dumbill E. What is Big Data? An introduction to the big data landscape. In: Strata 2012: making data work. O'Reilly, Santa Clara, CA O'Reilly. 2012.
23. Khoshgoftaar TM. Overcoming big data challenges. In: Proceedings of the 25th international conference on software engineering and knowledge engineering. Boston. ICSE. Invited Keynote Speaker. 2013.
24. Bengio Y. Learning deep architectures for AI. Hanover: Now Publishers Inc.; 2009.
25. Bengio Y. Deep learning of representations: looking forward. Proceedings of the 1st international conference on statistical language and speech processing. SLSP'13. Tarragona: Springer; 2013. p. 1–37. doi:10.1007/978-3-642-39593-2_1.
26. Hinton GE, Salakhutdinov RR (Science) Reducing the dimensionality of data with neural networks 313(5786):504–7.
27. Hinton GE, Zemel RS. Autoencoders, minimum description length, and helmholtz free energy. Adv Neural Inf Process Syst. 1994;6:3–10.
28. Smolensky P. Information processing in dynamical systems: foundations of harmony theory. Parallel distributed processing: explorations in the microstructure of cognition, vol. 1. Cambridge: MIT Press; 1986. p. 194–281.
29. Hinton GE. Training products of experts by minimizing contrastive divergence. Neural Comput. 2002;14(8):1771–800.
30. Garshol LM. Introduction to big data/machine learning. Online slide show. 2013. http://www.slideshare.net/larsga/introduction-to-big-datamachine-learning.
31. Grobelnik M. Big Data tutorial. European Data Forum. 2013. http://www.slideshare.net/EUDataForum/edf2013-bigdatatutorialmarkogrobelnik?related=1.

32. Salton G, Buckley C. Term-weighting approaches in automatic text retrieval. Inf Process Manag. 1988;24(5):513–23.
33. Robertson SE, Walker S. Some simple effective approximations to the 2-poisson model for probabilistic weighted retrieval. In: Proceedings of the 17th annual international ACM SIGIR conference on research and development in information retrieval. New York: Springer; 1994. p. 232–41.
34. Hinton G, Salakhutdinov R. Discovering binary codes for documents by learning deep generative models. Topics Cogn Sci. 2011;3(1):74–91.
35. Salakhutdinov R, Hinton G. Semantic hashing. Int J Approx Reason. 2009;50(7):969–78.
36. Ranzato M, Szummer M. Semi-supervised learning of compact document representations with deep networks. In: Proceedings of the 25th international conference on machine learning. ACM. 2008. p. 792–9.
37. Mikolov T, Chen K, Dean J. Efficient estimation of word representations in vector space. CoRR: Comput Res Repos. 2013;1–12. abs/1301.3781.
38. Dean J, Corrado G, Monga R, Chen K, Devin M, Le Q, Mao M, Ranzato M, Senior A, Tucker P, Yang K, Ng A. Large scale distributed deep networks. In: Bartlett P, Pereira FCN, Burges CJC, Bottou L, Weinberger KQ, editors. Advances in neural information processing systems, vol. 25. 2012. p. 1232–40. http://books.nips.cc/papers/files/nips25/NIPS2012_0598.pdf.
39. Mikolov T, Le QV, Sutskever I. Exploiting similarities among languages for machine translation. CoRR: Comput Res Repos. 2013;1–10. abs/1309.4168.
40. Li G, Zhu H, Cheng G, Thambiratnam K, Chitsaz B, Yu D, Seide F. Context-dependent deep neural networks for audio indexing of real-life data. In: Spoken language technology workshop (SLT), 2012 IEEE. IEEE. 2012. p. 143–8.
41. Zipern A. A quick way to search for images on the web. The New York Times. News Watch Article. 2001. http://www.nytimes.com/2001/07/12/technology/news-watch-a-quick-way-to-search-for-images-on-the-web.html.
42. Cusumano MA. Google: what it is and what it is not. Commun ACM Med Image Model. 2005;48(2):15–7. doi:10.1145/1042091.1042107.
43. Lee H, Battle A, Raina R, Ng A. Efficient sparse coding algorithms. Advances in neural information processing systems. Cambridge: MIT Press; 2006. p. 801–8.
44. Le Q, Ranzato M, Monga R, Devin M, Chen K, Corrado G, Dean J, Ng A. Building high-level features using large scale unsupervised learning. In: Proceeding of the 29th international conference in machine learning. Edingburgh. 2012.
45. Freytag A, Rodner E, Bodesheim P, Denzler J. Labeling examples that matter: relevance-based active learning with gaussian processes. In: 35th German conference on pattern recognition (GCPR). Germany: Saarland University and Max-Planck-Institute for Informatics; 2013. p. 282–91.
46. Socher R, Lin CC, Ng A, Manning C. Parsing natural scenes and natural language with recursive neural networks. In: Proceedings of the 28th international conference on machine learning. Madison: Omnipress; 2011. p. 129–36.
47. Kumar R, Talton JO, Ahmad S, Klemmer SR. Data-driven web design. In: Proceedings of the 29th international conference on machine learning..2012. icml.cc/Omnipress.
48. Le QV, Zou WY, Yeung SY, Ng AY. Learning hierarchical invariant spatio-temporal features for action recognition with independent subspace analysis. In: IEEE conference on computer vision and pattern recognition (CVPR) 2011 IEEE. 2011. p. 3361–8.
49. Zhou G, Sohn K, Lee H. Online incremental feature learning with denoising autoencoders. In: International conference on artificial intelligence and statistics. JMLR.org. 2012. p. 1453–61.
50. Vincent P, Larochelle H, Bengio Y, Manzagol P-A. Extracting and composing robust features with denoising autoencoders. In: Proceedings of the 25th international conference on machine learning. ACM. 2008. p. 1096–103.
51. Calandra R, Raiko T, Deisenroth MP, Pouzols FM. Learning deep belief networks from non-stationary streams. Artificial neural networks and machine learning–ICANN 2012. Berlin: Springer; 2012. p. 379–86.

52. Chen M, Xu ZE, Weinberger KQ, Sha F. Marginalized denoising autoencoders for domain adaptation. In: Proceeding of the 29th international conference in machine learning. Edingburgh; 2012.
53. Coates A, Ng A. The importance of encoding versus training with sparse coding and vector quantization. In: Proceedings of the 28th international conference on machine learning. Madison: Omnipress; 2011. p. 921–8.
54. Hinton GE, Srivastava N, Krizhevsky A, Sutskever I, Salakhutdinov R. Improving neural networks by preventing co-adaptation of feature detectors. CoRR: Comput Res Repos. 2012;1–18. abs/1207.0580.
55. Goodfellow IJ, Warde-Farley D, Mirza M, Courville A, Bengio Y. Maxout networks. In: Proceeding of the 30th international conference in machine learning. Atlanta. 2013.
56. Coates A, Huval B, Wang T, Wu D, Catanzaro B, Andrew N. Deep learning with Cots HPC systems. In: Proceedings of the 30th international conference on machine learning; 2013. p. 1337–45.
57. Glorot X, Bordes A, Bengio Y. Domain adaptation for large-scale sentiment classification: a deep learning approach. In: Proceedings of the 28th international conference on machine learning (ICML-11). 2011. p. 513–20.
58. Chopra S, Balakrishnan S, Gopalan R. Dlid: deep learning for domain adaptation by interpolating between domains. In: Workshop on challenges in representation learning, proceedings of the 30th international conference on machine learning. Atlanta. 2013.
59. Suthaharan S. Big data classification: problems and challenges in network intrusion prediction with machine learning. ACM sigmetrics: Big Data analytics workshop. Pittsburgh: ACM; 2013.
60. Wang W, Lu D, Zhou X, Zhang B, Mu J. Statistical wavelet-based anomaly detection in big data with compressive sensing. EURASIP J Wireless Commun Netw. 2013:269. http://www.bibsonomy.org/bibtex/25e432dc7230087ab1cdc65925be6d4cb/dblp.

Part II
LexisNexis Risk Solution to Big Data

Chapter 6
The HPCC/ECL Platform for Big Data

Anthony M. Middleton, David Alan Bayliss, Gavin Halliday, Arjuna Chala and Borko Furht

Introduction

As a result of the continuing information explosion, many organizations are experiencing what is now called the "Big Data" problem. This results in the inability of organizations to effectively use massive amounts of their data in datasets which have grown to big to process in a timely manner. Data-intensive computing represents a new computing paradigm [1] which can address the big data problem using high-performance architectures supporting scalable parallel processing to allow government, commercial organizations, and research environments to process massive amounts of data and implement new applications previously thought to be impractical or infeasible.

The fundamental challenges of data-intensive computing are managing and processing exponentially growing data volumes, significantly reducing associated data analysis cycles to support practical, timely applications, and developing new algorithms which can scale to search and process massive amounts of data. Researchers at LexisNexis believe that the answer to these challenges are:

(1) a scalable, integrated computer systems hardware and software architecture designed for parallel processing of data-intensive computing applications, and
(2) a new programming paradigm in the form of a high-level declarative data-centric programming language designed specifically for big data processing.

This chapter explores the challenges of data-intensive computing from a programming perspective, and describes the ECL programming language and the open source High-Performance Cluster Computing (HPCC) architecture designed for

© Springer International Publishing Switzerland 2016
B. Furht and F. Villanustre, *Big Data Technologies and Applications*,
DOI 10.1007/978-3-319-44550-2_6

data-intensive exascale computing applications. ECL is also compared to Pig Latin, a high-level language developed for the Hadoop MapReduce architecture.

Data-Intensive Computing Applications

High-Performance Computing (HPC) is used to describe computing environments which utilize supercomputers and computer clusters to address complex computational requirements or applications with significant processing time requirements or which require processing of significant amounts of data. Computing approaches can be generally classified as either *compute-intensive*, or *data-intensive* [2–4]. HPC has generally been associated with scientific research and compute-intensive types of problems, but more and more HPC technology is appropriate for both compute-intensive and data-intensive applications. HPC platforms utilize a high-degree of internal parallelism and tend to use specialized multi-processors with custom memory architectures which have been highly-optimized for numerical calculations [5]. Supercomputers also require special parallel programming techniques to take advantage of its performance potential.

Compute-intensive is used to describe application programs that are compute bound. Such applications devote most of their execution time to computational requirements as opposed to I/O, and typically require small volumes of data. HPC approaches to compute-intensive applications typically involves parallelizing individual algorithms within an application process, and decomposing the overall application process into separate tasks, which can then be executed in parallel on an appropriate computing platform to achieve overall higher performance than serial processing. In compute-intensive applications, multiple operations are performed simultaneously, with each operation addressing a particular part of the problem. This is often referred to as functional parallelism or control parallelism [6].

Data-intensive is used to describe applications that are I/O bound or with a need to process large volumes of data [2, 3, 7]. Such applications devote most of their processing time to I/O and movement of data. HPC approaches to data-intensive applications typically use parallel system architectures and involves partitioning or subdividing the data into multiple segments which can be processed independently using the same executable application program in parallel on an appropriate computing platform, then reassembling the results to produce the completed output data [8]. The greater the aggregate distribution of the data, the more benefit there is in parallel processing of the data. Gorton et al. [2] state that data-intensive processing requirements normally scale linearly according to the size of the data and are very amenable to straightforward parallelization. The fundamental challenges for data-intensive computing according to Gorton et al. [2] are managing and processing exponentially growing data volumes, significantly reducing associated data analysis cycles to support practical, timely applications, and developing new algorithms which can scale to search and process massive amounts of data.

Data-Parallelism

According to Agichtein [9], parallelization is considered to be an attractive alternative for processing extremely large collections of data such as the billions of documents on the Web [10]. Nyland et al. [8] define data-parallelism as a computation applied independently to each data item of a set of data which allows the degree of parallelism to be scaled with the volume of data. According to Nyland et al. [8], the most important reason for developing data-parallel applications is the potential for scalable performance, and may result in several orders of magnitude performance improvement. The key issues with developing applications using data-parallelism are the choice of the algorithm, the strategy for data decomposition, load balancing on processing nodes, message passing communications between nodes, and the overall accuracy of the results [8, 11]. Nyland et al. [8] also note that the development of a data-parallel application can involve substantial programming complexity to define the problem in the context of available programming tools, and to address limitations of the target architecture. Information extraction from and indexing of Web documents is typical of data-intensive processing which can derive significant performance benefits from data-parallel implementations since Web and other types of document collections can typically then be processed in parallel [10].

The "Big Data" Problem

The rapid growth of the Internet and World Wide Web has led to vast amounts of information available online. In addition, business and government organizations create large amounts of both structured and unstructured information which needs to be processed, analyzed, and linked. Vinton Cerf of Google has described this as an "Information Avalanche" and has stated "we must harness the Internet's energy before the information it has unleashed buries us" [12]. An IDC white paper sponsored by EMC estimated the amount of information currently stored in a digital form in 2007 at 281 exabytes and the overall compound growth rate at 57 % with information in organizations growing at even a faster rate [13]. In another study of the so-called information explosion it was estimated that 95 % of all current information exists in unstructured form with increased data processing requirements compared to structured information [14]. The storing, managing, accessing, and processing of this vast amount of data represents a fundamental need and an immense challenge in order to satisfy needs to search, analyze, mine, and visualize this data as information [15]. These challenges are now simple described in the literature as the "Big Data" problem. In the next section, we will enumerate some of the characteristics of data-intensive computing systems which can address the problems associated with processing big data.

Data-Intensive Computing Platforms

The National Science Foundation believes that data-intensive computing requires a "fundamentally different set of principles" than current computing approaches [16]. Through a funding program within the Computer and Information Science and Engineering area, the NSF is seeking to "increase understanding of the capabilities and limitations of data-intensive computing." The key areas of focus are:

- Approaches to parallel programming to address the parallel processing of data on data-intensive systems.
- Programming abstractions including models, languages, and algorithms which allow a natural expression of parallel processing of data.
- Design of data-intensive computing platforms to provide high levels of reliability, efficiency, availability, and scalability.
- Identifying applications that can exploit this computing paradigm and determining how it should evolve to support emerging data-intensive applications.

Pacific Northwest National Labs has defined data-intensive computing as "capturing, managing, analyzing, and understanding data at volumes and rates that push the frontiers of current technologies" [1, 17]. They believe that to address the rapidly growing data volumes and complexity requires "epochal advances in software, hardware, and algorithm development" which can scale readily with size of the data and provide effective and timely analysis and processing results. The ECL programming language and HPCC architecture developed by LexisNexis represents such an advance in capabilities.

Cluster Configurations

Current data-intensive computing platforms use a "divide and conquer" parallel processing approach combining multiple processors and disks configured in large computing clusters connected using high-speed communications switches and networks which allows the data to be partitioned among the available computing resources and processed independently to achieve performance and scalability based on the amount of data (Fig. 6.1). Buyya et al. [18] define a cluster as "a type of parallel and distributed system, which consists of a collection of inter-connected stand-alone computers working together as a single integrated computing resource." This approach to parallel processing is often referred to as a "shared nothing" approach since each node consisting of processor, local memory, and disk resources shares nothing with other nodes in the cluster. In parallel computing this approach is considered suitable for data processing problems which are "embarrassingly parallel", i.e. where it is relatively easy to separate the problem into a number of parallel tasks and there is no dependency or communication required between the tasks other than overall management of the tasks. These types of data processing

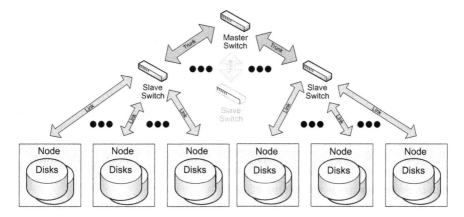

Fig. 6.1 Commodity hardware cluster [31]

problems are inherently adaptable to various forms of distributed computing including clusters and data grids and cloud computing.

Common Platform Characteristics

There are several important common characteristics of data-intensive computing systems that distinguish them from other forms of computing. First is the principle of collocation of the data and programs or algorithms to perform the computation. To achieve high performance in data-intensive computing, it is important to minimize the movement of data [19]. In direct contrast to other types of computing and high-performance computing which utilize data stored in a separate repository or servers and transfer the data to the processing system for computation, data-intensive computing uses distributed data and distributed file systems in which data is located across a cluster of processing nodes, and instead of moving the data, the program or algorithm is transferred to the nodes with the data that needs to be processed. This principle—"Move the code to the data"—is extremely effective since program size is usually small in comparison to the large datasets processed by data-intensive systems and results in much less network traffic since data can be read locally instead of across the network. This characteristic allows processing algorithms to execute on the nodes where the data resides reducing system overhead and increasing performance [2].

A second important characteristic of data-intensive computing systems is the programming model utilized. Data-intensive computing systems utilize a machine-independent approach in which applications are expressed in terms of high-level operations on data, and the runtime system transparently controls the scheduling, execution, load balancing, communications, and movement of programs and data across the distributed computing cluster [20]. The programming abstraction and language tools allow the processing to be expressed in terms of data flows and transformations incorporating new dataflow programming languages and shared libraries of common data manipulation algorithms such as sorting. Conventional high-performance computing and distributed computing systems typically utilize machine dependent programming models which can require low-level programmer control of processing and node communications using conventional imperative programming languages and specialized software packages which adds complexity to the parallel programming task and reduces programmer productivity. A machine dependent programming model also requires significant tuning and is more susceptible to single points of failure. The ECL programming language described in this chapter was specifically designed to address data-intensive computing requirements.

A third important characteristic of data-intensive computing systems is the focus on reliability and availability. Large-scale systems with hundreds or thousands of processing nodes are inherently more susceptible to hardware failures, communications errors, and software bugs. Data-intensive computing systems are designed to be fault resilient. This includes redundant copies of all data files on disk, storage of intermediate processing results on disk, automatic detection of node or processing failures, and selective re-computation of results. A processing cluster configured for data-intensive computing is typically able to continue operation with a reduced number of nodes following a node failure with automatic and transparent recovery of incomplete processing.

A final important characteristic of data-intensive computing systems is the inherent scalability of the underlying hardware and software architecture. Data-intensive computing systems can typically be scaled in a linear fashion to accommodate virtually any amount of data, or to meet time-critical performance requirements by simply adding additional processing nodes to a system configuration in order to achieve high processing rates and throughput. The number of nodes and processing tasks assigned for a specific application can be variable or fixed depending on the hardware, software, communications, and distributed file system architecture. This scalability allows computing problems once considered to be intractable due to the amount of data required or amount of processing time required to now be feasible and affords opportunities for new breakthroughs in data analysis and information processing.

HPCC Platform

HPCC System Architecture

The development of the open source HPCC computing platform by the Seisint subsidiary of LexisNexis began in 1999 and applications were in production by late 2000. The conceptual vision for this computing platform is depicted in Fig. 6.2. The LexisNexis approach also utilizes commodity clusters of hardware running the

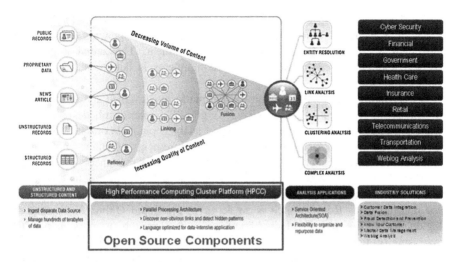

Fig. 6.2 LexisNexis vision for a data-intensive supercomputer

Linux operating system as shown in Figure 4.1. Custom system software and middleware components were developed and layered on the base Linux operating system to provide the execution environment and distributed filesystem support required for data-intensive computing. Because LexisNexis recognized the need for a new computing paradigm to address its growing volumes of data, the design approach included the definition of a new high-level language for parallel data processing called ECL (Enterprise Control Language). The power, flexibility, advanced capabilities, speed of development, and ease of use of the ECL programming language is the primary distinguishing factor between the LexisNexis HPCC and other data-intensive computing solutions. The following provides an overview of the HPCC systems architecture and the ECL language.

LexisNexis developers recognized that to meet all the requirements of data-intensive computing applications in an optimum manner required the design and implementation of two distinct processing environments, each of which could be optimized independently for its parallel data processing purpose. The first of these platforms is called a Data Refinery whose overall purpose is the general processing of massive volumes of raw data of any type for any purpose but typically used for data cleansing and hygiene, ETL processing of the raw data (extract, transform, load), record linking and entity resolution, large-scale ad hoc analysis of data, and creation of keyed data and indexes to support high-performance structured queries and data warehouse applications. The Data Refinery is also referred to as Thor, a reference to the mythical Norse god of thunder with the large hammer symbolic of crushing large amounts of raw data into useful information. A Thor system is similar in its hardware configuration, function, execution environment, filesystem, and capabilities to the Hadoop MapReduce platform, but offers significantly higher performance in equivalent configurations.

The Thor processing cluster is depicted in Fig. 6.3. In addition to the Thor master and slave nodes, additional auxiliary and common components are needed to implement a complete HPCC processing environment. The actual number of physical nodes required for the auxiliary components is determined during the configurations process.

The second of the parallel data processing platforms designed and implemented by LexisNexis is called the Data Delivery Engine. This platform is designed as an online high-performance structured query and analysis platform or data warehouse delivering the parallel data access processing requirements of online applications through Web services interfaces supporting thousands of simultaneous queries and users with sub-second response times. High-profile online applications developed by LexisNexis such as Accurint utilize this platform. The Data Delivery Engine is also referred to as Roxie, which is an acronym for Rapid Online XML Inquiry Engine. Roxie uses a special distributed indexed filesystem to provide parallel processing of queries. A Roxie system is similar in its function and capabilities to Hadoop with HBase and Hive capabilities added, but provides significantly higher throughput since it uses a more optimized execution environment and filesystem for high-performance online processing. Most importantly, both Thor and Roxie systems utilize the same ECL programming language for implementing applications, increasing continuity and programmer productivity. The Roxie processing cluster is depicted in Fig. 6.4.

The implementation of two types of parallel data processing platforms (Thor and Roxie) in the HPCC processing environment serving different data processing needs allows these platforms to be optimized and tuned for their specific purposes to provide the highest level of system performance possible to users. This is a distinct advantage when compared to Hadoop where the MapReduce architecture must be overlayed with additional systems such as HBase, Hive, and Pig which have different processing goals and requirements, and don't always map readily into the MapReduce paradigm. In addition, the LexisNexis HPCC approach

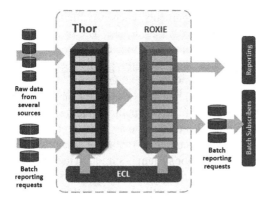

Fig. 6.3 HPCC Thor processing cluster

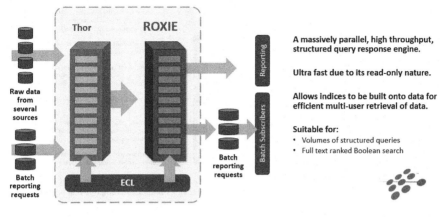

Fig. 6.4 HPCC Roxie processing cluster

incorporates the notion of a processing environment which can integrate Thor and Roxie clusters as needed to meet the complete processing needs of an organization. As a result, scalability can be defined not only in terms of the number of nodes in a cluster, but in terms of how many clusters and of what type are needed to meet system performance goals and user requirements. This provides significant flexibility when compared to Hadoop clusters which tend to be independent islands of processing. For additional information and a detailed comparison of the HPCC system platform to Hadoop, see [21].

HPCC Thor System Cluster

The Thor system cluster is implemented using a master/slave approach with a single master node and multiple slave nodes which provides a parallel job execution environment for programs coded in ECL. ECL is a declarative programming language, developed at LexisNexis, which is easy to use, data-centric and optimized for large-scale data management and query processing (Fig. 6.5). ECL is described in detail in "ECL Programming Language".

Each of the slave nodes is also a data node within the distributed file system for the cluster. Multiple Thor clusters can exist in an HPCC system environment, and job queues can span multiple clusters in an environment if needed. Jobs executing on a Thor cluster in a multi-cluster environment can also read files from the distributed file system on foreign clusters if needed. The middleware layer provides additional server processes to support the execution environment including ECL Agents and ECL Servers. A client process submits an ECL job to the ECL Agent which coordinates the overall job execution on behalf of the client process.

An ECL program is compiled by the ECL server which interacts with an additional server called the ECL Repository which is a source code repository and

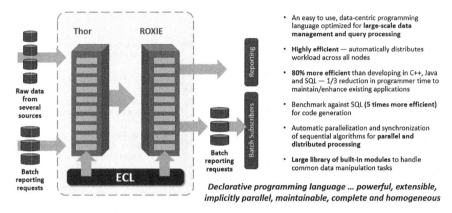

Fig. 6.5   ECL declarative programming language

contains shared, reusable ECL code. ECL code can also be stored in local source files and managed with a conventional version control system. ECL programs are compiled into optimized C++ source code, which is subsequently linked into executable code and distributed to the slave nodes of a Thor cluster by the Thor master node. The Thor master monitors and coordinates the processing activities of the slave nodes and communicates status information monitored by the ECL Agent processes. When the job completes, the ECL Agent and client process are notified, and the output of the process is available for viewing or subsequent processing. Output can be stored in the distributed filesystem for the cluster or returned to the client process.

The distributed filesystem (DFS) used in a Thor cluster is record-oriented which is somewhat different from the block format used in MapReduce clusters. Records can be fixed or variable length, and support a variety of standard (fixed record size, CSV, XML) and custom formats including nested child datasets. Record I/O is buffered in large blocks to reduce latency and improve data transfer rates to and from disk Files to be loaded to a Thor cluster are typically first transferred to a landing zone from some external location, then a process called "spraying" is used to partition the file and load it to the nodes of a Thor cluster. The initial spraying process divides the file on user-specified record boundaries and distributes the data as evenly as possible with records in sequential order across the available nodes in the cluster. Files can also be "desprayed" when needed to transfer output files to another system or can be directly copied between Thor clusters in the same environment. Index files generated on Thor clusters can also be directly copied to Roxie clusters to support online queries.

Nameservices and storage of metadata about files including record format information in the Thor DFS are maintained in a special server called the Dali server. Thor users have complete control over distribution of data in a Thor cluster, and can re-distribute the data as needed in an ECL job by specific keys, fields, or combinations of fields to facilitate the locality characteristics of parallel processing.

The Dali nameserver uses a dynamic datastore for filesystem metadata organized in a hierarchical structure corresponding to the scope of files in the system. The Thor DFS utilizes the local Linux filesystem for physical file storage, and file scopes are created using file directory structures of the local file system. Parts of a distributed file are named according to the node number in a cluster, such that a file in a 400-node cluster will always have 400 parts regardless of the file size. Each node contains an integral number of records (individual records are not split across nodes), and I/O is completely localized to the processing node for local processing operations. The ability to easily redistribute the data evenly to nodes based on processing requirements and the characteristics of the data during a Thor job can provide a significant performance improvement over the blocked data and input splits used in the MapReduce approach.

The Thor DFS also supports the concept of "superfiles" which are processed as a single logical file when accessed, but consist of multiple Thor DFS files. Each file which makes up a superfile must have the same record structure. New files can be added and old files deleted from a superfile dynamically facilitating update processes without the need to rewrite a new file. Thor clusters are fault resilient and a minimum of one replica of each file part in a Thor DFS file is stored on a different node within the cluster.

HPCC Roxie System Cluster

Roxie clusters consist of a configurable number of peer-coupled nodes functioning as a high-performance, high availability parallel processing query platform. ECL source code for structured queries is pre-compiled and deployed to the cluster. The Roxie distributed filesystem is a distributed indexed-based filesystem which uses a custom B+Tree structure for data storage. Indexes and data supporting queries are pre-built on Thor clusters and deployed to the Roxie DFS with portions of the index and data stored on each node. Typically the data associated with index logical keys is embedded in the index structure as a payload. Index keys can be multi-field and multivariate, and payloads can contain any type of structured or unstructured data supported by the ECL language. Queries can use as many indexes as required for a query and contain joins and other complex transformations on the data with the full expression and processing capabilities of the ECL language. For example, the LexisNexis Accurint® comprehensive person report which produces many pages of output is generated by a single Roxie query.

A Roxie cluster uses the concept of Servers and Agents. Each node in a Roxie cluster runs Server and Agent processes which are configurable by a System Administrator depending on the processing requirements for the cluster. A Server process waits for a query request from a Web services interface then determines the nodes and associated Agent processes that have the data locally that is needed for a query, or portion of the query. Roxie query requests can be submitted from a client application as a SOAP call, HTTP or HTTPS protocol request from a Web application, or through a direct socket connection. Each Roxie query request is associated with a specific deployed ECL query program. Roxie queries can also be executed from programs running on Thor clusters. The Roxie Server process that receives the request owns the processing of the ECL program for the query until it is completed. The Server sends portions of the query job to the nodes in the cluster and Agent processes which have data needed for the query stored locally as needed, and waits for results. When a Server receives all the results needed from all nodes, it collates them, performs any additional processing, and then returns the result set to the client requestor.

The performance of query processing on a Roxie cluster varies depending on factors such as machine speed, data complexity, number of nodes, and the nature of the query, but production results have shown throughput of 5000 transactions per second on a 100-node cluster. Roxie clusters have flexible data storage options with indexes and data stored locally on the cluster, as well as being able to use indexes stored remotely in the same environment on a Thor cluster. Nameservices for Roxie clusters are also provided by the Dali server. Roxie clusters are fault-resilient and data redundancy is built-in using a peer system where replicas of data are stored on two or more nodes, all data including replicas are available to be used in the processing of queries by Agent processes. The Roxie cluster provides automatic failover in case of node failure, and the cluster will continue to perform even if one or more nodes are down. Additional redundancy can be provided by including multiple Roxie clusters in an environment.

Load balancing of query requests across Roxie clusters is typically implemented using external load balancing communications devices. Roxie clusters can be sized as needed to meet query processing throughput and response time requirements, but are typically smaller that Thor clusters.

ECL Programming Language

Several well-known companies experiencing the big data problem have implemented high-level programming or script languages oriented toward data analysis. In Google's MapReduce programming environment, native applications are coded in C++ [22]. The MapReduce programming model allows group aggregations in parallel over a commodity cluster of machines similar to Figure 4.1. Programmers provide a Map function that processes input data and groups the data according to a key-value pair, and a Reduce function that performs aggregation by key-value on the output of the Map function. According to Dean and Ghemawat [22, 23], the processing is automatically parallelized by the system on the cluster, and takes care of details like partitioning the input data, scheduling and executing tasks across a processing cluster, and managing the communications between nodes, allowing programmers with no experience in parallel programming to use a large parallel processing environment. For more complex data processing procedures, multiple MapReduce calls must be linked together in sequence.

Google also implemented a high-level language named Sawzall for performing parallel data analysis and data mining in the MapReduce environment and a workflow management and scheduling infrastructure for Sawzall jobs called Workqueue [24]. For most applications implemented using Sawzall, the code is much simpler and smaller than the equivalent C++ by a factor of 10 or more. Pike et al. [24] cite several reasons why a new language is beneficial for data analysis and data mining applications: (1) a programming language customized for a specific problem domain makes resulting programs "clearer, more compact, and more expressive"; (2) aggregations are specified in the Sawzall language so that the programmer does not have to provide one in the Reduce task of a standard MapReduce program; (3) a programming language oriented to data analysis provides a more natural way to think about data processing problems for large distributed datasets; and (4) Sawzall programs are significantly smaller that equivalent C++ MapReduce programs and significantly easier to program.

An open source implementation of MapReduce pioneered by Yahoo! called Hadoop is functionally similar to the Google implementation except that the base programming language for Hadoop is Java instead of C++. Yahoo! also implemented a high-level dataflow-oriented language called Pig Latin and execution environment ostensibly for the same reasons that Google developed the Sawzall language for its MapReduce implementation—to provide a specific language notation for data analysis applications and to improve programmer productivity and reduce development cycles when using the Hadoop MapReduce environment. Working out how to fit many data analysis and processing applications into the MapReduce paradigm can be a challenge, and often requires multiple MapReduce jobs [25]. Pig Latin programs are automatically translated into sequences of MapReduce programs if needed in the execution environment.

Both Google with its Sawzall language and Yahoo with its Pig system and language for Hadoop address some of the limitations of the MapReduce model by providing an external dataflow-oriented programming language which translates language statements into MapReduce processing sequences [24, 26, 27]. These languages provide many standard data processing operators so users do not have to implement custom Map and Reduce functions, improve reusability, and provide some optimization for job execution. However, these languages are externally implemented executing on client systems and not integral to the MapReduce architecture, but still rely on the on the same infrastructure and limited execution model provided by MapReduce.

ECL Features and Capabilities

The open source ECL programming language represents a new programming paradigm for data-intensive computing. ECL was specifically designed to be a transparent and implicitly parallel programming language for data-intensive applications. It is a high-level, declarative, non-procedural dataflow-oriented language that allows the programmer to define what the data processing result should be and the dataflows and transformations that are necessary to achieve the result. Execution is not determined by the order of the language statements, but from the sequence of dataflows and transformations represented by the language statements. It combines data representation with algorithm implementation, and is the fusion of both a query language and a parallel data processing language.

ECL uses an intuitive syntax which has taken cues from other familiar languages, supports modular code organization with a high degree of reusability and extensibility, and supports high-productivity for programmers in terms of the amount of code required for typical applications compared to traditional languages like Java and C++. Similar to the benefits Sawzall provides in the Google environment, and Pig Latin provides to Hadoop users, a 20 times increase in programmer productivity is typical which can significantly reduce development cycles.

ECL is compiled into optimized C++ code for execution on the HPCC system platforms, and can be used for complex data processing and analysis jobs on a Thor cluster or for comprehensive query and report processing on a Roxie cluster. ECL allows inline C++ functions to be incorporated into ECL programs, and external programs in other languages can be incorporated and parallelized through a PIPE facility. External services written in C++ and other languages which generate DLLs can also be incorporated in the ECL system library, and ECL programs can access external Web services through a standard SOAPCALL interface.

The basic unit of code for ECL is called an attribute definition. An attribute can contain a complete executable query or program, or a shareable and reusable code fragment such as a function, record definition, dataset definition, macro, filter definition, etc. Attributes can reference other attributes which in turn can reference other attributes so that ECL code can be nested and combined as needed in a reusable manner. Attributes are stored in ECL code repository which is subdivided into modules typically associated with a project or process. Each ECL attribute added to the repository effectively extends the ECL language like adding a new word to a dictionary, and attributes can be reused as part of multiple ECL queries and programs. ECL can also be stored in local source files as with other programming languages. With ECL a rich set of programming tools is provided including an interactive IDE similar to Visual C++, Eclipse (an ECL add-in for Eclipse is available) and other code development environments.

The Thor system allows data transformation operations to be performed either locally on each node independently in the cluster, or globally across all the nodes in a cluster, which can be user-specified in the ECL language. Some operations such as PROJECT for example are inherently local operations on the part of a distributed file stored locally on a node. Others such as SORT can be performed either locally or globally if needed. This is a significant difference from the MapReduce architecture in which Map and Reduce operations are only performed locally on the input split assigned to the task. A local SORT operation in an HPCC cluster would sort the records by the specified key in the file part on the local node, resulting in the records being in sorted order on the local node, but not in full file order spanning all nodes. In contrast, a global SORT operation would result in the full

Fig. 6.6 ECL code example

distributed file being in sorted order by the specified key spanning all nodes. This requires node to node data movement during the SORT operation. Figure 6.6 shows a sample ECL program using the LOCAL mode of operation.

Figure 6.7 shows the corresponding execution graph. Note the explicit programmer control over distribution of data across nodes. The colon-equals ":=" operator in an ECL program is read as "is defined as". The only action in this program is the OUTPUT statement, the other statements are definitions.

An additional important capability provided in the ECL programming language is support for natural language processing (NLP) with PATTERN statements and the built-in PARSE function. The PARSE function cam accept an unambiguous grammar defined by PATTERN, TOKEN, and RULE statements with penalties or preferences to provide deterministic path selection, a capability which can significantly reduce the difficulty of NLP applications. PATTERN statements allow matching patterns including regular expressions to be defined and used to parse information from unstructured data such as raw text. PATTERN statements can be combined to implement complex parsing operations or complete grammars from BNF definitions. The PARSE operation function across a dataset of records on a specific field within a record, this field could be an entire line in a text file for example. Using this capability of the ECL language it is possible to implement parallel processing for information extraction applications across document files including XML-based documents or Web pages.

ECL Compilation, Optimization, and Execution

The ECL language compiler takes the ECL source code and produces an output with three main elements. The first is an XML representation of the execution graph, detailing the activities to be executed and the dependencies between those activities. The second is a C++ class for each of the activities in the graph, and the third contains code and meta information to control the workflow for the ECL program. These different elements are embedded in a single shared object that

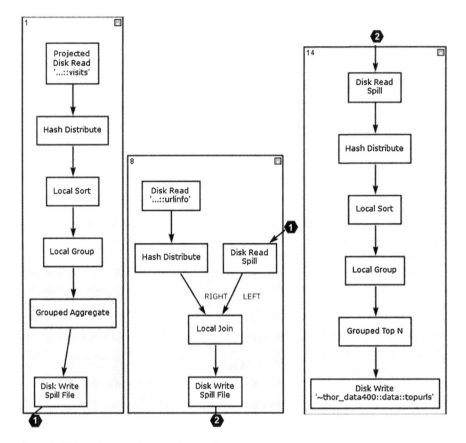

Fig. 6.7 ECL code example execution graph

contains all the information about the particular query. That shared object is passed to the execution engines, which take that shared object and execute the program it contains.

The process of compiling, optimizing, and executing the ECL is broken into several stages: (1) parsing, (2) optimization, (3) transforming, (4) generating, and (5) execution.

Parsing

The sources for an ECL program can come from a local directory tree, an external repository, or a single-source archive. The ECL compiler reads the ECL source, parses it, and converts it into an abstract graph representation of the program. The representation is then normalized to resolve ambiguities and ensure is it is suitable for subsequent processing. All of the subsequent operations within the compiler work on, and create, this same abstract representation.

Optimizations

The design of the ECL language provides abundant scope for optimizations. When reusable attributes are combined it often creates the scope for optimizations that would be hard, if not impossible, to be spotted by a programmer. Its declarative design allows many optimizations without the concerns about side-effects associated with imperative languages. Many different optimizations are performed on the program, some of the key ones are:

- *Constant folding.* This includes simple purely constant expressions like $12 * 3 => 36$, and more complex changes e.g. IF(a, 'b', 'c') IN ['a','c'] => NOT a
- *Tracking and propagating constant field values.* This can often lead to further constant folding, or reduce the lifetime of a field. Minimizing the fields in a row at each stage of the processing. This saves the programmer from unnecessary optimization, and often benefits from the other optimizations (e.g., constant propagation).
- *Reordering operations.* Sometimes changing the order of operations can significantly reduce the data processed by complex activities. Examples include ensuring a filter is done before a sort, or replacing a filter on a joined dataset with a filter on one (or both) of the inputs.
- *Tracking meta information including sort orders and record counts, and removing redundant operations.* This is an example of an optimization which often comes into play when reusable attributes are combined. A particular sort order may not be part of the specification of an attribute, but the optimizer can make use of the current implementation.
- *Minimizing data transferred between slave nodes.* There is sufficient scope for many additional optimizations. For example, a currently planned optimization would analyze and optimize the distribution and sort activities used in a program to maximize overlap and minimize data redistribution.

A key design goal is for the ECL programmer to be able to describe the problem, and rely on the ECL compiler to solve the problem efficiently.

Transforming

The ECL compiler needs to transform the abstract declarative ECL (what it should do) to a concrete imperative implementation (how it should do it). This again has several different elements:

- *Convert the logical graph into an execution graph.* This includes introducing activities to split the data stream, ensure dependencies between activities will be executed in the correct order, and resolving any global resourcing constraints.

- *Extracting context-invariant expressions to ensure they are evaluated a minimal number of times.* This is similar to spotting loop invariant code in an imperative language.
- *Selecting between different implementations of a sequence of activities.* For example generating either inline code or a nested graph of activities.
- *Common sub-expression elimination.* Both globally across the whole program, and locally the expressions used within the methods of the activity classes.
- *Mapping complex ECL statements into the activities supported by the target engine.* For instance a JOIN may be implemented differently depending on how the inputs are sorted, distributed, and the likely size of the datasets. Similarly an ECL DEDUP operation may sometimes be implemented as a local dedup activity followed by a global dedup activity.
- *Combining multiple logical operations into a single activity.* Compound activities have been implemented in the engines where it can significantly reduce the data being copied, or because there are likely to be expressions shared between the activities. One of the commonest examples is disk read, filter and project.

Generating

Following the transforming stage, the XML and C++ associated with the ECL program is generated. The C++ code is built using a data structure that allows peephole optimizations to be applied to the C++ that will be generated. Once the processing is complete, the C++ is generated from the structure, and the generated source files are passed to the system C++ compiler to create a shared object.

In practice, the optimization, transforming and generation is much more of an iterative process rather than sequential.

Execution

The details of executing ECL program vary depending on the specific HPCC system platform and its execution engine, but they follow the same broad sequence.

The engine extracts resources from the shared object that describe the workflow of the query. The workflow can include waiting for particular events, conditionally re-evaluating expressions, and executing actions in a particular order. Each workflow item is executed independently, but can have dependencies on other workflow items. A workflow item may contain any number of activity graphs which evaluate a particular part of the ECL program.

To execute a graph of activities the engine starts at the outputs and recursively walks the graph to evaluate any dependencies. Once the graph is prepared the graph of activities is executed. Generally multiple paths within the graph are executed in parallel, and multiple slave nodes in a cluster will be executing the graphs on different subsets of the data. Records are streamed through the graphs from the inputs to the outputs. Some activities execute completely locally, and others co-ordinate their execution with other slave nodes.

ECL Development Tools and User Interfaces

The HPCC platform includes a suite of development tools and utilities for data analysts, programmers, administrators, and end-users. These include ECL IDE, an integrated programming development environment similar to those available for other languages such as C++ and Java, which encompasses source code editing, source code version control, access to the ECL source code repository, and the capability to execute and debug ECL programs.

ECL IDE provides a full-featured Windows-based GUI for ECL program development and direct access to the ECL repository source code. ECL IDE allows you to create and edit ECL attributes which can be shared and reused in multiple ECL programs or to enter an ECL query which can be submitted directly to a Thor cluster as an executable job or deployed to a Roxie cluster. An ECL query can be self-contained or reference other sharable ECL code in the attribute repository. ECL IDE also allows you to utilize a large number of built-in ECL functions from included libraries covering string handling, data manipulation, file handling, file spray and despray, superfile management, job monitoring, cluster management, word handling, date processing, auditing, parsing support, phonetic (metaphone) support, and workunit services.

ECL Advantages and Key Benefits

ECL a heavily optimized, data-centric declarative programming language. It is a language specifically designed to allow data operations to be specified in a manner which is easy to optimize and parallelize. With a declarative language, you specify what you want done rather than how to do it. A distinguishing feature of declarative languages is that they are extremely succinct; it is common for a declarative language to require an order of magnitude ($10\times$) less code than a procedural equivalent to specify the same problem [28]. The SQL language commonly used for data access and data management with RDBMS systems is also a declarative language. Declarative languages have many benefits including conciseness, freedom from side effects, parallelize naturally, and the executable code generated can be highly optimized since the compiler can determine the optimum sequence of execution instead of the programmer.

ECL extends the benefits of declarative in three important ways [28]: (1) It is data-centric which means it addresses computing problems that can be specified by some form of analysis upon data. It has defined a simple but powerful data algebra to allow highly complex data manipulations to be constructed; (2) It is extensible. When a programmer defines new code segments (called attributes) which can include macros, functions, data definitions, procedures, etc., these essentially become a part of the language and can be used by other programmers. Therefore a new ECL installation may be relatively narrow and generic in its initial scope, but as new ECL code is added, its abilities expand to allow new problems and classes of problems to be stated declaratively; and (3) It is internally abstract. The ECL compiler generates C++ code and calls into many 'libraries' of code, most of which are major undertakings in their own right. By doing this, the ECL compiler is machine neutral and greatly simplified. This allows the ECL compiler writers to focus on making the language relevant and good, and generating highly-optimized executable code. For some coding examples and additional insights into declarative programming with ECL, see [29].

One of the key issues which has confronted language developers is to find solutions to the complexity and difficulty of parallel and distributed programming. Although high-performance computing and cluster architectures such have advanced to provide highly-scalable processing environments, languages designed for parallel programming are still somewhat rare. Declarative, data-centric languages because the parallelize naturally represent solutions to this issue [30]. According to Hellerstein, declarative, data-centric languages parallelizes naturally over large datasets, and programmers can benefit from parallel execution without modifications to their code. ECL code, for example can be used on any size cluster without modification to the code, so performance can be scaled naturally.

The key benefits of ECL can be summarized as follows:

- ECL is a declarative, data-centric, programming language which can expressed concisely, parallelizes naturally, is free from side effects, and results in highly-optimized executable code.
- ECL incorporates transparent and implicit parallelism regardless of the size of the computing cluster and reduces the complexity of parallel programming increasing the productivity of application developers.
- ECL enables implementation of data-intensive applications with huge volumes of data previously thought to be intractable or infeasible. ECL was specifically designed for manipulation of data and query processing. Order of magnitude performance increases over other approaches are possible.
- ECL provides a more than 20 times productivity improvement for programmers over traditional languages such as Java and C++. The ECL compiler generates highly optimized C++ for execution.
- ECL provides a comprehensive IDE and programming tools that provide a highly-interactive environment for rapid development and implementation of ECL applications.
- ECL is a powerful, high-level, parallel programming language ideal for implementation of ETL, Information Retrieval, Information Extraction, and other data-intensive applications.

HPCC High Reliability and High Availability Features

Thor and Roxie architectures of the HPCC system provide both high reliability and availability. The HPCC system in Fig. 6.8 shows the highly available architecture. In this architecture, Thor has several layers of redundancy:

1. Uses hardware RAID redundancy to isolate disk drive failure.
2. Two copies of the same data can exist on multiple nodes. This again is used to isolate against disk failure in one node or a complete node failure.
3. Multiple independent Thor clusters (as shown in Fig. 6.8) can be configured to subscribe to the same Job Queue. This is the highest form on redundancy available within Thor and this isolates you from disk failure, node failure and network failure within the same cluster.

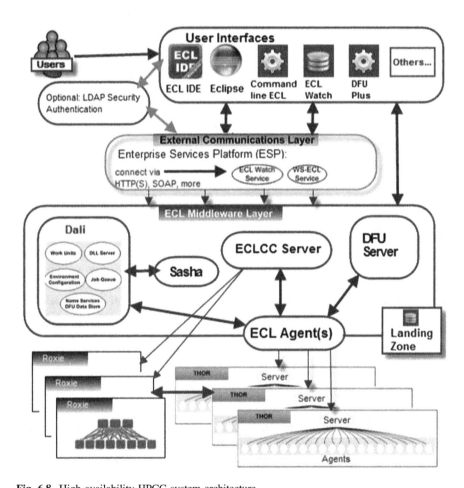

Fig. 6.8 High availability HPCC system architecture

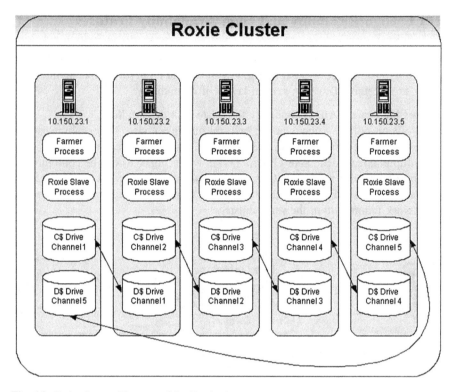

Fig. 6.9 Redundant architecture of the Roxie cluster

Thor cluster accepts jobs from a job queue. If there are two Thor clusters handling the queue, one will continue accepting jobs, if the other one fails. If a single component (Thorslave or Thormaster) fails, the other will continue to process requests. With replication enabled, it will be able to read data from the backup location of the broken Thor. Other components (such as ECL Server, or ESP) can also have multiple instances. The remaining components, such as Dali, or DFU Server, work in a traditional shared storage high availability fail over model.

The Roxie cluster has the highest form of redundancy, as illustrated in Fig. 6.9. Roxie will continue its operation even if half of the nodes are out of operation.

Conclusion

As a result of the continuing information explosion, many organizations are drowning in data and are experiencing the "Big Data" problem making it harder and harder to process and gain useful insights from their data. Data-intensive computing represents a new computing paradigm which can address the big data problem and allow government and commercial organizations and research environments to process massive amounts of data and implement applications previously thought to be impractical or infeasible. Several organizations developed new parallel-processing architectures using commodity computing clusters including Google who initially developed the MapReduce architecture and LexisNexis who developed the HPCC architecture and the ECL programming language. An open source version of MapReduce called Hadoop was developed with additional capabilities to enhance the platform including a data-oriented programming language and execution environment called Pig. The open source HPCC platform and the ECL programming language are described in this chapter, and a direct comparison of the Pig language of Hadoop to the ECL language was presented along with a representative benchmark. Availability of a high-level declarative, data-centric, dataflow-oriented programming language has proven to be a critical success factor in data-intensive computing.

The LexisNexis HPCC platform is at the heart of a premier information services provider and industry leader, and has been adopted by government agencies, commercial organizations, and research laboratories because of its high-performance cost-effective implementation. Existing HPCC applications implemented using the ECL language include raw data processing, ETL, and linking of enormous amounts of data to support online information services such as LexisNexis and industry-leading information search applications such as Accurint; entity extraction and entity resolution of unstructured and semi-structured data such as Web documents to support information extraction; statistical analysis of Web logs for security applications such as intrusion detection; online analytical processing to support business intelligence systems (BIS); and data analysis of massive datasets in educational and research environments and by state and federal government agencies.

There are many factors in choosing a new computer systems architecture and programming language, and usually the best approach is to conduct a specific benchmark test with a customer application to determine the overall system effectiveness and performance. A comparison of the Hadoop MapReduce architecture using a public benchmark for the Pig programming language to the HPCC architecture and ECL programming language on the same system hardware configuration in this chapter reveals significant performance advantages for the HPCC platform with ECL. Some additional advantages of choosing the LexisNexis HPCC platform with ECL include: (1) an open source architecture which implements a highly integrated system environment with capabilities from raw data processing to high-performance queries and data analysis using a common language; (2) a scalable architecture which provides equivalent performance at a much lower system cost based on the number of processing nodes required compared to other data-intensive computing architectures such as MapReduce; (3) an architecture which has been proven to be stable and reliable on high-performance data processing production applications for varied organizations over a 10-year period; (4) an architecture that uses a declarative, data-centric programming language (ECL) with extensive built-in capabilities for data-parallel processing, allows complex operations without the need for extensive user-defined functions, and automatically optimizes execution graphs with hundreds of processing steps into single efficient workunits; (5) an architecture with a high-level of fault resilience and language capabilities which reduce the need for re-processing in case of system failures; and (6) an architecture which is available in open source from and supported by a well-known leader in information services and risk solutions (LexisNexis) who is part of one of the world's largest publishers of information ReedElsevier.

References

1. Kouzes RT, Anderson GA, Elbert ST, Gorton I, Gracio DK. The changing paradigm of data-intensive computing. Computer. 2009;42(1):26–34.
2. Gorton I, Greenfield P, Szalay A, Williams R. Data-intensive computing in the 21st century. IEEE Comput. 2008;41(4):30–2.
3. Johnston WE. High-speed, wide area, data intensive computing: a ten year retrospective. In: Proceedings of the 7th IEEE international symposium on high performance distributed computing: IEEE Computer Society; 1998.
4. Skillicorn DB, Talia D. Models and languages for parallel computation. ACM Comput Surv. 1998;30(2):123–69.
5. Dowd K, Severance C. High performance computing. Sebastopol: O'Reilly and Associates Inc.; 1998.
6. Abbas A. Grid computing: a practical guide to technology and applications. Hingham: Charles River Media Inc; 2004.
7. Gokhale M, Cohen J, Yoo A, Miller WM. Hardware technologies for high-performance data-intensive computing. IEEE Comput. 2008;41(4):60–8.
8. Nyland LS, Prins JF, Goldberg A, Mills PH. A design methodology for data-parallel applications. IEEE Trans Softw Eng. 2000;26(4):293–314.
9. Agichtein E, Ganti V. Mining reference tables for automatic text segmentation. In: Proceedings of the tenth ACM SIGKDD international conference on knowledge discovery and data mining, Seattle, WA, USA; 2004. p. 20–9.
10. Agichtein E. Scaling information extraction to large document collections: Microsoft Research. 2004.
11. Rencuzogullari U, Dwarkadas S. Dynamic adaptation to available resources for parallel computing in an autonomous network of workstations. In: Proceedings of the eighth ACM SIGPLAN symposium on principles and practices of parallel programming, Snowbird, UT; 2001. p. 72–81.
12. Cerf VG. An information avalanche. IEEE Comput. 2007;40(1):104–5.
13. Gantz JF, Reinsel D, Chute C, Schlichting W, McArthur J, Minton S, et al. The expanding digital universe (White Paper): IDC. 2007.
14. Lyman P, Varian HR. How much information? 2003 (Research Report). School of Information Management and Systems, University of California at Berkeley; 2003.
15. Berman F. Got data? A guide to data preservation in the information age. Commun ACM. 2008;51(12):50–6.
16. NSF. Data-intensive computing. National Science Foundation. 2009. http://www.nsf.gov/funding/pgm_summ.jsp?pims_id=503324&org=IIS. Retrieved 10 Aug 2009.
17. PNNL. Data intensive computing. Pacific Northwest National Laboratory. 2008. http://www.cs.cmu.edu/~bryant/presentations/DISC-concept.ppt. Retrieved 10 Aug 2009.

18. Buyya R, Yeo CS, Venugopal S, Broberg J, Brandic I. Cloud computing and emerging it platforms: vision, hype, and reality for delivering computing as the 5th utility. Future Gener Comput Syst. 2009;25(6):599–616.
19. Gray J. Distributed computing economics. ACM Queue. 2008;6(3):63–8.
20. Bryant RE. Data intensive scalable computing. Carnegie Mellon University. 2008. http://www.cs.cmu.edu/~bryant/presentations/DISC-concept.ppt. Retrieved 10 Aug 2009.
21. Middleton AM. Data-intensive computing solutions (Whitepaper): LexisNexis. 2009.
22. Dean J, Ghemawat S. Mapreduce: simplified data processing on large clusters. In: Proceedings of the sixth symposium on operating system design and implementation (OSDI); 2004.
23. Dean J, Ghemawat S. Mapreduce: a flexible data processing tool. Commun ACM. 2010;53 (1):72–7.
24. Pike R, Dorward S, Griesemer R, Quinlan S. Interpreting the data: parallel analysis with sawzall. Sci Program J. 2004;13(4):227–98.
25. White T. Hadoop: the definitive guide. 1st ed. Sebastopol: O'Reilly Media Inc; 2009.
26. Gates AF, Natkovich O, Chopra S, Kamath P, Narayanamurthy SM, Olston C, et al. Building a high-level dataflow system on top of map-reduce: the pig experience. In: Proceedings of the 35th international conference on very large databases (VLDB 2009), Lyon, France; 2009.
27. Olston C, Reed B, Srivastava U, Kumar R, Tomkins A. Pig latin: a not-so_foreign language for data processing. In: Proceedings of the 28th ACM SIGMOD/PODS international conference on management of data/principles of database systems, Vancouver, BC, Canada; 2008. p. 1099–110.
28. Bayliss DA. Enterrprise control language overview (Whitepaper): LesisNexis. 2010b.
29. Bayliss DA. Thinking declaratively (Whitepaper). 2010c.
30. Hellerstein JM. The declarative imperative. SIGMOD Rec. 2010;39(1):5–19.
31. O'Malley O. Introduction to hadoop. 2008. http://wiki.apache.org/hadoop-data/attachments/HadoopPresentations/attachments/YahooHadoopIntro-apachecon-us-2008.pdf. Retrieved 10 Aug 2009.
32. Bayliss DA. Aggregated data analysis: the paradigm shift (Whitepaper): LexisNexis. 2010a.
33. Buyya R. High performance cluster computing. Upper Saddle River: Prentice Hall; 1999.
34. Chaiken R, Jenkins B, Larson P-A, Ramsey B, Shakib D, Weaver S, et al. Scope: easy and efficient parallel processing of massive data sets. Proc VLDB Endow. 2008;1:1265–76.
35. Grossman R, Gu Y. Data mining using high performance data clouds: experimental studies using sector and sphere. In: Proceedings of the 14th ACM SIGKDD international conference on knowledge discovery and data mining, Las Vegas, Nevada, USA; 2008.
36. Grossman RL, Gu Y, Sabala M, Zhang W. Compute and storage clouds using wide area high performance networks. Future Gener Comput Syst. 2009;25(2):179–83.
37. Gu Y, Grossman RL. Lessons learned from a year's worth of benchmarks of large data clouds. In: Proceedings of the 2nd workshop on many-task computing on grids and supercomputers, Portland, Oregon; 2009.
38. Liu H, Orban D. Gridbatch: cloud computing for large-scale data-intensive batch applications. In: Proceedings of the eighth IEEE international symposium on cluster computing and the grid; 2008. p. 295–305.
39. Llor X, Acs B, Auvil LS, Capitanu B, Welge ME, Goldberg DE. Meandre: semantic-driven data-intensive flows in the clouds. In: Proceedings of the fourth IEEE international conference on eScience; 2008. p. 238–245.
40. Pavlo A, Paulson E, Rasin A, Abadi DJ, Dewitt DJ, Madden S, et al. A comparison of approaches to large-scale data analysis. In: Proceedings of the 35th SIGMOD international conference on management of data, Providence, RI; 2009. p. 165–68.
41. Ravichandran D, Pantel P, Hovy E. The terascale challenge. In: Proceedings of the KDD workshop on mining for and from the semantic web; 2004.
42. Yu Y, Gunda PK, Isard M. Distributed aggregation for data-parallel computing: interfaces and implementations. In: Proceedings of the ACM SIGOPS 22nd symposium on operating systems principles, Big Sky, Montana, USA; 2009. p. 247–60.

Chapter 7
Scalable Automated Linking Technology for Big Data Computing

Anthony M. Middleton, David Bayliss and Bob Foreman

Introduction

The massive amount of data being collected at many organizations has led to what is now being called the "Big Data" problem, which limits the capability of organizations to process and use their data effectively and makes the record linkage process even more challenging [1, 2]. New high-performance data-intensive computing architectures supporting scalable parallel processing such as Hadoop MapReduce and HPCC allow government, commercial organizations, and research environments to process massive amounts of data and solve complex data processing problems including record linkage.

SALT (Scalable Automated Linking Technology), developed by LexisNexis Risk Solutions (a subsidiary of Reed Elsevier, one of the world's largest publishers of information), is a tool which automatically generates code in the ECL language for the open source HPCC scalable data-intensive computing platform based on a simple specification to address most common data integration tasks including data profiling, data cleansing, data ingest, and record linkage. Record linkage, one of the most complex tasks in a data processing environment, is the data integration process of accurately matching or clustering records or documents from multiple data sources containing information which refer to the same entity such as a person or business. A fundamental challenge of data-intensive computing is developing new algorithms which can scale to search and process big data [3].

SALT incorporates some of the most advanced technology and best practices of. LexisNexis, and currently has over 30 patents pending related to record linkage and other technology included in SALT, including innovative new approaches to approximate string matching (a.k.a. fuzzy matching), automated calculation of matching weights and thresholds, automated selection of blocking criteria,

This chapter has been developed by Anthony M. Middleton, David Bayliss, and Bob Foreman from LexisNexis.

automated calculation of best values for fields in an entity, propagation of field values in entities to increase likelihood of matching, automated calculation of secondary relationships between entities, automated splitting of entity clusters to remove bad links, automated cleansing of data to improve match quality, and automated generation of batch and online applications for entity resolution and search applications including an online search application including a *Uber* key which allows searches on any combination of field input data.

SALT is provided as an executable program that can be executed automatically from the ECL IDE (Integrated Development Environment), or run from a Windows command prompt. The input to the SALT tool is a user-defined specification stored as a text file with a *.spc* or *.salt* file extension which includes declarative statements describing the user input data and process parameters. The output of SALT is a text (.mod) file containing ECL code which can be imported and executed in an HPCC system environment. The SALT tool can be used to generate complete applications ready-to-execute for *data profiling*, *data hygiene* (also called data cleansing, the process of cleaning data), *data source consistency monitoring* (checking consistency of data value distributions among multiple sources of input), *data file delta changes*, *data ingest*, and *record linking and clustering*. SALT record linking and clustering capabilities include (1) *internal linking*—the batch process of linking records from multiple sources which refer to the same entity to a unique entity identifier; (2) *external linking*—also called *entity resolution*, the batch process of linking information from an external file to a previously linked base or authority file in order to assign entity identifiers to the external data, or an online process where information entered about an entity is resolved to a specific entity identifier, or an online process for searching for records in an authority file which best match entered information about an entity; and (3) *remote linking*, an online capability that allows SALT record matching to be incorporated within a custom user application.

This chapter describes how the SALT tool can be used to automatically generate executable code for the complete data integration process including record linkage. All data examples used in this chapter are fictitious and do not represent real information on any person, place, or business unless stated otherwise.

SALT—Basic Concepts

SALT is designed to run on the open source HPCC scalable data-intensive computing platform. It functions as an ECL code generator on the HPCC platform to automatically produce ECL code for a variety of applications. Although the primary use of SALT is for record linkage and clustering applications, SALT offers other capabilities including data profiling, data hygiene, data source consistency monitoring, data ingest and updating of base data files, data parsing, and file comparison to determine delta changes between versions of a data file.

SALT is an executable program coded in C++ which can be run from a Windows command prompt or built in directly from the ECL IDE in the HPCC environment. The SALT program reads as its input a text file containing user-defined specification statements, and produces an output file containing the generated ECL code to import into the user ECL code repository. SALT provides many command line options to control its execution, and to determine the type of ECL code to produce for a target application.

SALT offers many advantages when developing a new data-intensive application. SALT encapsulates a significant amount of ECL programming knowledge, experience, and best practices gained at LexisNexis for the types of applications supported, and can result in significant increases in developer productivity. It affords significant reductions in implementation time and cost over a hand-coded approach. SALT can be used with any type of data in any format supported by the ECL programming language to create new applications, or to enhance existing applications.

SALT Process

The SALT process begins with a user defined specification file for which an example is shown in Fig. 7.1. This is a text file with declarative statements and parameters that define the data file and fields to be processed, and associated processing options such as the module into which the generated code is imported.

```
1   OPTIONS:-ga
2   MODULE:MyModule
3   FILENAME:Sample
4   IDFIELD:EXISTS:BDID
5   RIDFIELD:rcid
6   RECORDS:9000000
7   POPULATION:1000000
8   NINES:3
9   FIELDTYPE:DEFAULT:LEFTTRIM:NOQUOTES("') :
10  FIELDTYPE:NUMBER:ALLOW(0123456789) :
11  FIELDTYPE:ALPHA:CAPS:ALLOW(ABCDEFGHIJKLMNOPQRSTUVWXYZ) :
12  FIELDTYPE:WORDBAG:CAPS:ALLOW(ABCDEFGHIJKLMNOPQRSTUVWXYZ0123456789') :SPACES( <>()[]-^~!+&,./) :ONFAIL(CLEAN) :
13  FIELDTYPE:CITY:LIKE(WORDBAG) :LENGTHS(0,4..) :ONFAIL(BLANK) :
14  FIELD:source:CARRY:
15  FIELD:vendor_id:CARRY:
16  FIELD:dt_first_seen:RECORDDATE(FIRST) :
17  FIELD:dt_last_seen:RECORDDATE(LAST) :
18  FIELD:company_name:BAGOFWORDS(MOST) :LIKE(WORDBAG) :TYPE(STRING120) :INITIAL:ABBR:FORCE(+10) :12,107
19  FIELD:prim_range:EDIT1:10,001
20  FIELD:predir:2,000
21  FIELD:prim_name:EDIT1:9,001
22  FIELD:addr_suffix:LIKE(ALPHA) :CONTEXT(prim_name) :3,000
23  FIELD:postdir:3,000
24  FIELD:unit_desig:1,012
25  FIELD:sec_range:7,022
26  FIELD:CITY:9,002
27  FIELD:state:LIKE(ALPHA) :4,000
28  FIELD:zip:LIKE(NUMBER) :11,001
29  FIELD:zip4:LIKE(NUMBER) :CONTEXT(zip) :11,007
30  FIELD:county:6,001
31  FIELD:msa:LIKE(NUMBER) :6,000
32  FIELD:phone:LIKE(NUMBER) :PROP:18,015
33  FIELD:fein:LIKE(NUMBER) :PROP:16,003
34  CONCEPT:locale:+:zip:state:city:msa:13,066
35  CONCEPT:address:prim_range+:sec_range:prim_name+:zip4:unit_desig:addr_suffix:10,027
36  SOURCEFIELD:source:CONSISTENT(company_name,prim_name,prim_range,sec_range,city,state)
37  ATTRIBUTEFILE:VEHICLES:NAMED(SALT_QA.File_Vehicle_Matches_Sample) :VALUES(vin) :IDFIELD(BDID) :16,375
38  ATTRIBUTEFILE:PROPERTY:NAMED(SALT_QA.File_Property_Matches_Sample) :VALUES(LN_FARES_id) :IDFIELD(BDID) :22,845
39  ATTRIBUTEFILE:BANKRUPTCY:NAMED(SALT_QA.File_Bankruptcy_Matches_Sample) :VALUES(court_case_number) :IDFIELD(BDID) :13,250
40
```

Fig. 7.1 SALT specification file example

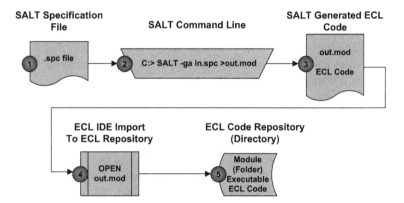

Fig. 7.2 SALT basic process

Figure 7.2 shows the basic steps in using SALT: (1) a specification file for the data and application is created by the user; (2) the SALT program is executed using a command line with specific options depending on the type of application for which the code is being generated and includes an input file with a .spc extension and an output file with a .mod extension; (3) the SALT program produces an output file in a special .mod format with the ECL coded needed for the application; (4) the generated code is imported into the ECL code repository; and (5) the ECL code is now available for execution using the ECL IDE.

Specification File Language

The SALT specification language is a declarative language which describes the input file data and the process parameters to be used in a SALT generated ECL language application. Each specification file language statement must appear on a single line of text in the specification file. The basic syntax for language statements is as follows:

KEYWORD:*parameter*:KEYWORD(*parameter*)[:OPTIONAL|WORD]

Keywords are not case-sensitive and Optional parameters can appear in any order within a specification file statement. Keywords are shown in caps for emphasis in all examples in this chapter. Although statements can generally appear in any order, definitions are usually ordered in a similar manner to the example in Fig. 7.2 for readability and consistency. A complete language reference is not presented here, but can be found in the SALT installation.

MODULE: *modulename[.submodule]*

The MODULE statement specifies a module name (folder) in the ECL repository (directory) where the source code generated by SALT will reside. The code generated by SALT uses the specified *modulename* with optional *submodule* as the base for the ECL code generated and is used for external references the code.

OPTIONS: *option_switches*

The OPTIONS statement allows the .spc file to override or add in command line options normally specified on the SALT command line when using SALT directly from the ECL IDE.

FILENAME: *name]*

The FILENAME statement allows a logical name for the input file to be specified and processed by the code generated by SALT. The *name* parameter is incorporated into various attribute names including attributes which identify the input dataset and the input record layout for the process, and additional temporary and output filenames in the ECL code generated by SALT.

PROCESS: *processname[:*UBER(ALWAYS|REQUIRED|NEVER)]

The PROCESS statement specifies an overall name for an external linking or remote linking process generated by SALT, but is not required for other processes. The *processname* is arbitrary and used for symbol naming in the generated code. The UBER option defines how the UBER key is used in an external linking process. The default is the UBER key is used if searching using all of the LINKPATHs specified for external linking could satisfy the query.

IDFIELD:
IDFIELD:EXISTS:*fieldname*

The IDFIELD identifies the field to be used as the entity ID for record linkage. If IDFIELD: is specified with nothing following, then it is assumed that no ID exists and the generated code will be used to cluster the input file records and assign a clustering ID based on the record id field specified in the RIDFIELD statement. If IDFIELD:EXISTS:*fieldname* is specified, then the input file is assumed to have a field previously defined identifying matching records for entity clusters. When used in an record linkage process, this allows additional records to be clustered with the existing IDs.

IDNAME:*fieldname*

The IDNAME statement specifies the *fieldname* to be used for the ID field in the output of a record linkage process. If an ID field does not already exist in the input data, then IDFIELD: is used with IDNAME:*fieldname* which specifies the name of the output field for the ID.

RIDFIELD: *fieldname*

The RIDFIELD statement specifies the name of the numeric field containing the record identifier or RID. Each record in the input dataset should have a unique RID value. The RIDFIELD is used as the basis for the record linkage process when no IDFIELD:EXISTS is specified. The entity cluster ID for each matched set of records will be the lowest value RID in the group at the end of the record linkage process.

RECORDS: *record_count*

The RECORDS statement specifies the expected number of records at the end of a record linkage process. The *record_count* value is the expected number of records at the end of the process which initially can be specified as the input record count. The RECORDS statement in combination with the NINES and POPULATION statements in a specification file allow SALT to compute a suitable matching score threshold for record linkage as well as a block size for the number of records to compare for the blocking process.

POPULATION: *entity_count*

The POPULATION statement specifies the expected number of entities at the end of a record linkage process. When the matching process is complete, entity clusters or records are formed, each identified by a unique entity ID. The *entity_count* value is the expected number of entities or unique entity IDs that will be generated by the matching process.

NINES: *precision_value*

The NINES statement specifies the precision required for a SALT generated record linkage process. The *precision_value* parameter specifies the precision required expressed as a number of nines such that a value of 2 means 2 nines or a precision of 99 %. A value of 3 means 3 nines or 99.9 %.

FIELDTYPE: *typename*: **[ALLOW(** *chars* **):]   [SPACES(** *chars* **):]**
[IGNORE(*chars* **):] [LEFTTRIM:] [CAPS:] [LENGTHS**
(*length_list* **):]**
[NOQUOTES():] [LIKE(*fieldtype* **):] [ONFAIL(IGNORE|CLEAN|**
BLANK|
REJECT):] [CUSTOM(*functionname* **[<|>n] [,** *funcparam1,*
funcparam2,
...funcparamn] **:]**

The FIELDTYPE statement allows field editing and validity checking requirements used for data hygiene processing to be defined and grouped into common definitions which can then be associated with any field. A FIELDTYPE field does not really exist; it is used to assign editing constraints to a field. The FIELDTYPE parameters are essentially assertions defining what the given field must look like. The LIKE parameter specifies a base or parent for the field type allowing

FIELDTYPEs to be nested. All of the restrictions of the parent field type are then applied in addition to those of the field type being specified. The ONFAIL parameter allows the user to select what occurs when an editing constraint is violated. These include ignoring the error, cleaning the data according to the constraint, blanking or zeroing the field, or rejecting the record. The CUSTOM parameter allows a user defined function to be referenced to perform validity checking.

BESTTYPE:*name*:*BASIS(fixed_fields:[?|!]:optional_fields)*: *construction_method*:*construction_modifiers*: *propagation_method*

The BESTTYPE statement is used to define a best value computation for a field or concept for a given basis for an entity. The calculated best value can be used for propagation during record linkage, and is available for external application use. The basis is typically the entity identifier specified by the IDFIELD, but a more complex basis can be specified consisting of multiple fields. Multiple BESTTYPEs can be associated with a field or concept, and all are evaluated, but the leftmost non-null best value is considered the overall best value for the field. SALT generates code for calculating the best values in the Best module and exported dataset definitions are provided which allow output of a dataset of best values for each field or concept and associated BESTYPE definitions. In addition SALT provides several aggregate files using whatever fields are defined in the basis.

BESTTYPE construction methods provided are COMMONEST (most frequently appearing value), VOTED (a user-defined function is provided to weight the field value by source type), UNIQUE (best value is produced if there is only one unique value for the field in the entity cluster), RECENT (uses the most recent value specified by a date field parameter), LONGEST (picks the longest value for a field). Construction modifiers include MINIMUM (candidates must have a minimum number of occurrences in an entity cluster), FUZZY (specifies that the fuzzy matching criteria of the target field are used to allow less common values to support candidates for best value), and VALID (specifies that only those values considered valid will be considered available for BEST computation). Propagation methods include PROP (copy the best value into null fields with a matching basis), EXTEND (copy the best value into null fields and those that are partial exact matches to the best value), FIX (copy the best value onto null fields and overwrite those fields which are fuzzy matches to the best value), and ENFORCE (copy the best value into the field regardless of the original data content).

Note that the BESTTYPE statement is a powerful capability and interested readers are referred to the SALT User's Guide for a more in-depth explanation.

FIELD:*fieldname*[:PROP] [:CONTEXT(*context_fieldname*)] [:BAGOFWORDS[(MANY|ALL|ANY|MOST|TRIGRAM)]] [:CARRY] [:TYPE(datatype)] [:LIKE(*fieldtype*)] [:EDIT1] [:EDIT2] [:PHONETIC] [:INITIAL] [:ABBR] [:HYPHEN1[(*n*)]] [:HYPHEN2 [(*n*)]]

```
[:fuzzy_function...[:fuzzy_function]]
[:MULTIPLE][:RECORDDATE(FIRST|LAST[,YYYYMM])]
[:besttype...[:besttype]] [:FLAG][:OWNED]
[:FORCE[(+|-[n])]:specificity,switch_value1000
```

The FIELD statement defines a data field in the input file record including its type and other characteristics which affect hygiene, validity, and matching. The PROP parameter specifies a default propagation for the field if there is no associated BESTTYPE. If the CONTEXT parameter is specified, then a match occurs only if both the values in *fieldname* and the *context_fieldname* match. If the BAGOFWORDS parameter is specified then the string field is treated as a sequence of space delimited tokens. The LIKE parameter specifies additional editing characteristics of the field defined by the named FIELDTYPE statement. EDIT1 and EDIT2 specify edit-distance fuzzy matching, PHONETIC specifies phonetic fuzzy matching, INITIAL allows a partial string to match the first characters of another string, ABBR allows the first character of tokens in one string appended together to match another string, HYPHEN1 and HYPHEN2 provide for partial and reverse matching of hyphenated fields, MULTIPLE allows multiple values to be specified for entity resolution, RECCORDATE allows a date field to be specified as FIRST or LAST in context and YYYYMM allows dates to be year and month only. fuzzy_function specifies the name of a custom fuzzy matching function defined by the FUZZY statement.

The *besttype* parameters refer to BESTTTYPE definitions associated with the field, FLAG allows statistics to be calculated about the fields when using BESTTYPE, OWNED with FLAG implies the best value should only appear in a single entity cluster. The FORCE parameter is used to require a match on the field for a record match, or specify the minimum field match score needed for a record match, and can also specify that no negative contribution to the record score is allowed. The specificity and switch_value1000 are computed by SALT and added to the FIELD statements prior to record linkage. Specificity is the weighted average field score for matching and the switch_value1000 is the average variability of field values across all entity clusters (fraction * 1000).

FUZZY: name:RST:TYPE(FuzzyType):CUSTOM(FunctionName)

The FUZZY statement specifies a custom user-supplied fuzzy matching function for a FIELD. SALT automatically handles other requirements such as scaling of the field value specificity. The *name* parameter associates a name with the custom fuzzy processing. Once defined, the *name* can be used as a parameter of a FIELD definition. The *FuzzyType* parameter allows the return type of the fuzzy function to be specified as a valid ECL datatype. The *FunctionName* parameter defines an ECL function which performs the fuzzy processing.

DATEFIELD:fieldname[:PROP][:SOFT1][:YEARSHIFT][:MDDM]
[:CONTEXT(context_fieldname)][:FORCE[(+|-[n]
[,GENERATION])]:specificity,switch_value1000

The DATEFIELD statement specifies a numeric string field in the format YYYYMMDD. It functions in an identical manner to the FIELD statement except for requiring the specific date format. The FORCE parameter includes a special option GENERATION which applies only to a DATEFIELD. If used the YEAR portion of the date has to be within 13 years of the other (or null). The SOFT1, YEARSHIFT, and MDDM options provide some fuzzy matching capabilities for dates.

SOURCEFIELD:*fieldname*[**:CONSISTENT[(checkfieldname,checkfieldname, …)]]**

The SOURCFIELD statement specifies the name of the field containing the input data source type. The source field is not processed as a normal field definition for matching, but is used for the data source consistency checking process. If the CONSISTENT parameter is provided then SALT generates code into the hygiene module to check for consistency of field values between the various sources represented in the input file.

SOURCERIDFIELD: *fieldname*

The SOURCERIDFIELD statement specifies the name of a field in the input file which contains a unique identifier for a corresponding record in source or ingest file which has been merged into the base file. This value in combination with the value of the SOURCEFIELD provides a link to the original source record for the data.

LATLONG: *name***:LAT(***latitude_field***):LONG(***longitude_field***):** **[DISTANCE(***n***)][DIVISIONS(***n***)]**

The LATLONG statement specifies a geo-point field for the location associated with a record based on latitude and longitude fields included in the specification file. If a LATLONG is specified, the geo-point is made up of the combined latitude field and longitude field and is treated as one single 'pin-point' location instead of the two separate measures during a record linkage process. LATLONG field values are treated fuzzily for matching records. The LATLONG geo-points must also be within DISTANCE(n) as defined above from each other to make a positive contribution to the match score, otherwise it can make a negative contribution. The population density of entities in the grid as defined by the DISTANCE and DIVISIONS parameters for the grid around all geo-points is calculated giving the field match score for a given distance from a geo-point.

CONCEPT:fieldname[:+]:*child1***[+]:***child2***[+]:***childn***[+]…** **[:FORCE[(+|−[n])]]:[:SCALE(NEVER|ALWAYS|MATCH)][:** **BAGOFWORDS]:** **specificity,switch_value1000**

The CONCEPT statement allows a group of related or dependent fields to be defined and is used so that dependent fields are not over weighted in the record linkage process. SALT makes an implicit assumption of field independence which can lead to under or over weighting during the matching process when the fields only really have meaning in the context of other fields. This can be corrected by appropriately defining CONCEPT fields. A CONCEPT replaces the child fields *only if matched* between records during the record matching process. If the Concept field does not match, the child fields are independently evaluated in the record matching and scoring process. A Concept field is a computed field and does not appear in the input file.

ATTRIBUTEFILE: *name* **[:NAMED(** *modulename.filename* **)] :IDFIELD (** *id_field_name* **) :VALUES(** *attribute_field_name* **[,LIST]) [:KEEP (** *n* **/ALL)] [:WEIGHT(** *value* **)] [SEARCH(** *list_of_fields* **)] [:speci- ficity, switch_value1000]**

An ATTRIBUTEFILE statement defines a special type of field which provides a set of values for matching from an external file, child dataset which is part of the main input file, or a child dataset which is part of the external file. Each matching value must be paired with an ID value of the same type as defined for the input file in the IDFIELD or IDNAME statement. During the matching process, if attribute values match between records being compared, the match will contribute to the overall score of the record match. The VALUES field list allows additional fields to be included which can then be used in search applications. The KEEP parameter allows the user to specify how many matching attribute values are allowed to contribute to a record match.

INGESTFILE: *name* **:NAMED(** *module.attribute_name* **)**

The INGESTFILE statement specifies the *name* to be used for an ingest file to be appended/merged with the base file as part of a SALT record linkage process. The *module.attribute_name* specified in the NAMED() parameter specifies the module and attribute name of a dataset attribute. The dataset is assumed to be in the same format as the base file. Ingest files are appended to and merged with the base file specified in the FILENAME attribute for a record linkage process. Typically these files are generated from external source files or base files for other types of entities.

LINKPATH: *pathname* **[:** *fieldname:fieldname:fieldname...:* *fieldname* **]**

The LINKPATH statement specifies the name of a search path for an external linking entity resolution process generated by SALT. The *pathname* is arbitrary and used for symbol naming. A *fieldname* references either a field defined in the specification file, an ATTRIBUTEFILE value field, or is a '?' or '+' character separating groups of fields. A linkpath can be divided into 3 groups: *required* fields which immediately follow the *pathname* and must match, *optional* fields which follow the '?' character used as a *fieldname* and must match if data is present in both records for the field, and *extra credit* fields which follow a '+' character used as a *fieldname* and are not required to match but will add to the match score if they do. The fieldnames used in a linkpath typically correspond to field combinations used frequently in user queries.

RELATIONSHIP: *relationshipname***:BASIS(***FieldList***):DEDUP (***FieldList***)[:SCORE(***FieldList***)][:MULTIPLE(***n***)][:SPLIT(***n***)] [:THRESHOLD(***n***)]**
RELATIONSHIP: *relationshipname***:** *RelationshipList***)][: MULTIPLE(***n***)] [:THRESHOLD(***n***)]**

SALT record linkage provides the capability to cluster together records to form an entity. In some situations, the objective is not to determine that two records or clusters are close enough to become part of the same entity, but to determine if a statistically significant link exists between the two clusters and to record this relationship. The RELATIONSHIP statement provides this function. Relationships provide a way to record instances when multiple occurrences of specific set of fields (the BASIS field list) matching between clusters provide information that a specific relationship exists or evidence that the clusters may need to be linked. The second form of the RELATIONSHIP statement definition above allows a relationship to be formed as the sum of other relationships.

THRESHOLD: *threshold_value*

The THRESHOLD statement overrides the default record matching threshold calculated by the SALT code generation process. The *threshold_value* specifies a new value for the specificity matching threshold which is the minimum amount of total specificity needed for a record match.

BLOCKLINK: *NAMED(modulename.attribute)*

The BLOCKLINK statement is used to define a file which will be used to block linking of specific matching records during an internal linking process. BLOCKLINK provides a user-specified unlink capability which prevents certain records from being combined in an entity cluster. This may be required as part of a linking process for compliance or other reasons.

SALT—Applications

The starting point for utilizing the SALT tool is creating a specification file which defines your input data file, the fields in your input data file to be used, as well as additional statements and parameters to direct the ECL code generation process including the module name to be used for the generated ECL. The content of your specification file varies depending on the specific type of process for which you need ECL code generated by SALT.

SALT can be used for processes including:

- Internal linking/record matching/clustering
- External record matching
- Data hygiene
- Data profiling
- Data source consistency checking
- Data source cluster diagnostics
- Delta file comparisons
- Data Parsing and Classification
- Generation of inverted index records for Boolean search.

Figure 7.3 shows the SALT user data integration process and application flow.

Data Profiling

Data profiling or exploratory data analysis [2] is a step usually performed by data analysts on raw input data to determine the characteristics of the data including type, statistical, and pattern information as well as field population counts. The goal of profiling is to fully understand the characteristics of the data and identify any bad data or validity issues and any additional cleansing, filtering, or de-duplication that may be needed before the data is processed further. Data profiling can also provide information on the changing characteristics of data over time as new data is linked. Data profiling can occur prior to the parsing step if needed to identify raw data fields which need to be parsed, but is usually performed once the input data has been projected into a structured format for the record linkage process.

SALT data profiling is a process which provides important type, statistical, and pattern information on the data fields and concepts and their contents in any input data file. This information is essential in analyzing the content and shape (patterns) of the data in the source data files and facilitates important decisions concerning data quality, cleansing, de-duping, and linking of records, and to provide information on the changing characteristics of data over time. Data profiling is a task usually performed by data analysts as exploratory data analysis [2], and is an important preparatory step for the record linkage process.

SALT data profiling provides by field breakdowns of all the characters, string lengths, field cardinality (the number of unique values a field contains), top data values, and word counts for every data field or concept (dependent groups of data fields) defined in the specification file. In addition, SALT calculates and displays the top data patterns to help analyze the shape of the data. The data profiling capability also provides summary statistical data such as the number of records in the input file, and the percentage of non-blank data, maximum field length, and average field length for every field and concept. This summary information provides a quick view

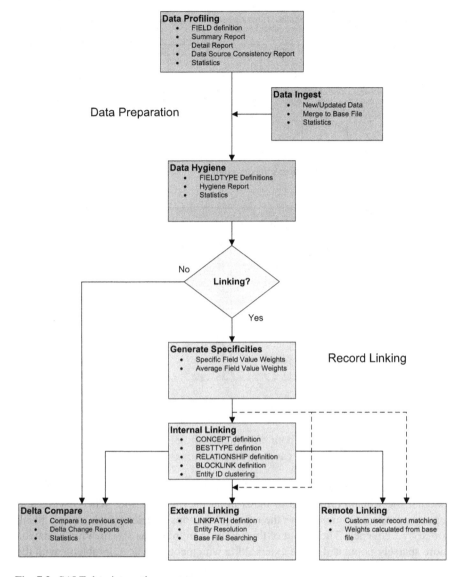

Fig. 7.3 SALT data integration process

which can be compared with previous versions of a data file to identify anomalies or to verify anticipated changes in the content of a data file.

The data profiling information can also be used as input data to a change tracking system. If any of the data profiling information is not consistent with expectations, it may be an indication of bad data in the source file which may need further cleansing. Figure 7.4 shows a partial data profiling summary report produced by SALT for a sample input file of business data.

	txt	numberofrecords	populated source pcnt	maxlength source	avelength source	populated vendor id pcnt	maxlength vendor id	avelength vendor id
1	Data Profiling Summa	7959271	100.000000	2	1.787971	81.667705	34	15.146526

populated msa pcnt	maxlength msa	avelength msa	populated phone pcnt	maxlength phone	avelength phone	populated fein pcnt	maxlength fein	avelength fein
85.654164	4	4.000000	37.854836	10	9.957977	4.412515	9	8.923079

Fig. 7.4 SALT data profiling summary report example

fldno	fieldname	cardinality	len		words		characters		patterns		frequent terms	
			len	cnt	words	cnt	c	cnt	data pattern	cnt	val	cnt
1	19	phone	970479	1	4946302	1	7959271	0	8485069	9	4946302	0	4946302
				10	2970565			2	3325862	9999999999	2970565	8002882020	979
				7	42044			3	3182462	9999999	42044	9146408100	833
				9	238			7	3028430	999999999	238	2076248603	661
				8	122			5	2971896	99999999	122	5618334573	591
								8	2957289			8004387325	565
								4	2929773			8007721213	505
								6	2836442			8008291040	469
								1	2799816			8002758777	453
								9	2432339			9086386462	443
												8666070317	408
												2125732323	384
												5617500369	363

Fig. 7.5 SALT data profiling field detail report example

Figure 7.5 shows a partial data profiling detail field report for the phone field in the same sample file. SALT can run data profiling field profiles on all fields or selected fields.

SALT data profiling also provides the capability to analyze fields existing in an external file which correspond to fields in an internal base file. When executed, a report is produced which shows the top combinations of fields which are non-blank sorted by frequency. The output of this report is an indicator of which fields may be more sparsely populated in the external file which could indicate problems with the data source and can also help identify the type of data represented in the external file. If the external file is representative of typical data requiring entity resolution using the base file, this report helps determine the best field combinations to use to define the linkpaths required, and SALT can automatically generated suggested linkpaths using the data from this report. Figure 7.6 shows partial sample output from this report.

	fields	cnt
1	:predir:prim_name:postdir:unit_desig:zip:zip4:county:msa:locale	239398
2	:predir:prim_name:postdir:unit_desig:zip:zip4:county:msa	236589
3	:predir:prim_name:addr_suffix:postdir:unit_desig:zip:zip4:county:msa:locale	99631
4	:predir:prim_name:addr_suffix:postdir:unit_desig:zip:zip4:county:msa	80250
5	:predir:zip:zip4	55903
6	:predir:locale	40800
7	:predir:prim_name:postdir:unit_desig:CITY:state:zip:zip4:county:msa:locale	24342
8	:predir:prim_name:postdir:unit_desig:sec_range:zip:zip4:county:msa:locale	22128
9	:predir:postdir:zip:zip4:county:msa:locale	18771
10	:predir:prim_name:postdir:unit_desig:sec_range:zip:zip4:county:msa	17454
11	:predir:prim_name:postdir:unit_desig:zip:zip4:county:msa:address	16296
12	:predir:prim_name:postdir:zip:zip4:county:msa	12038
13	:predir:prim_name:addr_suffix:postdir:unit_desig:CITY:state:zip:zip4:county:msa:locale	11745
14	:predir:prim_name:postdir:zip:zip4:county:msa:locale	11670
15	:predir:prim_name:postdir:unit_desig:CITY:state:zip:zip4:county:msa	9883

Fig. 7.6 SALT field combination analysis report example

Data Hygiene

Data cleansing, also called data hygiene, is the process of cleaning the raw input data so that it can be used effectively in a subsequent process like record linkage. The cleanliness of data is determined by whether or not a data item is valid within the constraints specified for a particular field. For example, if a particular data field is constrained to the numeric characters 0–9, then any data item for the field which contains characters other than 0–9 would fail a cleansing validity check for the data field. So 5551212 would be a valid value for a data item, but 555-1212 would not. Data which has not been cleansed properly can have adverse effects on the outcome of the record linkage process [4]. Some data issues can be identified and corrected through the cleansing process, for others such as misspellings or character transpositions or deletions, the record linkage process will need to support comparison methods such as edit-distance, phonetic, and other forms of fuzzy matching to allow for common typographical errors and then scale match weights appropriately.

Once the initial data profiling process is complete, SALT can be used to check the cleanliness of the data. SALT uses the term data hygiene to refer to both the cleanliness of the data and the process by which data is cleansed so that it can be used effectively in a subsequent data integration process such as record linkage. Cleanliness of data is determined by whether or not a data item is valid within the constraints specified for a particular data field. For example, if a particular data field is constrained to the numeric characters 0–9, then any data item for the field which contains characters other than 0–9 would fail a hygiene validity check for the data field.

SALT includes capabilities to define hygiene constraints on its input data, identify invalid data in fields, and cleanse the data if needed. However, by default, no error checking will occur unless specified for field definitions in the specification file. SALT includes standard syntax using the FIELDTYPE statement to support most common types of validity checks on data in fields. Custom user-defined functions which perform user-specific validity checks can also be included.

SALT data hygiene can be used as an independent process to check the input data, and if appropriate, the user can correct any problems identified to create a cleansed input file before continuing with other SALT processes like record linkage. SALT can also automatically cleanse bad data before proceeding in which is controlled by the ONFAIL parameter of the FIELDTYPE statement. If the value in a field is not valid according to the editing constraints imposed, ONFAIL actions include: IGNORE (data is accepted as is), BLANK (the value in the data field is changed to blank or zero depending on the type of the field), CLEAN (removes any invalid characters), or REJECT (removes/filter out records with the invalid field data).

The following are sample FIELDTYPE statements:

```
FIELDTYPE:DEFAULT:LEFTTRIM:NOQUOTES('''):
FIELDTYPE:NUMBER:ALLOW(0123456789):
FIELDTYPE:ALPHA:CAPS:ALLOW(ABCDEFGHIJKLMNOPQRSTUVWXYZ):
FIELDTYPE:WORDBAG:CAPS:ALLOW
(ABCDEFGHIJKLMNOPQRSTUVWXYZ0123456789'):SPACES(<>{}[]-
^=!+&,./):ONFAIL(CLEAN):
FIELDTYPE:CITY:LIKE(WORDBAG):LENGTHS(0,4..):ONFAIL
(BLANK):
```

The DEFAULT fieldtype applies to all fields unless overridden, and the LIKE parameter allows fieldtypes to be nested in a hierarchical manner. If the name of a FIELDTYPE also matches the name of a field like CITY, then the field automatically assumes the hygiene constraints of the FIELDTYPE with the same name. This facilitates building a library of FIELDTYPE statements which can be used to

	source	fieldname	errormessage	cnt	sourcegroupcount
1	BR	company_name	String is quoted using one of:'''	1	333821
2	C)	company_name	String is quoted using one of:'''	3	65110
3	C,	company_name	String is quoted using one of:'''	2	27322
4	C-	company_name	String is quoted using one of:'''	5	13575
5	CO	company_name	String is quoted using one of:'''	1	38720
6	CI	company_name	String is quoted using one of:'''	1	1450
7	D	company_name	String is quoted using one of:'''	1	541244
8	FL	company_name	String is quoted using one of:'''	1	14134
9	LC	company_name	String is quoted using one of:'''	1	11989
10	SP	company_name	String is quoted using one of:'''	1	8659
11	WF	company_name	String is quoted using one of:'''	1	286949
12	Y	company_name	String is quoted using one of:'''	4	362739
13	+E	company_name	Contains characters not in:ABCDEFGHIJKLMNOPQRSTUVWXYZ0123456789' <>{}[]-^=! +&,./	1	228
14	.E	company_name	Contains characters not in:ABCDEFGHIJKLMNOPQRSTUVWXYZ0123456789' <>{}[]-^=! +&,./	4	1178

Fig. 7.7 SALT data hygiene report example

insure consistency across data models. Figure 7.7 shows a partial example of the SALT data hygiene report.

Data Source Consistency Checking

SALT has the capability to check the field value consistency between different sources of data for an input file. This capability requires that the input file being checked has a field on each record designating from which unique source the data was provided for the record. The SALT specification file includes a special SOURCEFIELD statement which provides the information needed to perform the consistency checking. Consistency checking can be specified for all fields in the record, or only specific fields. Typically consistency checking is used only on specific fields where a consistent distribution of data values across all sources is expected. For example, for an input file containing person names, we expect data values in the last name field would generally be consistently represented in terms of its distribution within sources. For example, the last name Smith would be represented in all sources and no source would have this data value in abnormally high numbers compared to the average across all sources.

The output of the data source consistency checking process is a list of outliers, data values whose distribution is not consistently represented across all sources. This list contains the name of the data field(s) being checked, the data value of the outlier, the unique source identifier, and the number of records containing the outlier. These outliers could represent bad data values being introduced from a specific source, missing data, or other anomalies and inconsistencies related to the data source containing the outliers. Some outliers may be legitimate, for example if the source field contains a geographic identifier, there may be high concentrations of a particular last names in certain geographical areas which could be flagged by the consistency checking. Figure 7.8 shows partial sample output from a data source consistency report.

	fieldname	fieldvalue	src	c
1	prim_name	PO BOX 1349	BA	953
2	prim_name	PERSON	BA	820
3	prim_name	LA POSADA	BA	805
4	prim_name	PO BOX 1785	BA	757
5	prim_name	PO BOX 272749	UT	744
6	prim_name	GLENWAY	BA	743
7	prim_name	PO BOX 615	BA	733
8	prim_name	MEANS	W	733
9	prim_name	KABLER	GG	677
10	prim_name	COLISEUM CENTER 2730 TYVOLA	JI	645

Fig. 7.8 SALT data source consistency report example

Delta File Comparison

SALT includes the capability to compare two versions of a file and provides two reports showing the differences. A *differences summary report* which outputs five records similar to the data profiling summary report for the records in the new file, records in the old file, updated/changed records in the new file, records added to the new file, and records deleted from the old file. The differences summary provides the number of records for each of these categories (New, Old, Updates, Additions, Deletions), and the percentage of non-blank data, maximum field length, and average field length for every field for each of the categories. The Changed category is only available if an RIDFIELD statement is included in the specification file. A *differences detail report* which outputs any record (Added, Deleted, Changed) which is different in the new file from the old file with additional columns to flag the type of change. Added and Changed records are shown from the new file, and Deleted records are shown from the old file. The Changed category is only available if an RIDFIELD statement is included in the specification file and otherwise a change is shown as an addition and deletion.

The delta difference reports show the differences between two versions of the same file which has been updated through an ETL type of process, for example a monthly update of a data source. Even though summary statistics are normally generated in a typical ETL update process, the statistics for the delta difference reports may highlight smaller errors that may be obscured by the statistics on the full files. Figure 7.9 shows partial sample output from delta difference summary and detail reports.

Data Ingest

The data ingest step is the merging of additional standardized input data source files with an existing base file or with each other to create the base file on which the record linkage process will be performed. If a linked base or authority already file exists, the data ingest process functions as an update and merges the new or updated record information into the base file. The subsequent record linkage process can add any new records to existing entity clusters, form new entity clusters, splitting and collapsing entity clusters as required based on the matching results and new information included in the input files to create a new linked version of the base file.

Data processing applications which maintain a base or authority file with information on an entity typically require periodic updates with new or updated information. The reading and processing of new information to add or update the base file is usually referred to as a *data ingest* process. The SALT data ingest process applies the ingest records to the base file and determines which records are: *new*, never seen before; *updates*, identical record to an existing record in the base file but with newer record dates; *unchanged*, identical to an existing record in the

SALT—Applications

	txt	numberofrecords	populated rcid pcnt	maxlength rcid	avelength rcid	populated bdid pcnt	maxlength bdid	avelength bdid	populated source pcnt	maxlength source	avelength source
1	New	151553	100.000000	10	9.164708	100.000000	10	9.164160	100.000000	2	1.916683
2	Old	151475	100.000000	10	9.164159	100.000000	10	9.164159	100.000000	2	1.916627
3	Updates	44	100.000000	10	9.181818	100.000000	10	9.181818	100.000000	2	1.909091
4	Additions	100	100.000000	10	10.000000	100.000000	10	9.170000	100.000000	2	2.000000
5	Deletions	22	100.000000	10	9.181818	100.000000	10	9.181818	100.000000	2	1.909091

	rcid	bdid	source	vendor id	dt first seen	dt last seen	company name	prim range	predir	prim name	addr suffix
1	1967797467	1967797467	FA	OOFRA0660106157	0	0	VALENARNI MEDICAL	3031903165	N	NORDART	AVE
2	829704187	829704187	U2	S-DNB410766020001120	20060100	20090700	DONNSTON BANK NATIONAL	3125497287		TELLIN	ST
3	883387279	883387279	U2	S-DNB0679300628219930824	20060100	20090700	PAULMORE CORP	722179063	W	CHORZAIMES	BLVD
4	2293585932	2293585932	PF	CP3520282754	20010301	20010301	DESUR INSURANCE AGENCY	350	N	NORDART	AVE
5	2380337696	829465110	++	24	20060100	20090700	HOLKIMER	100		TELLIN	ST

postdir	unit desig	sec range	city	state	zip	zip4	county	msa	phone	fein	added	deleted	changed
	STE	202	GAGLEPARK	MI	96939	1337	125	2160	0	0	false	false	true
			BELLFORT	MA	6810	1802	025	1120	0	0	false	false	true
			GARBMONT	MI	59742	3037	163	2160	0	0	false	false	true
	STE	100	GAGLEPARK	MI	96939	5388			0	0	false	true	false
			BELLFORT	MA	6810	1802	025	1120	0	0	true	false	false

Fig. 7.9 SALT delta difference summary and detail reports

base file but not altering a record date; and *old*, records exist in the base file but not in the ingest file. SALT can generate code which will automatically perform data ingest operations as an independent process, or as part of and combined with an internal record linking process described later in this chapter.

The SALT data ingest process requires the ingest file format to match the record layout of the base file. The base file record must include a numeric record id field specified by the RIDFIELD statement which uniquely identifies any record in the base file. The GENERATE option on the RIDFIELD statement allows fresh record IDs to be automatically generated by the data ingest process. The base file may also include a field which indicates the external source file type for a record and a field which is the unique identifier of the record from the data ingest file identified by the source type specified by the SOURCEFIELD and SOURCERIDFIELD statements in the specification file. Including these fields allows SALT to provide additional functionality including enhanced statistics. The base file and ingest file records may also include specific date fields which indicate the first date and the last date that the data meets some condition such as being valid for the specified source, or when the data first added entered and last entered the base file for the specified source.

	rcid	bdid	source	vendor id	dt first seen	dt last seen	company name	prim range	predir
1	1001283731	42704331	C2	15-F1-18002	20061117	20080502	JOHANOVA FINANCE INC	260	E
2	1001283731	42704331	C2	15-F1-18002	20061117	20080902	JOHANOVA FINANCE INC	260	E
3	882151981	43761025	U2	S-DNB00-67359920000317	20060100	20060100	GAGLEMILLE FINANCE INC	877	S
4	1050567253	43898400	C-	08-20031095258	20070821	20091027	CRIMETTA FINANCE SERVICE INC	770	S
5	420241466	420241466	C=	26-B73295	20030917	20091027	TEMURMORE EQUITY	322	N
6	1242558670	42438	ZM	995362027C92832084	20080418	20080418	NORDSTON LLP	425	

prim name	addr suffix	postdir	unit desig	sec range	city	state	zip	zip4	county	msa	phone	fein	tpe
DONNONEN	ST		STE	200	GAGLEPARK	MI	96939	6231	125	2160	0	0	3
DONNONEN	ST		STE	200	GAGLEPARK	MI	96939	6231	125	2160	0	0	3
CHORZAIMES	RD				GAGLEPARK	MI	96939	7026	125	2160	0	0	4
CHORZAIMES	RD		STE	300	GAGLEPARK	MI	96939	6949	125	2160	0	0	2
NORDART	AVE				GAGLEPARK	MI	96939	5321	125	2160	0	0	2
DOMOST	AVE				GERFOREST	NY	89107	3903	061	5600	1855767802	0	4

Fig. 7.10 SALT data ingest sample updated base file

Three reports are produced by the data ingest process in addition to the updated base file: (1) statistics by ingest change type and source defined by the SOURCEFIELD statement with record counts where type indicates *old*, *new*, *updated*, or *unchanged* as described previously; (2) field change statistics between old and new records where the source field as defined by the SOURCEFIELD statement and the unique id as defined by the SOURCERIDFIELD statement (vendor_id for the sample data example shown below) match between old and new records; and (3) record counts by ingest file source defined by the SOURCEFIELD statement. The updated base file will be identical in format to the previous base file but can include an additional field which will contain a numeric value corresponding to the ingest change type: 0-unknown, 1-Unchanged, 2-updated, 3-old, 4-new.

Figure 7.10 shows a partial sample of an updated base file for a data ingest operation.

Record Linkage—Process

Record linkage fits into a general class of data processing known as *data integration*, which can be defined as the problem of combining information from multiple heterogeneous databases [5]. Data integration can include data preparation [6] steps such as parsing, profiling, cleansing, normalization, and parsing and standardization of the raw input data prior to record linkage to improve the quality of the input data [2] and to make the data more consistent and comparable [1, 7] (these data preparation steps are sometimes referred to as ETL or extract, transform, load). The data preparation steps are followed by the actual record matching or clustering process which can include probability and weight computation, data ingest of source data, blocking/searching, weight assignment and record comparison, and weight aggregation and match decision to determine if records are associated with the same entity [4–6, 8–11]. Figure 7.11 shows the phases typical in a data integration processing model.

The record linking approach used by SALT is similar to the classic probabilistic record linkage approach. However, the SALT approach has some significant advantages over the typical probabilistic record linkage approach. The amount of specificity added per field for a match is variable, based on the actual matching field value. This effectively assigns higher weights automatically to the more rare values which have higher specificity. This in turn allows record matches to occur even when the data in a record is sparse or inconsistent (i.e. fields with missing values) increasing recall significantly, when the remaining matching field values are sufficiently rare. In addition, field specificities are automatically scaled for fuzzy matches and other editing constraints specified for a field improving overall precision. Since specificities are also effectively trained on all the available data, and

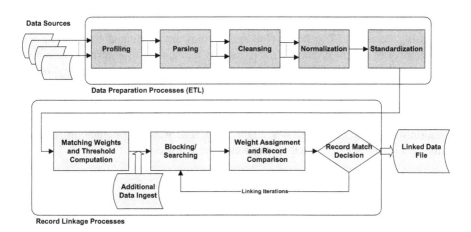

Fig. 7.11 Data integration process model

not just a hand-labeled sample of the data, the SALT approach can provide higher precision and recall than other machine learning approaches.

Record Matching Field Weight Computation

SALT calculates record matching field weights based the concept of *term specificity* and matching weights are referred to within SALT as *specificities*. The measure of term specificity for documents was first proposed by Karen Spärk Jones in 1972 in a paper titled "A Statistical Interpretation of Term Specificity and its Application in Retrieval" [12], but later became known as *inverse document frequency* (IDF) [13]. It is based on counting the documents in a collection or set of documents which contain a particular term (indexed by the term). The basic idea is that a term that occurs in many documents is less specific as an index term and should be given less weight than a term which occurs in only a few documents. The IDF is frequently used in combination with the *term frequency* or TF which is the frequency of a term within a single document. The combination called TF-IDF is used as a weight or statistical measure to evaluate how important a term is to a document that is part of a set of documents. The use of frequencies in calculating weights for record matching was first proposed by Newcombe et al. [14], formalized in a mathematical model by Fellegi and Sunter [15], and extended by Winkler [16]. TF-IDF calculations for matching weights have also been used by Cohen [8], Cohen et al. [17], Koudas et al. [18], Bilenko and Mooney [19], and Gravano et al. [7].

SALT applies the concept of term specificity to the unique field values for a field defined for a record in the input dataset(s) to be matched to calculate a *field value specificity* for each unique value contained in the field across all records in the dataset. The rarer a field value is in the input dataset, the higher the specificity value. SALT also calculates a weighted average field specificity taking into account the distribution of unique values for each field which is used when individual field values are not available during processing and for internal code generation and processing decisions. The field value specificities are calculated by dividing the total number of unique entities in the dataset by the number of entities containing a non-null unique field value in a field and taking the logarithm base 2 ($\log_2$) of the quotient. Note that initially in an unlinked dataset the number of entities is equal to the number of records. SALT recalculates field value specificities and the weighted average field value specificity which are used directly as matching weights for each iteration of linking based on all available data. The weight computation for field value specificity is represented by the following equation:

$$wf_{vs} = \log_2 \frac{n_{ent}}{n_{val}}$$

where wf_{vs} is the field value specificity, n_{ent} is the number of unique entities in the dataset, and n_{val} is the number of entities containing a non-null unique field value for the field for the current iteration of linking. The average field specificity is calculated by the following equation:

$$w_{avg} = \log_2 \frac{\left(\sum_{i=1}^n n_{val_i}\right)^2}{\sum_{i=1}^n n_{val_i^2}}$$

SALT uses the field value specificities as weights for determining record matches in a record linking/clustering process. For example, when two separate records are being matched, SALT compares each field in the two records for similarity based on the definition of the field in the SALT specification file. If the field values match between the two records, the specificity for the field value (scaled for fuzzy matches and otherwise adjusted based on the editing options for the field) is added to a total specificity to help determine a record match. Each field defined in the specification file for the record can make a positive, negative, or no contribution to the total specificity. If the total specificity exceeds a pre-determined record matching threshold, then the two records are considered a match. The SALT record linking/clustering technology operates on a dataset of records containing information about a specific entity type. As records are linked in an iterative process to form entity clusters, specificities are recalculated based on the number of entities that have a field value as a proportion of the total entities represented in the dataset. As clustering occurs specificities converge to a more accurate value based on the number of entities represented. Figure 7.12 shows an example of the specificity

fieldname	field specificity	val	cnt	specificity	fieldname	field specificity	val	cnt	specificity	fieldname	field specificity	val	cnt	specificity
state	4.486855	MH	3	21.496943			SC	86406	6.683062			FL	590933	3.909271
		AS	6	20.496943			KY	93459	6.569860			TX	678783	3.709315
		AE	6	20.496943			KS	94496	6.553940			IL	738654	3.587367
		MP	10	19.759978			LA	98611	6.492445			CA	967544	3.197938
		GU	195	15.474576			MS	103093	6.428319					
		VI	358	14.598090			OK	104315	6.411319					
			362	14.582060			IA	105273	6.398130					
		PR	4918	10.818050			AL	106175	6.385921					
		WY	18218	9.928829			CT	115563	6.263586					
		VT	18842	8.880241			AZ	117683	6.237360					
		RI	22809	8.604590			TN	140490	5.981798					
		HI	23947	8.534349			MD	148264	5.904097					
		AK	24888	8.490384			MA	151475	5.873224					
		WV	27936	8.312068			CO	164337	5.755608					
		ND	30704	8.175767			WA	168722	5.717617					
		SD	38743	7.840258			VA	169259	5.713033					
		DC	39678	7.805854			IN	172179	5.688356					
		NM	39821	7.900664			MO	177523	5.644260					
		MT	40180	7.787716			NJ	199331	5.477099					
		ME	45376	7.612264			OR	200419	5.469246					
		DE	45557	7.606521			NC	203731	5.445600					
		NH	48939	7.503209			MN	221901	5.322349					
		NV	53032	7.387330			MI	244718	5.181145					
		ID	64190	7.112070			OH	262240	5.081378					
		AR	76418	6.860281			WI	264963	5.066475					
		UT	77458	6.840779			GA	295923	4.907044					
		NE	82401	6.751532			PA	307005	4.854003					
							NY	580383	3.985852					

Fig. 7.12 Example average and field value specificities for state

values for state codes calculated on a large dataset. Note that the state code with the largest count of records (CA—California) has the lowest specificity, and the state code with the fewest records (MH—Marshall Islands) has the highest specificity.

Generating Specificities

The first step in running a SALT record linking process is to generate the field value and average field specificities described in the last section that will be used as weights for matching during the linking process. There are two different modes which can be used: (1) a single-step mode in which specificity values are stored in persisted files on the HPPC processing cluster, and (2) a two-step mode in which specificity values are stored in key/index files. Specificities can take a large amount of time to calculate when the base data is extremely large depending on the size (number of nodes) of the processing cluster. The two-step mode allows the option of not recalculating specificities each time a process like internal or external linking is run based on updates to the base data. This can save a significant amount of processing time when data is updated and linked on a processing cycle such as a monthly build to add new or changed data.

Intially in the SALT specification file, the specificity and switch value information is unknown for the FIELD, CONCEPT, and ATTRIBUTEFILE stetements (refer to the description of the FIELD statement in "Specification File Language' section for a further description). Once specificities have been calculated using the SALT generation process, the average field specificity and switch values can be added to the specification file. This information allows SALT to generate optimized code and set various thresholds appropriately for the record linkage processes. SALT produces two reports when specificities are generated: (1) the specificites report displays an average field specificity value, maximum specificity value, and switch value for each FIELD, CONCEPT, and ATTRIBUTE statement in the specification file. In addition, SALT shows which values if any for each field will also be treates as nulls (other than blanks and zeros) by SALT in the matching process. The specificities shift report shows the change (positive or negative) in specificity from the previous value in specification file. The field value specificities are stored in either persisted data files or index/key files depending on the generation mode selected. Persisted files are an HPCC and ECL feature that allow datasets generated by ECL code to be stored, and if a process is run again, and the code or other data affecting the persisted file has not changed, it will not be recomputed. Figure 7.13 shows a partial specificities and specificities shift report example for a sample data file.

dummy	company name specificity	company name switch	company name max	nulls company name company cnt id name		prim range specificity	prim range switch	prim range max	nulls prim range prim cnt id range		predir specificity	predir switch	predir max
1	0	14.044804	0.000000	22.924206	0	0	10.493466	0.000000	22.924206	0	0	2.282632	0.000000	7.975655

company name shift0	company name shift0	prim range shift0	prim range shift0 switch	predir shift0	predir shift0 switch	prim name shift0	prim name shift0 switch	addr suffix shift0	addr suffix shift0 switch	postdir shift0	postdir shift0 switch	unit desig shift0	unit desig shift0 switch	sec range shift0	sec range shift0 switch	city shift0	city shift0 switch
1	14	0	10	0	2	0	10	0	3	0	3	0	1	0	7	0	9	0

Fig. 7.13 SALT specificities and specificity shift sample reports

Internal Linking

SALT includes three types of record linkage processes: Internal, External, and Remote.

Internal linking is the classic process of matching and clustering records that refer to the same entity and to assign entity identifiers to create a base or authority file. An entity is typically a real-world object such as a person or business, but can be anything about which information is collected in fields in a record where each record refers to a specific entity. The goal is to identify all the records in a file that are related to the same entity. This process is useful in many information processing applications including identifying duplicate records in a database and consolidating account information for example. Input records are matched using the fields and process parameters defined by FIELD, CONCEPT, ATTRIBUTEFILE and other statements in the specification SALT specification file.

The goal of the internal linking process is SALT is to match records containing data about a specific entity type in an input file and to assign a unique identifier to records in the file which refer to the same entity. For example, in a file of records containing customer information such as a customer order file, internal linking could be used to assign a unique customer identifier to all the records belonging to each unique customer. Internal linking can also be thought of as clustering, so that records referring to the same entity are grouped into clusters, with each cluster having a unique identifier.

SALT uses the field value specificities as weights for determining record matches in the internal linking process. For example, when two separate records are being matched, SALT will compare each field, concept, and attribute file in the two records for similarity based on the definition of the field specified by the FIELD, CONCEPT, and ATTRIBUTEFILE statements in the SALT specification file. If the values match between the two records, the specificity for the value (scaled for fuzzy matches and otherwise adjusted based on the editing options for the field) will be added to a total specificity to help determine a record match. Each field defined in the specification file for the record can make a positive, negative, or no contribution to the total specificity. If the total specificity exceeds the pre-determined matching threshold, then the two records are considered a match.

The tendency is to think of the record match decision as a yes/no question as in many rule-based systems. However, since SALT uses specificity values for match scores based on every field value available in the input data, a record match score of $n + 1$ denotes a link which is $2\times$ less likely to be false than a score of n. In addition, during an iteration of SALT internal linking, entity links are only generated (a) if they are above the calculated threshold (either the default automatically calculated by SALT or user-specified); and (b) are the highest scoring linkage for both records involved in the link.

The internal matching process is iterative beginning with the input base file and any additional ingest files which are merged with the input base file, with each processing iteration attempting additional matches of records to records and entity clusters formed in the previous iteration. As new entity clusters are formed or expanded during each iteration, more information becomes available about an entity. In a successive iteration, this may allow additional records or entire clusters to be merged with an existing cluster. The output of each iteration effectively becomes the training set for the next iteration, effectively learning from the previous iteration, as new entity clusters are formed or extended and matching weights are recalculated. Multiple iterations are usually required for convergence (no additional matches occur) and to achieve high levels of precision and recall for a given population of entities. A typical SALT-generated record linkage system will be iterated quite extensively initially, but may only need additional iterations once or twice a month as new or updated data is ingested.

The results from each iteration should be reviewed to determine if the record matching results have met precision and recall goals or if under-matching or over-matching has occurred. Adjustments may need to be made to field and concept definitions or the specificity matching threshold and the entire process repeated. If the goals of the linking process have been met, the result of final iteration becomes the new linked base file. This result will contain the same number of records as the original input file, but the entity identifier field specified by the IDFIELD or IDNAME statement on each record will now contain a unique identifier for the entity cluster to which the record belongs.

SALT produces a wealth of information to assess the quality of the results for each iteration of the internal linking process. This information includes match sample records, field specificities used in the current iteration, pre- and post-iteration field population stats, pre- and post-iteration clustering stats showing the number of clusters formed by record count for the cluster, the number of matches that occurred, rule efficacy stats showing how many matches occurred as a result of each blocking/matching rule (each rule is implemented as an ECL join operation), confidence level stats showing total specificity levels for matches and how many matches for each level, percentages of records where propagation assisted or was required for a match, validity error flags which can indicate an internal problem with the process or data, a match candidates debug file which contains all the records in the input file with individual field value specificities appended and propagation flags appended, a match_sample_debug file which contains a record for each match attempted with both left and right field data and

Results: (21)				
Match Sample Records	[3000 rows]	.gz	.xls	
Slice Out Candidates	[0 rows]	.gz	.xls	
Specificities	[1 rows]	.gz	.xls	
SPC Shift	[1 rows]	.gz	.xls	
Pre Clusters	[1 rows]	.gz	.xls	
Post Clusters	[487 rows]	.gz	.xls	
Matches Performed				3984751
Slices Performed				0
Rule Efficacy	[26 rows]	.gz	.xls	
Confidence Levels	[90 rows]	.gz	.xls	
Propagation Assisted Pcnt				0.000000
Propagation Required Pcnt				0.000000
Pre Pop Stats	[1 rows]	.gz	.xls	
Post Pop Stats	[1 rows]	.gz	.xls	
Validity Statistics	[1 rows]	.gz	.xls	
Result 16	[7959271 rows]	.gz	.xls	temp::bdid::tmm_test::match_candidates_debug
Result 17	[3996215 rows]	.gz	.xls	temp::bdid::tmm_test::match_sample_debug
Result 18	[1 rows]	.gz	.xls	temp::bdid::tmm_test::specificities_debug
Result 19	[102 rows]	.gz	.xls	temp::bdid::tmm_test::attribute_matches
Result 20	[7959271 rows]	.gz	.xls	temp::bdid::tmm_test::it1
Result 21	[7959271 rows]	.gz	.xls	temp::bdid::tmm_test::patched_candidates

Fig. 7.14 SALT internal linking output results

scores for field matches and the total match score, an iteration result file sorted in order of the entity identifier, and a patched match candidates file with the entity identifier appended to each record. SALT also produces various debug key files and a ID compare online service which can be deployed to a HPCC Roxie cluster (refer to Chap. 4 for more information on Roxie and the HPCC technology) that allows you to compare the data for two entity identifiers to debug matches and non-matches. Figure 7.14 shows an example of the output results produced for each iteration of linking.

Figure 7.15 shows a partial result example of the post iteration cluster statistics after the first iteration of internal linking.

The input base file for the internal linking process is specified by the FILENAME statement in the specification file. Other data ingest files can also be included which will be appended and merged with the base file prior to the linking process. The INGESTFILE statement allows you to define a dataset which provides records to be ingested in the same format of the base file. Typically these files are generated from external source files or base files for other types of entities. The data ingest is executed automatically if the specification file includes INGESTFILE statements.

After analyzing match sample records generated by the internal linking process on each iteration, the results may indicate the system is overmatching (too many false positives), or under matching (too many false negatives). False positive matches are evidenced by entity clusters that have records which should not have been included. False negative matches are evidenced by records which should have been matched and included in an entity cluster, but were not. There are many reasons why either of these conditions could be occurring including the need or adjustment of parameters in the specification file such as the FORCE on the FIELD statements, and the overall definition of statements and concepts. If the matching

Fig. 7.15 SALT internal
linking cluster statistics

incluster	numberofclusters	
1	1	2291293
2	2	898524
3	3	400952
4	4	164832
5	5	77980
6	6	44730
7	7	26067
8	8	17149
9	9	11691
10	10	8287
11	11	5917
12	12	4486
13	13	3556
14	14	2786
15	15	2394

criteria in you specification file appears to be correct, then the match threshold value may need to be adjusted manually using the THRESHOLD statement.

The match sample records generated by the internal linking process include samples of record matches at and above the match threshold, and also matches in the range of the match threshold value within 3 points of specificity. If matches below the threshold appear to actually be valid, then the match threshold may need to be lowered. If records above the current match threshold appear to be invalid, then you may need to raise the match threshold. A sufficient number of records needs to be examined at the match threshold, below, and above before making a decision. It is not uncommon to have some false positives and false negatives in a linking process.

SALT automatically generates an ID Compare Service for use with internal linking. Once the service has been deployed to a HPCC Roxie cluster, the query can be accessed manually through the WsECL interface. The query allows you to look at all the data associated with two identifiers to see if they should be joined. It is also useful for looking at all the data associated with a specific entity identifier if you only enter one identifier. SALT also automatically generates an ID sliceout service for use with internal linking. This query allows examination of records which the internal linking process has identified as sliceouts.

External Linking

External linking is the process of matching an external file or an online query to an existing, linked base or authority file which has been previously linked by an internal linking process, or some other linking process. The goal of external linking is to determine if a record in the external file is a match to an entity cluster in the internal base file and assign it the unique identifier for the matching entity. This process is also referred to as *entity resolution*. External linking is useful in establishing foreign key relationships between an external file and an existing file based on the unique entity identifier. For example, a person may have a unique identifier in a base file that contains general information about the person entity, and the external file may have information on vehicles which are or have been owned or leased by a person entity. SALT external linking also supports a base file search mode in which all records which are similar to the search criteria are returned.

The goal of the external linking process of SALT is to match records containing data about a specific entity type in an external file or online query to a previously linked base file of entities and to assign a unique entity identifier from the base file to records in the external file or to the query which refer to the same entity. External linking is also useful in establishing foreign key relationships between an external file and an existing file based on the unique entity identifier. For example, in an external file of records containing property information for people, external linking could be used to assign a unique person entity identifier to all the property records associated with a base file of people. External linking can also be thought of as *entity resolution*, so that records or onlilne queries containing information about an entity are resolved by matching the records to a specific entity in an authority file, and assigning the corresponding unique entity identifier.

The external linking capability requires a previously linked input file in which all the records have been clustered for a specific entity type. The linked input file is used to build keys required for external matching. The linked file is a single flat file that functions as the authority or base file to be used for matching corresponding fields from an external file to perform the entity resolution process. The records in this authority file should contain all the fields to be used for matching with an entity identifier (unique ID for the associated entity cluster) assigned to each record. The authority file can be the output of a previous SALT internal linking process.

External linking in SALT requires the definition of fields to be used for searching for candidate records for matching. In SALT, these search definitions are called *linkpaths* and defined in the SALT specification file using the LINKPATH statement. Linkpaths define various combinations of fields which are used to inform the external linking process how the internal data should be searched for a potential match. Linkpaths are analogous to defining indexes on a data base, and result in the generation of an index on the base file data to support the external linking process.

User-defined linkpaths are specified using the LINKPATH statement with a name and a field list. The field list can be grouped into required fields, optional fields, and extra-credit fields. The required fields defined in a linkpath must match exactly during external linking. Optional fields must match if provided, and fuzzy matches are acceptable. Extra-credit fields do not need to match, but add to the total matching score if they do and can also include any of the fuzzy matching edit characteristics. Each linkpath defined results in the creation of an HPCC ECL index (key) file which is used in the matching process.

Although the user is primarily responsible for defining appropriate linkpaths based on knowledge of the data and user query patterns, SALT includes a capability to suggest possible linkpaths based on the data in the base file and a sample external file or files. The output of the Data Profiling Field Combination Analysis report on an external file can be used as an additional input file to the SALT tool along with a specification file defining the fields in the base file to create a Linkpath Generation Report with suggested linkpaths.

The key to implementing an efficient external linking capability with high precision and recall using SALT is the choice of linkpaths defined by the LINKPATH statement in the specification file. Figure 7.16 shows an example of LINKPATH statements used for external linking of a base file of person entities.

Each LINKPATH statement will result in the creation of an ECL Index (key) file which is used in the external matching process. The ultimate responsibility for choosing linkpaths to be used for external linking entity resolution rests with the developer. Linkpath definitions in the specification file can be divided into required (compulsory for a match) and non-required fields. User-defined linkpaths are specified using the LINKPATH statement beginning with a linkpath name, followed by a specific field list with the required fields first, followed by optional fields and then extra-credit fields as described earlier in this section.

Each field in the authority file to be used for external linking is defined in the specification file using either the FIELD or CONCEPT statement, or can be a value field in an attribute file specified in an ATTRIBUTEFILE statement, and the entity identifier is defined using the IDFIELD statement. The specificity of each field, concept, or attribute file value field must be included, so specificities on the authority file need to be generated if the specificities are not already known from a previous internal linking process. If the field definition includes a BESTTYPE definition with a propagation method, propagation of fields within entity clusters in the authority file will be automatically handled to improve matching results. Field definitions used for external linking can include the MULTIPLE parameter which specifies that the external file matching field contains multiple values. FIELDTYPE statements can also be used in a specification file used for external linking, and if

```
1   // User-defined Linkpath Definitions
2   LINKPATH:FLCTS:FNAME:LNAME:COUNTY:STATE:?:MNAME:+:CITY:PRIM_RANGE:PRIM_NAME:SEC_RANGE:ZIP
3   LINKPATH:ADDRESS:PRIM_RANGE:PRIM_NAME:STATE:ZIP:?:SEC_RANGE:FNAME:LNAME:MNAME:+:CITY
4   LINKPATH:S:SSN:FNAME:+:LNAME:MNAME:CITY:STATE
5   LINKPATH:DO:DOB:LNAME:FNAME:+:STATE:CITY:SSN:MNAME
6   LINKPATH:ZPR:ZIP:PRIM_RANGE:?:FNAME:+:LNAME:PRIM_NAME:SEC_RANGE:CITY:STATE
7
```

Fig. 7.16 SALT LINKPATH definitions example

included are used to clean the data for the external linking keybuild process, and also to clean external file data or queries for the search process.

The required fields defined in a linkpath must match exactly during external linking. Optional fields must match if provided, and if the field is not defined as MULTIPLE, then fuzzy matches are adequate. Extra-credit fields do not need to match, but add to the total matching score if they do and can also included any of the fuzzy matching edit characteristics.

SALT also automatically creates an additional key called the UBER key using all the fields and concepts defined in your specification file. By default, UBER key is not used unless an external record or query fails to match any records using the linkpaths you have defined, essentially providing a "fallback" alternative for searching. The default behavior can be changed by using a parameter on the PROCESS statement in your specification file. The parameters include ALWAYS (the search process will always use the UBER key as well as any other linkpath specified), REQUIRED (same as the default, the UBER key will be used if none of the other linkpaths could satisfy the query), and NEVER (the UBER key is not used for external linking or searching). The UBER key can provide recall lift when the data in the external record or query does not match any existing linkpath, but at a higher processing cost. The UBER key does not support any form of fuzzy matching, all fields provided much match exactly for a search to be successful.

Another interesting feature of the UBER key which can raise recall significantly is that it works entirely at the entity level. Thus if *any* entity record has a particular middle name and *any* entity record has a particular address, then the entity will be returned; even if both did not originally appear on the same record. This feature allows an UBER key search to work with many multiple fields. You can search, for example, for someone with two different last names who have lived in two different counties.

The SALT external linking process will mandate that some fields defined for a linkpath become required for a link to occur based on the total specificity required for a match. The SALT external linking process will also automatically divide the non-required fields in a linkpath into optional and extra-credit fields if the specification file has not done that already.

Before the SALT external linking capability can be used, a keybuild process on the internal base file must be run. The specification file must be edited to ensure that all FIELD, CONCEPT, ATTRIBUTEFILE, and LINKPATH statements required for the matching process are defined and field specificities are included. Figure 7.17 shows the results of a keybuild process. Figure 7.18 is a partial sample of a key file built for a LINKPATH which begins with CITY and STATE as required fields and COMPANY_NAME as an optional field.

Once the external linking keybuild is completed, record matching using an external file to your internal/base file can be processed. Batch mode external linking allows you to perform the external linking function on a HPCC Thor cluster as a batch process. SALT automatically generates a macro which can be used in ECL code implemented to perform the actual matching process.

Results: (14)

Result 1	[7959271 rows]	.gz .xls	key::salt em test20b3 sample
Result 2	[19341991 rows]	.gz .xls	key::salt em test20b3samplerefs
Result 3	[4780199 rows]	.gz .xls	key::salt em test20b3samplewords
Result 4	[126294 rows]	.gz .xls	salt em test20b3::key::meow xadl headerprepped0
Result 5	[3095638 rows]	.gz .xls	salt em test20b3::key::meow xadl headerprepped1
Result 6	[9498 rows]	.gz .xls	salt em test20b3::key::meow xadl headerprepped2
Result 7	[2673592 rows]	.gz .xls	key::salt em test20b3samplecsrefs
Result 8	[477820 rows]	.gz .xls	key::salt em test20b3sampleferefs
Result 9	[2231117 rows]	.gz .xls	key::salt em test20b3samplephrefs
Result 10	[2371324 rows]	.gz .xls	key::salt em test20b3sampleaddrrefs
Result 11	[660910 rows]	.gz .xls	key::salt em test20b3samplevehrefs
Result 12	[29927480 rows]	.gz .xls	key::salt em test20b3sampleproprefs
Result 13	[22034 rows]	.gz .xls	key::salt em test20b3samplebkrptrefs
Result 14	[301390 rows]	.gz .xls	key::values::salt em test20b3 bdid company name

Fig. 7.17 SALT external linking keybuild results

	city	state	company name	bdid	zip	county	prim range	prim name	sec range	msa	city weight100	state weight100	company name weight100	zip weight100	county weight100	prim range weight100	prim range e1 weight100
1	ALPHAINGTON	TX	1332 ARDERO 12 HOLDING 7 INC 3	120146742	48802	113	2300	JOHANSON		1920	398	182	1332	398	866	363	956
2	ALPHAINGTON	TX	1332 ARDERO 12 HOLDING 7 INC 3	120146742	48802	113	2300	JOHANSON	300	1920	398	182	1332	398	866	314	975
3	ALPHAWATER	MI	1332 CRIMETTA 9 CITY 8 NATIONAL 8	578792192	95939						0	0	1332	0	0	0	0
4	AMSTROCK	FL	1100 PERSETTA 9 PRINTING 10	64390722	81471	121		PETERERO	110	8280	357	162	1100	357	1075	0	0
5	AMSTROCK	FL	1100 PERSETTA 9 PRINTING 10	87499922	81471	103				8280	357	162	1100	357	816	0	0
6	AMSTROCK	FL	1332 BORDARZI 8 TIES 14	27808067	81471	121		PETERERO	110	8280	357	162	1332	357	1075	0	0
7	AMSTTOWN	TX	1173 GREENSTON 10 LLC 3	452688274	29476	339	4	RASELLIN		3360	338	184	1173	338	1075	311	766

Fig. 7.18 SALT City, State, Company_Name LINKPATH key example file

The output dataset from the external linking batch process contains a corresponding record for any external file record which contained sufficient data for matching to a defined linkpath. This is determined by filtering the external file records to ensure that the records contained data in the required fields in the linkpath. Each record in the output dataset contains a parent record with a *reference* field corresponding to a unique id assigned to the external input file prior to the external linking process, a set of Boolean result flags, and a child dataset named *results* containing the results of the matching process. Resolved records (successful linking to an entity in the base file) in the output dataset are indicated by the Boolean *resolved* flag set to true. The *reference* field for each record in the child dataset is the same as the reference on the parent record. The matching process will return one or more result records with scores in the child dataset depending on how many viable matches to different entities in the internal base are found. The identifier specified by the IDFIELD statement in your specification file will contain the matching entity identifier. The output recordset can be used to append resolved entity identifiers to the external input file based on the reference field, or for other application uses such as to display the candidate matches for a query when the record is not resolved.

SALT external linking automatically generates two deployable Roxie services to aid in debugging the external linking process which also can be used for manual examination of data to evaluate linkpaths, as well as to support the online mode external linking capability described later in this section. These services also provide an example for incorporating online external linking and searching the base file into other online queries and services.

Online mode external linking allows the external linking function to be performed as part of an online query on a HPCC Roxie cluster. This capability can be utilized to incorporate external linking into other Roxie-based online queries and applications or you can use the provided online service for batch mode linking from a Thor. SALT automatically generates a macro can be used in the ECL code implemented to perform the actual matching process for an online mode batch external linking application. Figure 7.19 shows an example of the automatically generated online service and manual query for entity resolution. Figure 7.20 shows the entity resolution result for this query.

The same Boolean flags used for batch mode external linking including the resolved flag are displayed along with the weight field which contains the score for the match and ID field for the resolved entity (bdid in this example).

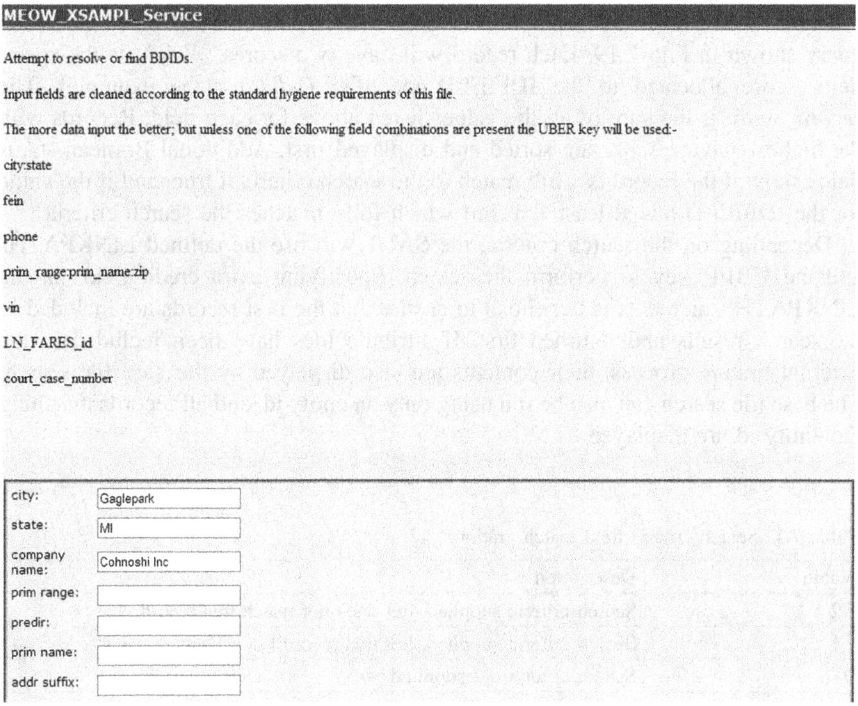

Fig. 7.19 SALT external linking online query example input

TM M TES T2. MEO W XSAMP L SERVICE Response

Dataset: Result 1

	verified	ambiguous	shortlist	handful	resolved	bdid	weight	reference	city	cityweight	state	stateweight	company name	company nameweight
1	true	false	true	true	true	40652012	17	0	GAGLEPARK	0	MI	0	1332 COHNOSHI 14 INC 3	17

Fig. 7.20 SALT external linking entity resolution result example

Base File Searching

SALT provides an additional query which displays all the entity records from the internal base file matching the input information. This query is useful in debugging the external linking process to assess how a particular record was resolved or not resolved to an entity. The ECL function called by this service provides a base file search capability that can be incorporated into other HPCC online applications.

The base file search is intended to return records organized with the records which best match the search criteria first. All data returning from the search is graded against the search criteria, and for each field in the data a second field is appended which will contain one of the following values (Table 7.1).

Figure 7.21 shows an example of the base file search results using the same query shown in Fig. 7.19. Each record will have two scores. *Weight* is the specificity score allocated to the IDFIELD identifier (*bdid* for the example). The *record_score* is the sum of all the values listed above for each field. Records with the highest *record_score* are sorted and displayed first. Additional Boolean status fields show if the record is a full match to the search criteria if true, and if the value for the IDFIELD has at least 1 record which fully matches the search criteria.

Depending on the search criteria, the SALT will use the defined LINKPATHs and the UBER key to perform the search. Specifying extra credit fields in the LINKPATH statements is beneficial to ensure that the best records are included in the search results and returned first. If attribute files have been included in the external linking process, their contents are also displayed by the base file search. The base file search can also be run using only an entity id, and all records matching the entity id are displayed.

Table 7.1 Search criteria field match grading

Value	Description
−2	Search criteria supplied, but does not match this record
−1	Search criteria supplied, but this record has a blank
0	Search criteria not supplied
1	Search criteria is fuzzy match to this record
2	Search criteria is a match to this record

TRAININGTONYMIDDLETO N. XSAMP L HEADE R SERVICE Response

Dataset: Header_Data

	uniqueid	weight	keysused	bdid	rcid	source	vendor id	dt first seen	dt last seen	company name	prim range	predir	prim name	addr suffix	postdir	unit desig	sec range
1	0	24	2	40652012	51926691	W	MINIVANTATHLON.COM	20040416	20040504	COHNOSHI INC	135	N	NORDART	AVE			
2	0	24	2	40652012	71203891	W	LIEMAC.COM	20030528	20040504	COHNOSHI INC	135	N	NORDART	AVE			
3	0	24	2	40652012	107342291	W	MINIVANTATHALON.COM	20040416	20040504	COHNOSHI INC	135	N	NORDART	AVE			
4	0	24	2	40652012	513220257	U	MID884724	20020315	20040715	COHNOSHI INC	135	N	NORDART	AVE			
5	0	24	2	40652012	513321534	U	MID884724	20020315	20050309	COHNOSHI INC	135	N	NORDART	AVE			
6	0	24	2	40652012	513373798	U	MI31494C	20011205	20011205	COHNOSHI INC	135	N	NORDART	AVE			
7	0	24	2	40652012	881907187	U2	D-DNB31494C20011205	20060100	20090700	COHNOSHI INC	135	N	NORDART	AVE			

city	state	zip	zip4	county	msa	phone	fein	fullmatch required	has fullmatch	recordsonly	is fullmatch	record score	match city	match state	match company name	match prim range	match predir	match prim name	match addr suffix
GAGLEPARK	MI	96939	3313	261	2160	0	0	false	true	false	true	6	2	2	2	0	0	0	0
GAGLEPARK	MI	96939	3313	261	2160	0	0	false	true	false	true	6	2	2	2	0	0	0	0
GAGLEPARK	MI	96939	3313	261	2160	0	0	false	true	false	true	6	2	2	2	0	0	0	0
GAGLEPARK	MI	96939	3372	125	2160	0	0	false	true	false	true	6	2	2	2	0	0	0	0
GAGLEPARK	MI	96939	3372	125	2160	0	0	false	true	false	true	6	2	2	2	0	0	0	0
GAGLEPARK	MI	96939	3372	125	2160	0	0	false	true	false	true	6	2	2	2	0	0	0	0
GAGLEPARK	MI	96939	3372	125	2160	0	0	false	true	false	true	6	2	2	2	0	0	0	0

match postdir	match unit desig	match sec range	match zip	match zip4	match county	match msa	match phone	match fein	match locale	match address
0	0	0	0	0	0	0	0	0	0	0
0	0	0	0	0	0	0	0	0	0	0
0	0	0	0	0	0	0	0	0	0	0
0	0	0	0	0	0	0	0	0	0	0
0	0	0	0	0	0	0	0	0	0	0
0	0	0	0	0	0	0	0	0	0	0
0	0	0	0	0	0	0	0	0	0	0

Fig. 7.21 SALT sample base file search results

Remote Linking

SALT can be used to generate code to perform record matching and scoring and link together records that are completely independent from a base file without directly using the base file during the linking process. This capability is called *remote linking*. For remote linking, SALT still generates statistics from the base file data which can be used to significantly improve the quality of record to record matching/linking for any application assuming the records contain fields with the same type of data in the base file. The remote linking capability is implemented as a compare service, which compares the fields in two records and generates scoring information similar to SALT internal linking.

For remote linking, SALT still generates specificity weights from the base file data which can be used to significantly improve the quality of record to record matching/linking assuming the records contain fields with the same type of data in the base file.

The remote linking capability is implemented as an online compare service for the HPCC Roxie cluster, which compares the fields in two records and generates scoring information similar to SALT internal linking. This allows user-defined matching to be implemented in a Roxie query, using the power of SALT generated statistics, specificity weights, and field editing features on the independent records to improve the matching result. Remote linking requires the definition of a specification file for the fields that will be matched from the base file. The base file is used only for calculating the specificities needed for remote matching, the base is not actually used during the remote linking process.

The remote linking code works by constructing two input records from input data to the service which are then passed to the internal linking process to determine if they would link using the following steps: (1) the normal cleaning process is performed as required on input data for fields defined with editing constraints using FIELDTYPE statements in the specification file; and (2) the weighting and scoring is done exactly as if an internal linking process was executed without any propagation. In this manner, remote linking can be added to a conventional record linking application to provide improved matching decisions.

Attribute Files

Sometimes there are additional fields related to an entity identifier which may help in record linkage except these fields do not exist in the input file being linked. Examples from the LexisNexis public records are properties, vehicles, and bankruptcies which contain information relating to person entities. These are files external to the linking process that contain a person entity identifier and some form of data or attribute that is associated with that entity identifier. For example, a unique property id, vehicle identification number (VIN), or bankruptcy filing number. SALT refers to these external files as attribute files and they are defined in the SALT specification file using an ATTRIBUTEFILE statement.

The properties needed for these external fields are that they have high specificity (usually a unique identifier about something like a vehicle which could be associated with more than one entity) and low variability (some variability in value for a given entity is permissible, i.e., one person entity could be associated with multiple vehicles). This implies looking for things which are associated with an entity and which are shared by relatively few entities (one vehicle hasn't had too many owners), and where a single entity doesn't have too many. By default only the best of the matching entity identifiers from each attribute file is allowed to score towards matching one pair of entity identifiers in the input file. Attribute files can contain additional fields from the external file which can be used by SALT in search applications. For example if appropriate fields are included, a search for persons who own or have owned red Corvette convertibles living in Florida could be done.

Summary and Conclusions

Data integration and data analysis are fundamental data processing requirements for organizations. Organizations now collect massive amounts of data which has led to the Big Data problem and the resulting need for data-intensive computing architectures, systems, and application solutions. Scalable platforms such as Hadoop and

HPCC which use clusters of commodity processors are now available which can address data-intensive computing requirements. One of the most complex and challenging data integration applications is record linkage [2]. Record linkage allows information from multiple sources that refer to the same entity such as a person or business to be matched and identified or linked together. The record linkage process is used by organiztions in many types of applications ranging from maintaining customer files for customer relationship management, to merging of all types of data into a data warehouse for data analysis, to fraud detection.

This chapter introduced SALT, a code generation tool for the open source HPCC data-intensive computing platform, which can automatically generate executable code in the ECL language for common data integration applications including data profiling, data hygiene, record linking and entity resolution. SALT provides a simple, high-level, declarative specification language to define the data and process parameters in a user-defined specification file. From the specification file, SALT generates ECL code which can then be executed to perform the desired application. SALT encapsulates some of the most advanced technology and best practices of LexisNexis Risk Solutions, a leading aggregator of data and provider of information services significantly increasing programmer productivity for the applications supported. For example, in one application used in LexisNexis Risk Solutions for processing insurance data, a 42-line SALT specification file generates 3980 lines of ECL code, which in turn generates 482,410 lines of C++. ECL code is compiled into C++ for efficient execution on the HPCC platform.

SALT specific record linking capabilities presented in this chapter include internal linking, a batch process to link records from multiple sources which refer to the same entity to a unique entity identifier; external linking, the batch process of linking information from an external file to a previously linked base or authority file in order to assign entity identifiers to the external data (entity resolution), or an online process where information entered about an entity is resolved to a specific entity identifier, or an online process for searching for records in an authority file which best match entered information about an entity; and remote linking, an online capability that allows SALT record matching to be incorporated within a custom user application. The key benefits of using SALT can be summarized as follows:

- SALT automatically generates executable code for the open source HPCC data-intensive computing platform to address the Big Data problems of data integraton.
- SALT provides important data preparation applications including data profiling, data hygiene, and data source consistency checking which can significantly reduce bugs related to data cleanliness and consistency.
- SALT provides record linking applications to support clustering of data referring to the same entity, entity resolution of external data to a base or authority file, and advanced searching capabilties to find data related to an entity, and generates code for both batch and online access.

- SALT automatically generates field matching weights from all the available data, and calculates default matching thresholds and blocking criteria for record linking applications.
- SALT incorporates patent-pending innovations to enhance all aspects the record linkage process including new approaches to approximate string matching such as BAGOFWORDS which allows matching to occur with no order dependency of word tokens and using the specificity of the individual words contained in the field as weights for matching.
- SALT data hygiene supports standard and custom validity checking and automatic cleansing of data using field editing constraints defined by FIELDTYPE statements which can be standardized for specific data fields..
- SALT record linking applications are data neutral and support any data type available in the ECL programming language, support both real-world and abstract entity types, can provide higher precision and recall than hand-coded approaches in most cases, can handle relationships and dependencies between individual fields using CONCEPT statements, support calculation of best values for a field in an entity cluster using the BESTTYPE statement which can be used to propagate field values to increase matching precision and recall, support additional relationship detection for non-obvious relationships between entity clusters using the RELATIONSHIP statement, provide many built-in fuzzy matching capabilities, and allow users to define custom fuzzy-matching funtions using the FUZZY statement.
- SALT applications are defined using a simple, declarative specification language edited in a standard text file, significantly enhancing programmer productivity for data integration applications.
- SALT automatically generates statistics for processes which can be utilized to analyze cyclical changes in data for repeating processes and quickly identify problems.
- SALT is provided and supported by LexisNexis Risk Solutions, a subsidiary of Reed Elsevier, one of the largest information companies in the world.

Using SALT in combination with the HPCC high-performance data-intensive computing platform can help organizations solve the complex data integration and processing issues resulting from the Big Data problem, helping organizations improve data quality, increase productivity, and enhance data analysis capabilities, timeliness, and effectiveness.

References

1. Christen P. Automatic record linkage using seeded nearest neighbor and support vector machine classification. In: Proceedings of the KDD '08 14th ACM SIGKDD international conference on knowledge discovery and data mining, Las Vegas, NV; 2008. p. 151–9.
2. Herzog TN, Scheuren FJ, Winkler WE. Data quality and record linkage techniques. New York: Springer Science and Business Media LLC; 2007.

3. Middleton AM. Data-intensive technologies for cloud computing. In: Furht B, Escalante A, editors. Handbook of cloud computing. New York: Springer; 2010. p. 83–136.
4. Winkler WE. Record linkage software and methods for merging administrative lists (No. Statistical Research Report Series No. RR/2001/03). Washington, DC: US Bureau of the Census; 2001.
5. Cohen W, Richman J. Learning to match and cluster large high-dimensional data sets for data integration. In: Proceedings of the KDD '02 Eighth ACM SIGKDD international conference on knowledge discovery and data mining, Edmonton, Alberta, Canada; 2002.
6. Cochinwala M, Dalal S, Elmagarmid AK, Verykios VV. Record matching: past, present and future (No. Technical Report CSD-TR #01–013): Department of Computer Sciences, Purdue University; 2001.
7. Gravano L, Ipeirotis PG, Koudas N, Srivastava D. Text joins in an rdbms for web data integration. In: Proceedings of the WWW '03 12th international conference on world wide web, Budapest, Hungary, 20–24 May; 2003.
8. Cohen WW. Data integration using similarity joins and a word-based information representation language. ACM Trans Inf Syst. 2000; 18(3).
9. Gu L, Baxter R, Vickers D, Rainsford C. Record linkage: current practice and future directions (No. CMIS Technical Report No. 03/83): CSIRO Mathematical and Information Sciences; 2003.
10. Winkler WE. Advanced methods for record linkage. In: Proceedings of the section on survey research methods, American Statistical Association; 1994. p. 274–9.
11. Winkler WE. Matching and record linkage. In: Cox BG, Binder DA, Chinnappa BN, Christianson MJ, Colledge MJ, Kott PS, editors. Business survey methods. New York: Wiley; 1995.
12. Jones KS. A statistical interpretation of term specificity and its application in information retrieval. J Doc. 1972;28(1):11–21.
13. Robertson S. Understanding inverse document frequency: on theoretical arguments for IDF. J Doc. 2004;60(5):503–20.
14. Newcombe HB, Kennedy JM, Axford SJ, James AP. Automatic linkage of vital records. Science. 1959;130:954–9.
15. Fellegi IP, Sunter AB. A theory for record linkage. J Am Stat Assoc. 1969;64(328):1183–210.
16. Winkler WE. Frequency-based matching in fellegi-sunter model of record linkage. In: Proceedings of the section on survey research methods, American Statistical Association; 1989. p. 778–8.
17. Cohen WW, Ravikumar P, Fienberg SE. A comparison of string distance metrics for name matching tasks. In: Proceedings of the IJCAI-03 workshop on information integration, Acapulco, Mexico, August; 2003. p. 73–8.
18. Koudas N, Marathe A, Srivastava D. Flexible string matching against large databases in practice. In: Proceedings of the 30th VLDB Conference, Toronto, Canada; 2004. p. 1078–86.
19. Bilenko M, Mooney RJ Adaptive duplicate detection using learnable string similarity measures. In: Proceedings of the KDD '03 Ninth ACM SIGKDD International Conference on Knowledge Discovery and Data Mining, Washington, DC, 24–27 August; 2003. p. 39–48.
20. Branting LK. A comparative evaluation of name-matching algorithms. In: Proceedings of the ICAIL '03 9th international conference on artificial intelligence and law, Edinburgh, Scotland; 2003. p. 224–32.
21. Cohen W, Richman J. Learning to match and cluster entity names. In: Proceedings of the ACM SIGIR'01 workshop on mathematical/formal methods in IR; 2001.
22. Dunn HL. Record linkage. Am J Public Health. 1946;36:1412–5.
23. Maggi F. A survey of probabilistic record matching models, techniques and tools (No. Advanced Topics in Information Systems B, Cycle XXII, Scientific Report TR-2008-22): DEI, Politecnico di Milano; 2008.
24. Newcombe HB, Kennedy JM. Record linkage. Commun ACM. 1962;5(11):563–6.
25. Winkler WE. The state of record linkage and current research problems. U.S. Bureau of the Census Statistical Research Division; 1999.

Chapter 8
Aggregated Data Analysis in HPCC Systems

David Bayliss

Introduction

The HPCC (High Performance Cluster Computing) architecture is driven by a proprietary data processing language: Enterprise Control Language (ECL). When considered briefly, the proprietary nature of ECL may be perceived as a disadvantage when compared to a widespread query language such as SQL.

The following chapter compares and contrasts the traditional Relationship Database Management System (DBMS)/Structured Query Language (SQL) solution to the one offered by the HPCC ECL platform. It is shown that ECL is not simply an adjunct to HPCC, but is actually a vital technological lynchpin then ensures that the HPCC offering achieves performance levels that an SQL system is not even capable of as a theoretical ideal. While many of the points made are applicable to data processing in general, the particular setting for this paper is the integration of huge amounts of heterogeneous data. It will be argued that the relational data model is excellent for data which is generated, collected and stored under relational constraints. However for data which is not generated or collected under relational constraints, the attempt to force the data into the relational model involves crippling compromises. The model-neutral nature of ECL obviates these concerns.

The capabilities of the HPCC ECL platform are sufficiently disruptive in that certain algorithms can be considered that are not realistic using a traditional RDBMS/SQL solution. The paper ends by considering a couple of case studies illustrating the new horizons that are opened by the HPCC and ECL combination.

The relational database is the most prevalent database management system available today; however, it is not the most suitable system for the integration and analysis of massive amounts of data from heterogeneous data sources. This unsuitability does not stem from a defect in the design of the RDBMS but is instead

---

This chapter has been adopted from the white paper authored by David Bayliss, Lexis Nexis.

© Springer International Publishing Switzerland 2016
B. Furht and F. Villanustre, *Big Data Technologies and Applications*,
DOI 10.1007/978-3-319-44550-2_8

a feature of the engineering priorities that went into their creation. The object of this paper is to contrast an idealized RDBMS processing model with the model employed within the HPCC platform in the context of the integration and analysis of massive volumes of disparate data.

The RDBMS as a theoretical concept is distinct from SQL which is just one manifestation of an RDBMS. However, the reality is that almost every RDBMS out there supports SQL either natively or indirectly. Thus in the following SQL will be used interchangeably with RDBMS.

The RDBMS Paradigm

Relational databases are built upon the principle that the physical data representation should be entirely disjointed from the way that the data is viewed by people accessing the database. An SQL system has a notion of a table which is an unsorted bag of records where each record contains columns. An SQL query can then return a sub-set of that table and may perform simple operations to alter some of the columns of that table. An SQL system then, at least theoretically, allows any two tables to be compared or joined together to produce a new composite table. The values in those fields are pure strings or pure numbers with an SQL defined behavior. Thus the application coder can use data in an extremely neat and clean format without any thought to the underlying data structures.

The underlying data structures are then managed by the database architect/administrator. It is the architect's job to map the underlying physical data within the warehouse to the logical view of the data that has been agreed between the architect and the programmers. In particular, the architect chooses the keys that are needed to allow data to be filtered and to allow tables to be joined together.

The idealized model is shown in Fig. 8.1. Please note: in this model it is assumed that the visualization logic is separate from the application (or analysis)

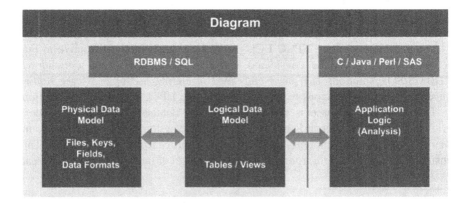

Fig. 8.1 The idolized model of the relational database system

logic. However the visualization logic may not have been placed on this diagram. It would naturally be located to the right-hand side of the Application Logic discussed in this chapter.

The other significant factor in this design which is not mandated but is extremely common is the notion of normal form. The premise behind normal form is that no scrap of data should appear in the system twice and that the database should be able to maintain and enforce its own integrity. Thus, for example, the 'city' column of a relational database will not typically contain the characters of the city name. Instead they contain a foreign key that is a link into a city file that contains a list of valid city names. Most RDBMS systems allow logic to be defined in the physical layer to enforce that only valid records enter the system. The beauty of removing the burden of data validation from the application logic is that the programmers cannot 'infect' the database with bad data.

The Reality of SQL

Even within the domains for which the SQL paradigm was designed there are some flaws in the Fig. 8.1 model that render the system practically unusable. For example, the SQL system allows for any columns in a table to act as a filter upon the result set. Thus theoretically, every query coming into an SQL system requires the system to read every one of the records on the systems hard disk. For a large system that would require terabytes of data to be read hundreds of times a second.

The database administrator (DBA) therefore produces keys for those filters he or she believes are likely to be required. Then as long as the coders happen to use the filters the administrator happens to have guessed they wanted the query will execute rapidly. If either side misguesses then a table scan is performed and the system grinds to a halt which is not unacceptable in the business world. So what actually happens is that the encapsulation which is supposed to exist between the three boxes in Fig. 8.1 is actually circumvented by a series of memos and design meetings between the programmers and the DBAs.

It is important to realize that modern, significant SQL applications are not independent of the underlying physical data architecture; they just don't have a systematic way of specifying the dependence that exists.

A consequence of Fig. 8.1 is that many SQL vendors have extended the SQL system to allow the application codes to specify 'hints' or 'methods' to the underlying database to try to enforce some of the semantic correlations that really need to exist. Unfortunately, these extensions are not covered by the SQL standard and thus the implementation of them differs from SQL system to SQL system and sometimes even within different releases of a given SQL system.

This can result in a tie between the application logic and the SQL system itself that would prevent database portability. Thus a layer of logic has been inserted to recreate that database independence. Two famous examples of this logic are Open Database Connectivity (ODBC) and Java Database Connectivity (JDBC). However, there is a tradeoff. While an application using ODBC can port between different SQL vendors, it can only take advantage of those extensions that are supported by all, or most, of the SQL vendors. Furthermore, while some SQL vendors support certain extensions the quality of that support can vary from feature to feature. Even when the support is high quality there is usually some performance overhead involved in the insertion of another layer of data processing.

It should be noted that ODBC and JDBC interface come in different forms:

- Those in which the ODBC is a native interface of the SQL system and executes on the database servers, and
- Those where the ODBC layer is actually a piece of middleware acting either upon dedicated middleware servers or in the application layer itself.

Normalizing an Abnormal World

A premise of the RDBMS concept is that the data is generated, stored and delivered according to the same data model. For those in the business of collecting data from external sources, this premise is fundamentally broken.

Each data source that is collected will, at best, have been generated according to a different data model. Far more commonly, the data has not really been collected according to a data model at all. The procedures in place to ensure a RDBMS has integrity simply do not apply for the majority of data that is available today.

Here are some examples of constraints placed upon a well-designed RDBMS that are violated by most data that is ingested:

(1) **Required fields**—When ingesting real world data, you cannot assume ANY of the fields will always be populated.
(2) **Unique fields are unique**—Consider the SSN on a customer record. Often these will be mis-typed or a person will use a family member's SSN, resulting in duplications.
(3) **Entity can be represented by a single foreign key**—Many of the fields relating to a person can have multiple valid values meaning the same thing. Therefore, if you wish to store not just what was referenced but how it was referenced you need at least two Tables
(4) **A single foreign key can refer to only one entity**—Consider the city name. A city can be replicated in many different states.
(5) **A field can take one of a discrete set of values**—Again misspellings and variations between different systems mean that the standard field lookup is invalid.

A result of the above is that it is impossible to construct a normalized relational model that accurately reflects the data that is being ingested without producing a model that will entirely destroy the performance of the host system.

The above assertion is very strong and there seems to be wars between teams of data architects that will argue for or against the above. However the following example has convinced anyone that has spent the time to look at the issue.

Consider the fields: city, zip and state. Try to construct a relational model for those three fields that accurately reflects

(a) A single location can validly have multiple city names (vanity city and postal city).
(b) A vanity city can exist in multiple postal cities.
(c) A postal city contains multiple vanity cities.
(d) A zip code can span between vanity cities and postal cities even if a given vanity city of which the zip code is a part does not span the postal cities.
(e) There are multiple valid abbreviations for given vanity cities and postal cities.
(f) The range of valid abbreviations for a given city name can vary, dependent upon the state of which the city is a part.
(g) The same city can span states, but two states can also have different cities with the same name.
(h) The geographical mapping of zip codes has changed over time.
(i) City names are often misspelled.
(j) A single collection of words could be a misspelling of multiple different cities.

From our experience, the best seen anyone has been able to tackle the above still took eight tables, and it relied upon some extensions in one particular data vendor.

There are a number of pragmatic solutions that are usually adopted:

(a) Normalize the data fully, investing in enough hardware and manpower to get the required performance. This is the theoretically correct solution. However, it can result in a single but large file of ingested data producing multiple terabytes of data into tens or even hundreds of sub-files. Further the data architecture team potentially has to alter the model for every new ingested file.
(b) Abandon normalization and move the data manipulation logic down into the application layer. With this approach, the fields contain the data as collected and the task of interpreting the data is moved down to the programmers. The application typically has to fetch a lot of data in multiple steps for a process that should have been executed atomically on the database server.
(c) Insert a significant data ingest phase where the data is 'bashed' into a format that has been predefined by the data architects. This is the best in terms of performance of the query system but has the twin downsides of creating a significant delay during the data ingest phase and also throwing away potentially vital data that was not compatible with the pre-defined ingest data architecture.
(d) Hybridize the above three approaches on a largely ad hoc file by file basis dependent upon the particular restrictions that were uppermost in the programmers mind at the point the data came in through the door.

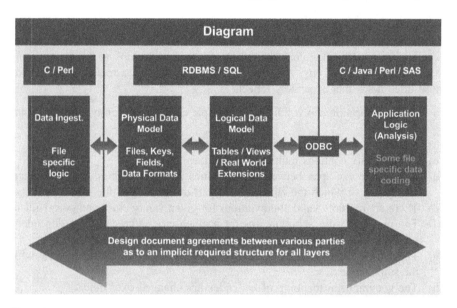

Fig. 8.2 Hybrid RDBMS model

While each of the first three of the above solutions has been heralded in design documents, most functioning systems evolve towards the fourth solution—the hybrid approach, as illustrated in Fig. 8.2.

A Data Centric Approach

In 1999, Seisint (now a LexisNexis Company) and LexisNexis independently conducted an evaluation of existing database technology and both concluded that the RDBMS was an unsuitable solution for large scale, disparate data integration.

After coming to this conclusion, in 2000, a team of Seisint employees that had been world leaders in the field of utilizing RDBMS technology was handed a blank sheet of paper and asked to construct a model to handle huge amounts of real-world data. The result was almost diametrically opposed to the diagram presented above.

Remember, the basic premise of the RDBMS is that the physical and logical data models are entirely disjoint. This was necessary in the late eighties as programming languages lacked the ability to adequately support multiple programmers from cooperating upon a task. Therefore processes had to be produced that allowed for teams to operate. By 2000 procedural encapsulation was well understood; these layers did not need to be kept distinct for programmatic reasons. In fact, as has been discussed previously, the separation of these two layers would routinely be violated by implicit usage agreements to boost performance to acceptable levels.

Another observation is that a key factor in the integration of disparate datasets is the disparity of the datasets. Attempting to fix those disparities in three different layers with three different skill sets is entirely counterproductive. The skills need to be developed around the data, not specifically around the processes that are used to manipulate the data. The layers in a data process should represent the degree of data integration not the particular representation used by the database.

The next decision was that compiler optimization theory had progressed to the point where a well specified problem was more likely to be correctly optimized automatically than by hand. This is especially true in the field of parallel execution and sequencing. Therefore a commitment was made to invest whatever resources were required to ensure that performance could be tuned by an expert system; leaving the data programmer with the responsibility to correctly specify the correct manipulation of the data.

In particular implicit agreements between execution layers are not tolerated; they are explicitly documented in the code to allow the optimizer to ensure optimal performance.

The final piece of the puzzle was a new programming language: **Enterprise Control Language (ECL)**. This was designed to have all of the data processing capabilities required by the most advanced SQL or ETL systems but also to have the code encapsulation mechanisms demanded by systems programmers.

There some advantages of the system in Fig. 8.3 that may not be immediately obvious from the diagram.

(1) The data sources are stored unmodified, even though they are modified as part of delivery. Thus there is never any "loss" of information or significant pain in re-mapping the incoming files to the target formats

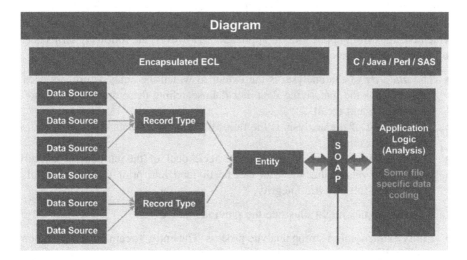

Fig. 8.3 Data centric RDBMS model

(2) The data teams can be segmented by data type rather than language skill. This allows for every file type to be handled by individuals skilled in that field.
(3) If required, a storage point between a batch ingest facility and a real-time delivery mechanism is available without a need to significantly recode the processing logic.
(4) Introducing parallelism is natural and can even be done between remote processing sites.

Data Analysis

The model in Fig. 8.3 essentially allows for huge scale data ingest and integration. It allows for scaling of hardware and personnel. It encourages deep data comprehension and code reuse, which enhances development productivity and improves the quality of the results. However, it does not significantly improve either the quality or the performance of any analysis that is performed. This should not be surprising; the data analysis is being performed outside of the ECL system.

Data analysis is one of those terms that everyone claims to perform, although very few can define what it really is. Most will claim that it somehow involves the translation of data into knowledge, although exactly what that translation means is ill defined. In particular, some would suggest that analysis is occurring if data is simply presented on screen for an analyst to review. Others would say analysis has occurred if the data is aggregated or summarized. Others may consider data analysis if it has be represented in some alternative format.

For the purposes of this document, a much simpler definition will be used:

Analysis is the process of concentrating information from a large data stream with low information content to a small data stream with high information content.

Large, small, low and high are deliberately subjective terms the definition of which changes from application to application. However, the following terms need to be defined that are crucial to the following.

- The *integrity* of the analysis is the extent to which the analysis process accurately reflects the underlying data. For data searching these are often measured by precision and recall.
- The *strength* of the analysis is the ratio of the about of data considered to the size of the result.
- The *complexity* of the analysis is the reciprocal of the number of entirely independent pieces that the data can be divided into prior to analysis whilst maintaining full analytic integrity.

A few examples might illustrate the previous terms:

- Entity extraction is a strong analytic process. The entity stream from a document is typically very small. As implemented today, entity extraction has very low complexity; typically every document is extracted independently of every other one.

- The LexisNexis process that ascertains whether or not people are using their correct SSN is not a strong process. The number of SSNs coming out is very similar to the number going in. Yet the process is complex in that every record in the system is compared to every other.
- The LexisNexis process that computes associates and relatives for every individual in the US is both strong and complex.
- Result summarization (such as sorting records by priority and counting how many of each kind you have) is a weak, non-complex process.
- Fuzzy matching is a strong, non-complex process.
- Pattern matching is a complex, weak process.
- Non-obvious relationship finding is a strong, complex process.

If the application data analysis being performed by the system is weak and simple then the fact that the data analysis has not been improved by this architecture is insignificant.[1] In such a scenario, well over 90 % of the system performance is dominated by the time taken to retrieve the data.

A system where the application analysis is strong will probably find that a bottleneck develops in the ability of the application server to absorb the data generated from the supercomputer. If the application code is outside of the control of the developer, then this problem may be insurmountable. If the application code is within the developer's control, then movement of the strength of the analysis down into the ECL layer will produce a corresponding improvement in system performance.

Case Study: Fuzzy Matching

A classic example of the above occurs millions of times each day on our LexisNexis servers. We allow sophisticated fuzzy matching of the underlying records. This includes edit distance, nick-naming, phonetic matching, zip-code radius, city-aliasing, street aliasing and partial address matching—all of which can happen at the same time. The integrity of the fuzzy match is directly proportional to the percentage of the underlying data that you perform the fuzzy scoring up. The LexisNexis HPCC system will often scan hundreds of records for every one that is returned. By moving this logic down into the ECL layer the optimizer is able to execute this code with negligible performance degradation compared to a hard match. Had the filtering logic not been moved down into the ECL layer then the fuzzy fetch would have been hundreds of times slower than a simple fetch because the application logic (on a different server) would have been receiving hundreds of times as much data.

---

[1]In such a situation, the main strength of the analysis will have been performed during the record selection process.

A complex system will suffer performance degradation that is exponential in the complexity of the query. This is because complexity implicitly strengthens a query. If the database is one terabyte in size, then a one-record result that requires all data to be considered requires the full terabyte to be exported from the system. Of course this proves unacceptable, so the application typically settles upon reducing the integrity of the system by using less than the full dataset for the analysis.

The LexisNexis HPCC system is designed to completely remove the barriers to high complexity, strong data analysis. However to leverage this capability the analysis code has to be entirely ported into the ECL layer so that the optimizer can move the code down into the data servers. This fact actually produces an extremely simple and accurate way to characterize the engineering decision involved:

You either have to move the code to the data or move the data to the code.

The LexisNexis HPCC system, utilizing ECL, has been designed on the premise that the dataset is huge and the code is relatively simplistic. Therefore, moving code into the ECL system generally results in performance improvement that is measured in orders of magnitude. The cost is that the algorithms have to be ported out of their existing format. For algorithms that don't yet exist, the modular, structured and data centric ECL language will actually speed up the development of the algorithm.

Case Study: Non-obvious Relationship Discovery

The original implementation of relatives within the LexisNexis Accurint product was performed in the application layer. Going to three levels of relationship involved hundreds and sometimes thousands of fetches per query. The result, even on high performance hardware, was that many relationship trees would fail the systems default timeout. The logic was moved into an ECL process that simultaneously computes the associates and relatives for every person in the US. The process at one point is evaluating seven hundred and fifty billion simultaneous computations stored in a sixteen terabyte data file. The result is presented across the SOAP layer as a simplistic relatives table which can now be delivered up with a sub-second response time. Further these relationships now exist as facts which can themselves be utilized in other non-obvious relationship computations.

The work of LexisNexis in the fields of law enforcement and anti-terrorism has all been achieved using the processing model shown in Fig. 8.4.

It should be noted that some third-party data analysis tools are low complexity by this definition. In this situation LexisNexis has an integration mechanism whereby we can actually execute within the ECL environment. In this situation, HPCC acts as an operating environment to manage the parallelization of the third-party software. The lift provided by HPCC will be directly proportional to the number of blades the system is running upon. For those third parties that do not offer a parallel blade solution, this is a significant win. For those that do have a server farm solution, the HPCC architecture will represent an improvement in the ease of system management for a modest reduction in performance.

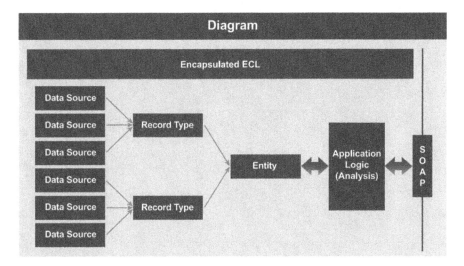

Fig. 8.4 Processing model applied by LexisNexis in the fields of law enforcement and anti-terrorism

Conclusion

This chapter has described, in stages, the premises behind the RDBMS engines, the data architectures of those engines and the problems associated with deploying either to the problem of high volume, real world, disparate data integration and analysis. It then detailed a platform architecture that allows teams to integrate large volumes of data quickly and efficiently while retaining third-party tools for analysis. Finally, it documented the pitfalls of external data analysis that is either strong or complex and outlined with case studies and architecture that solves this problem in a maintainable and efficient manner.

Chapter 9
Models for Big Data

David Bayliss

The principal performance driver of a Big Data application is the data model in which the Big Data resides. Unfortunately most extant Big Data tools impose a data model upon a problem and thereby cripple their performance in some applications.[1] The aim of this chapter is to discuss some of the principle data models that exist and are imposed; and then to argue that an industrial strength Big Data solution needs to be able to move between these models with a minimum of effort.

As each data model is discussed various products which focus upon that data model will be described and generalized pros and cons will be detailed. It should be understood that many commercial products when utilized fully will have tricks, features and tweaks designed to mitigate some of the worst of the cons. This chapter attempts to show that those embellishments are a weak substitute for basing the application upon the correct data model.

Structures Data

Perhaps the dominant format for data throughout the latter part of the twentieth century and still today the dominant format for corporate data is the data table.[2] Essentially structured data has records of columns and each column has a value the

---

This chapter has been adopted from the LexisNexis' white paper authored by David Bayliss.

---

[1] In fact, the performance is **so** crippled that the application just "doesn't happen". It is impossible to 'list all the things that didn't happen' although our paper "Math and the Multi-Component Key" does give a detailed example of a problem that would take 17 h in a key-value model and which runs at about 60 transactions per second in a structured model. It is easy to imagine that the key-value version would not happen.

[2] This is changing very rapidly in some areas. The other models are tackled later; but it is probably still true that today this one deserves to be tackled first.

© Springer International Publishing Switzerland 2016
B. Furht and F. Villanustre, *Big Data Technologies and Applications*,
DOI 10.1007/978-3-319-44550-2_9

meaning of which is consistent from record to record. In such a structure, vehicles could be represented as:

Make	Model	Color	Year	Mileage
Dodge	Caravan	Blue	2006	48,000
Toyota	Camry	Blue	2009	12,000

Many professional data modelers would immediately wince when handed the above. This is because the proposed structure did not allow for a very common feature of structured databases: normalization. Normalization is a topic which can (and does) fill many university Computer Science courses. For brevity[3] we shall define normalization as occurring when two or more columns are found to be dependent and the combinations of field values occurring are placed into a separate table that is accessed via a foreign key.

Thus looking at the above we might decide that the Make and Model of a car are related facts (only Toyota makes a Camry)—the other three columns are not. Thus the above could be defined as:

Vehicle type	Color	Year	Mileage
1	Blue	2006	48,000
2	Blue	2009	12,000

Key	Make	Model
1	Dodge	Caravan
2	Toyota	Camry

The advantage of the above may not be immediately apparent. But we have reduced the width of the vehicle file by two strings at the cost of one fairly small integer column. There is also a secondary table but that will have relatively few entries (hundreds) compared to the millions of entries one may have in a vehicle file. Better yet we can add new information about each vehicle type (such as Weight, Horsepower et cetera) in one place and all the other data referring to it is 'automatically right'.

The real power of structured data becomes apparent at the point you wish to **use** it. For example; if you wanted to find out how many miles travelled by all vehicles you simply need to sum the fourth column of the first table. If you want to find the ratio of vehicle colors in any given year—or trend it over time—a simple aggregate on the second and third table suffice.

---

[3]A rather longer and more formal treatment is given here: http://en.wikipedia.org/wiki/Database_normalization.

Things start to get a little more interesting if you want to count the number of miles driven in cars for a given manufacturer for a given model year. The issue is that the key in the vehicle file no longer identifies manufacturer; rather it identifies manufacturer AND model. Put another way the required result wants information spanning two tables. To get this result efficiently some clever tricks need to be performed. If you use a high level language such as SQL[4] and if the database administrators predicted this query then the cleverness should be hidden from you. However, under the hood one of the following will be happening:

1. The Tables can be JOINed back together to create a wider table (the first table) and then the required statistic is an aggregate on three columns.
2. The aggregate can be formed on the larger table to produce summaries for manufacturer and model and then that table can be joined to the vehicle type table and a second aggregate performed
3. The vehicle type table can be scanned to produce SETs of vehicle-type ID for each manufacturer, these sets can then be used to produce a temporary 'manufacturer' column for the main table which is then aggregated.
4. Perhaps something else that an SQL vendor found that works in some cases

As can be seen, aggregating and querying data across the entire database can get very complicated even with two files; as the number of data files grows then the problem becomes exponentially[5] more complex. It is probably fair to say that one of the key indicators of a quality of an SQL optimizer is the way it handles cross-file queries.

The other key feature of an SQL system to keep in mind is that in its purest form the performance is dismal. The reason is simply that each query, however simple, requires the entire dataset to be read. As the data becomes larger this becomes prohibitive. For this reason almost every SQL system out there has the notions of KEYs (or Indexes) built in. A key is an access path that allows records matching on one or more columns to be retrieved without reading the whole dataset. Thus, for example, if our vehicle dataset had been indexed by Color it would be possible to retrieve all of the red vehicles without retrieving vehicles of any other Color.

A critical extra feature of most SQL systems is the concept of multi-component keys; this is simply a key that contains multiple columns. For example, one could construct a key on Color/Year. This allows for two extremely fast access paths into

---

[4]Structured data does NOT need to imply SQL—but SQL is, without doubt, the leading method through which structured data is accessed.

[5]Many people use 'exponentially' as an idiom for 'very'. In this paper the term is used correctly to denote a problem that grows as a power of the scaling factor. In this case if you have three choices as to how to perform a join between two files, then between 10 files you have at least $3^{10} = 59,049$ choices. In fact you have rather more as you can also choose the order in which the files are processed; and there are 3,628,800 orders of 10 files giving a total of 2.14×1011 ways to optimize a query across 10 files.

the vehicle file: Color and Color/Year.[6] The criticality of this feature comes from an understanding of how the Color/Year value is found. The index system does NOT have to fetch all of the records with a matching color looking for a matching year. Rather the index itself is able to resolve the match on both color and year in one shot. This can give a performance lift of many orders of magnitude.[7]

Naturally a performance lift of orders of magnitude is always going to come at a price; in this case the price is the double whammy of build cost and flexibility. For these multiple-component keys to be useful they have to exist; in order to exist you have to **predict** that they are going to be useful and you have to have spent the time building the key. The visible manifestations of this cost are the high price attached to good system DBAs and also the slow data ingest times of most SQL systems.

The final, though obvious, flaw in the structured data model is that not all data is structured. If data is entered and gathered electronically then it has probably been possible to force the data to be generated in a model that fits the schema. Unfortunately there are two common reasons for data not existing easily in this model:

1. The data is aggregated. In the examples I have used vehicles as an example of a structured data file. In fact vehicle data is gathered, at least in the USA, independently by each state. They have different standards for field names and field contents. Even when they have the same column-name that is supposed to represent the same thing, such as color, they will have different words they use to represent the same value. One state might categorize color using broad strokes (Blue, Green, Red) others can be highly detailed (Turquoise, Aqua, Cyan). Either the data has to be brutalized into a common schema (at huge effort and losing detail), or left unintegrated (making genuine cross-source queries infeasible) or the schema has to get extremely complicated to capture the detail (making queries extremely complex and the performance poor).
2. Data cannot be readily split into independent columns. Consider any novel, white paper or even email. It doesn't represent a readily quantified transaction. It can contain detail that is not known or understood a priori; no structured model exists to represent it. Even if a structured model were available for one particular record (say Pride and Prejudice) then it would not fit other records (say War and Peace) in the same file.

In closing the structured data section we would like to mention one important but often overlooked advantage of SQL—it allows the user of the data to write (or generate) queries that are related to the problem (question being asked) rather than to the vagaries of the underlying data. The user only has to know the declared data

---

[6]Because of the way most Key systems work it does not in general provide a fast access path for Year only. Some good structured systems can access year quickly if the earlier component has low cardinality.

[7]Again, this is not an idiomatic expression. If a field is evenly distributed with a cardinality of N then adding it as a component of a key in the search path reduces the amount of data read by a factor of N. Thus if you add two or three fields each with a cardinality of 100 then one has produced a system that will go 4–6 orders of magnitude ($10{,}000$–$1{,}000{,}000\times$) faster.

model—the data has already been translated into the data model by the time the user is asking questions.

Text (and HTML)

At the complete opposite end of the scale you have text. Within text the data has almost no explicit structure and meaning.[8] A slightly more structured variant is HTML which theoretically imposes some form of order upon the chaos but in reality the generations of evolving specifications and browsers which do not comply to any specifications mean that HTML documents cannot even be assumed to be well formed. The beauty of text is that anyone can generate it and it often contains useful information. The World Wide Web is one example of a fairly large and useful text repository.

The downside of text, at least at the moment, is that it has no computer understandable semantic content. A human might be able to read it, and a thought or impression or even sets of facts might be conveyed quite accurately—but it is not really possible to analyze the text using a computer to answer meaningful questions.[9]

For this reason query systems against Text tend to be very primitive analytically. The most famous interface is that presented by the internet search engines. Typically these will take a list of words, grade them by how common the words are, and then search for documents in which those words are prevalent.[10] This work all stems from term and document frequency counts first proposed in 1972.[11]

An alternative approach, used in fields where precision is more important and the users are considered more capable, is the Boolean search. This method allows sophisticated expressions to be searched for. The hope or presumption is that the query constructor will be able to capture 'all the ways' a given fact might have been represented in the underlying text.

Text databases such as Lucene are extremely flexible and require very little knowledge of the data to construct. The downside is that actually extracting

---

[8]Many people refer to text as 'unstructured data'. I have generally avoided that term as good text will usually follow the structure of the grammar and phonetics of the underlying language. Thus text is not genuinely unstructured so much as 'structured in a way that is too complex and subtle to be readily analyzed by a computer using the technology we have available today.' Although see the section on semi-structured data.

[9]Of course, people are researching this field. Watson is an example of a system that appears to be able to derive information from a broad range of text. However if one considers that 'bleeding edge' systems in this field are correct about 75 % of the time it can immediately be seen that this would be a very poor way to represent data that one actually cared about (such as a bank balance!).

[10]Google pioneered a shift from this model; the 'page ranking' scheme effectively places the popularity of a page ahead of the relevance of the page to the actual search. Notwithstanding the relevance ranking of a page is still computed as discussed.

[11]Of course one can build multi-billion dollar empires by 'tweaking' this formula correctly.

aggregated information from them is the field of data-mining; which is generally a PhD level pursuit.

Semi-structures Data

Almost by definition it is impossible to come up with a rigorous and comprehensive definition of semi-structured data. For these purposes we shall define Semi-Structured data as that data which **ought** to have been represented in a structured way but which wasn't.

Continuing our vehicle example; consider this used car classified:

2010 Ferrari 599 GTB HGTE, FERRARI APPROVED, CERTIFIED PRE-OWNED WITH WARRANTY, Don't let this exceptional 599 GTB HGTE pass you by. This car is loaded with desirable options such as black brake calipers, carbon ceramic brake system, heat insulating windscreen, front and rear parking sensors, full recaro seats, and a Bose hifi system., The HGTE package comes with a retuned suspension consisting of stiffer springs, a thicker rear anti-roll bar, returned adjustable shocks, and wider front wheels. In addition, the car sits 0.4 of an inch lower to the ground and has a retuned exhaust note, and is fitted with the stickier Pirelli P Zero rubber. Inside, the HGTE package includes every possible carbon-fiber option., This vehicle has been Ferrari Approved as a Certified Pre Owned vehicle. It has passed our 101 point inspection by our Ferrari Factory trained technicians., 100 % CARFAX, CERTIFIED!!!

All of the text is free form and without pre-processing it would need to be stored as a variable length string. None of the queries of our structured section could be applied to this file. As it stands one would be limited to one or more 'string find' or 'search' statements in the hope of keying to text in the description. In an aggregated database this restriction results in a crucial shift of responsibility. The person asking the question (or writing the query) now has to understand all the possible ways a given piece of information might have been represented.[12] Of course, many internet search engines operate exactly this way.

However, with even some simplistic pre-processing it should be possible to turn the text above into:

Make	Model	Year	Other
Ferrari	599 GTB HGTE	2010	FERRARI APPROVED, CERTIFIED PRE-OWNED WITH WARRANTY, Don't let this exceptional 599 GTB HGTE pa...

Of course if one were clever and dedicated then even more information could be taken. In general, one could write a suite of routines to parse car classifieds and

---

[12]Or not care; if one is just 'surfing the web' then as long as the page offered is 'interesting enough' then one is happy—whether or not it was the 'best' response to the question is immaterial.

build up a fairly strong data model. This process of transforming from unstructured to semi-structured text requires up-front processing but the pay-off is that structured techniques can be used upon those facts that have been extracted; with all of the inherent performance and capability advantages that have been mentioned. Exactly how much data is 'parsed out' and how much is left unparsed is (of course!) ill defined; however for the process of classification we shall define semi structured data as data in which one or more (but not all) of the fields have been parsed into columns.

Thus far the presumption has been that the data is starting unstructured and that we are 'creating' some fields from nothing. In fact many structured databases contained semi-structured components. At the most extreme these are represented as BLOBs inside an SQL database. Rather more usefully there may be text strings which appear free-format but which realistically have a limited range of values. These are often the 'comment' fields stored in relational databases and used as a catch-all for fields that the modelers omitted from the original data model and which no-one has ever fixed.

Bridging the Gap—The Key-Value Pair

Whilst the processing paradigm of Map-Reduce is much vaunted, the equally significant enabler, the Key-Value pair, has gone relatively unheralded. In the context of the foregoing it may be seen that it is the data model underlying Map-Reduce (and thus Hadoop) that is actually the fundamental driver of performance. In fact those with a sharp eye may notice that key-value pairs derive their power from their position part-way between semi-structured and textual data.

A file of key value pairs has exactly two columns. One is structured—the KEY. The other, the value, is unstructured—at least as far as the system is concerned. The Mapper then allows you to move (or split) the data between the structured and unstructured sections at will. Thus our vehicle table could be:

Key	Value
Dodge | Caravan	Blue | 2006 | 48,000
Toyota | Camry	Blue | 2009 | 48,000

The reducer then allows data to be collated and aggregated **provided** it has an identical key. Thus with the Mapper I used above it would be very easy to perform statistics aggregated by Make and Model. Aggregating by Color and Year would be extremely painful; in fact the best idea would probably be to re-map the data so that Color and Year were the key and then perform a different reduce. The advantage of this system over a keyed and structured SQL system is that the sort-order required is defined at query time (no prediction is required) and until that sort order is used then there is no overhead: in other words data ingest is quick.

In pure map reduce any query requires at least one full table scan of every file involved; Hadoop has no keys. Of course, a number of add-ons have been written that allow Key-Value pairs to be stored and quickly retrieved. These function in a very similar way to SQL keys except that the keys can only have one component, they must contain the payload of the record, and there is no existing support for normalization or the concept of foreign keys. In other words, if you have one or two access paths there is no problem—but you cannot access your data using a wide range of queries.

It should now be evident why Key-Value (and Map-Reduce) has been able to achieve general purpose popularity. It can read data in either structured, text or semi-structured form and dynamically translate it into a semi-structured data model for further processing. The restrictions within the current implementations are:

1. A single map-reduce only supports one semi-structured data model
2. Multiple map-reduces require multiple-maps (and thus reprocessing of the data)
3. Either: all queries require a full data scan (Hadoop) OR all data-models require their own (potentially slimmer) copy of the data (Hive etc.)
4. The work of constructing the semi-structured model is entirely the responsibility of the programmer
5. If the data is at all normalized then optimizing any joins performed is entirely the responsibility of the programmer

XML—Structured Text

It is really incorrect to speak of XML as a data model. It is really a data transmission format that allows a plethora of different data models to be defined and utilized. In particular many if not all of the preceding data models can be expressed and utilized from within XML. That said, XML databases are usually thought of in the context of the storage of hierarchical databases where each record has a different but conformant format. An example might be a personnel database where the person is the outer container and various nested details are provided about that person, some of which themselves have nested details. Another very common usage of an XML database is document storage where the annotations upon and structure of the document is expressed in XML. In this model it is very common for the XML to also have an HTML rendering generated via XSLT.

From the perspective of data access and summarization the key to an XML document is epitomized by the name of the most common query language for them: XPATH.[13] Each element of an XML document is identified by a path which defined

---

[13]XQuery has probably surpassed XPATH in more modern installations.

from the root; the 'navigation path' to the individual element or elements. Along the route various filters and conditions based upon the elements and tags encountered along the route may be executed; it is even possible to include data derived from routes into other parts of the data within the expressions.

Thus the pain of any given query; both in terms of encoding and execution is directly proportional to the extent to which that query can be expressed as a linear path. Simple fetches of one or more columns are easy; queries that rely upon the relationship between the columns of a record, or between columns of different records are harder to express and execute.[14] In short, whilst XML supports almost any data model; its features, search syntax, and performance footprint encourage, if not mandate, a hierarchical data model. Hierarchical data models are 'great if it is true'. If the data naturally fits a model whereby every element has a single parent and that the children of different parents are logically independent then the model is efficient in both usage and expression. If, however, it is the relationship between child elements that is interesting then the imposition of a rigid hierarchy will simply get in the way.[15] One might conclude that XML/XPATH provides most of the features of SQL but with an antithetical data model.

A hierarchical view of vehicles might be:

Parent	Child	
Vehicle one	FactType	Value
	Make	Dodge
	Model	Caravan
	Year	2006
	Color	Blue
	Mileage	48,000
Vehicle two	FactType	Value
	Make	Toyota
	Model	Camry
	Year	2009
	Color	Blue
	Mileage	12,000

[14]A good XML database such as MarkLogic will allow for optimization of complex queries provided the access paths can be predicted and declared to the system.

[15]Within the academic literature there have been numerous attempts to extend XPATH towards more relational or graph-based data.

RDF

Given the characterization of XML data as 'typically' hierarchical it is probably wise to mention one extremely nonhierarchical 'data model' based upon XML which is RDF (Resource Description Framework). Like many things designed by committee this standard has many features and purposes; however at the data model level it is a method of describing data entirely using triples. Put another way RDF is a method of describing data as a collection of typed relationships between two objects. Our vehicle file could be represented using:

Object 1	Relationship	Object 2
Vehicle 1[a]	MadeBy	Dodge
Vehicle 1	ModelName	Caravan
Vehicle 1	Color	Blue
Vehicle 1	Mileage	48,000
Vehicle 1	Year	2006
Vehicle 2	MadeBy	Toyota
Vehicle 2	ModelName	Camry
Vehicle 2	Color	Blue
Vehicle 2	Mileage	12,000
Vehicle 2	Year	2009

[a]The labels I am using here are for illustration; RDF contains a range of specifications to ensure unique naming, integrity etc

Viewed at the level of a single entity this appears to be a painful way of expressing a simple concept. The power comes from the fact that the objects in the third column are 'first class' objects; exactly the same as those on the left. Therefore this table, or another, could be expressing various facts and features of them. Viewed holistically data is no longer a series of heterogeneous tables that may or may not be linked; rather it is a homogenous web of information.

While not mandated by the data model RDF is often queried using declarative languages such as SPARQL. These languages take advantage of the simplicity and homogeneity of the data model to produce elegant and succinct code. The challenge here is that every query has been transformed into a graph query against a graph database. These queries represent some of the most computationally intensive algorithms that we know. To return all 'Blue Toyota Camry' we need to retrieve three lists—all blue things, all Toyota things, all Camry things and then look for any objects appear on all three huge lists. Compared to a structured data multi-component fetch, we have turned one key fetch into three; each of which would be at least two orders of magnitude slower.

Data Model Summary

Big Data owners typically adopt one of three approaches to the above issue:

(1) Adopt one model that excels in their core need and either accept that alternative forms of application run slowly; or simply ignore alternative forms of data usage. As long as the 'other uses' weren't really that important; this is probably optimal.

(2) Adopt a model that isn't **too** bad across a range of potential applications—typically this results in 'lowest common denominatorism'—but it allows the data to be used to its fullest extent. This gets the most out of your data—but the cost of hardware—or lack of performance may be an issue.

(3) Replicate some or all of the data from a core system into one or more data-marts; each in a slightly different data model. If the cost of maintaining multiple systems, and training staff for multiple systems can be justified then at an 'external' data level this appears optimal.

Data Abstraction—An Alternative Approach

A driving goal behind the HPCC[16] system was the abstraction of the data model from both the language and the system. As each of the previous data models was discussed the 'typical query language' was discussed alongside the model. The HPCC system language, ECL, allows all of the supported data models to be queried using the same language. Further most, if not all, of the previous data models have an engine built around them that is designed for one particular data model. HPCC offers two different execution engines each of which is designed to support each of the data models. Finally, and most importantly, much of the technology built into the HPCC is focused upon the translation of one data model to another as efficiently as possible—in terms of both programmer and system resources.

Of course, in a whitepaper such as this, it is always easy to conclude 'we do everything better than everyone without any compromises'; it is much harder to develop a system that renders the claim true. The truth is that HPCC has been used to implement all of the above data models and in some aspects[17] we beat all the existing systems we have been measured against; of course any system that specializes in only one data model will usually have some advantage within that data model.[18] Where we believe HPCC is untouchable is in its ability to natively support a broad range of models and in the speed with which we can translate between models.

---

[16]http://en.wikipedia.org/wiki/HPCC.

[17]Usually including performance.

[18]Usually speed of update or standards conformance.

As an aid to further research this paper will now review some of the system features available for each of the stated data models, the extent of the work done, and some features available for translation in and out of the data model.

Structured Data

The ECL[19] handling of structured data is based upon the RECORD structure and corresponding DATASET declarations. ECL has a huge range of data types from INTEGERs of all sizes from 1 to 8 bytes, packed decimals, fixed and variable length strings, floating point numbers and user defined types. ECL also supports variant records and arrays of structured types.[20] Like SQL, columns are accessed by name and type conversions occur to allow columns to be compared and treated at a logical level (rather than low-level). Unlike SQL it is also possible to define an ordering of records within a file; this allows for records to be iterated over in sequence. ECL also has NORMALIZE and DENORMALIZE functions built in so that questions regarding 'just how much to normalize data' can be changed and modified even once the system is in place.

By default ECL files are flat and they have to be fully scanned to obtain data (like Hadoop). However ECL also has the ability to create indexes with single or multiple components and it also allows for variant degrees of payload to be imbedded within the key. Thus an ECL programmer can choose to have a large number of highly tuned keys (similar to SQL) or one or two single component keys with a large payload (like Key-Value) whichever they prefer (including both[21]).

There are many, many ECL features to support structured data but some of the most important are PROJECT, JOIN (keyed and global variants) and INDEX. Simple ingest of structured data is handled via DATASET which can automatically translated from flat-file, CSV[22] and well-structured XML. More complex (and partial) ingest of other data models will be handled under the section 'Semi-Structured Data'.

---

[19]http://en.wikipedia.org/wiki/ECL,_data-centric_programming_language_for_Big_Data.

[20]For those familiar with COBOL this was a method of having narrower records whereby collections of fields would only exist based upon a 'type' field in the parent record.

[21]Our premier entity resolution system uses multi-component keys to handle the bulk of queries and falls back to a system similar to key value if the multi-components are not applicable.

[22]Referred to as 'comma separated variable' although there are many variants; most of which don't include commas!.

Text

ECL handles Text in two different ways depending upon requirements. At the most basic level variable length strings[23] are native within ECL and a full set of string processing functions are available in the Std.Str module of the ECL standard library. For text which needs to be 'carried around' or 'extracted from' this is usually adequate.

For text that needs to be rigorously searched, an ECL programmer typically utilizes an inverted index. An inverted index is a key that has a record for every word in a document and a marker for where in the document the word occurs.[24] This index can then be accessed to perform searches across documents in the manner described in the text section. Our 'Boolean Search' add-on module exemplifies this approach. At a low level ECL has two technologies, global smart stepping and local smart stepping,[25] to improve the performance of this form of search.

Text can be brought into ECL as a 'csv' dataset; if it currently exists as a series of independent documents then a utility exists to combine that into one file of records. It will be covered in greater depth in the semi-structured section but it should be noted that ECL has an extremely strong PATTERN definition and PARSE capability that is designed to take text and extract information from it.

Semi-structured Data

We firmly believe that the next big 'sweet spot' in Big Data exists in Semi-Structured data. The action of turning data which is otherwise inaccessible into accessible information is a major competitive differentiator. ECL has extensive support at both a high and low level for this process.

Firstly, at a high level, ECL is a data **processing** language, not a storage and retrieval language. The two execution engines (one batch, one online), coupled with PERSIST, WORKFLOW and INDEX capability all push towards a model where data goes through an extensive and important processing stage **prior** to search and retrieval.

---

[23]In standard ASCII and UNICODE formats.

[24]There may be other flags and weights for some applications.

[25]When accessing inverted indexes naively you need to read every entry for every value that is being searched upon; this can require an amount of sequential data reading that would cripple performance. 'Smart stepping' is a technique whereby the reading of the data is interleaved with the merging of the data allowing, on occasions, vast quantities of the data for one or more values to be skipped (or stepped) over. The 'local' case is where this is done on a single machine; 'global' is the case where we achieve this even when the merge is spread across multiple machines.

Secondly, at a lower level, the ECL PARSE statement makes available two different text parsing capabilities that complement each other to extract information in whichever form it is in. One capability is essentially a superset of regular expressions; wrapped in an easy-to-read and re-use syntax. This is ideal where the data is highly unstructured[26] and particular patterns are being 'spotted' within in. Examples applications include screen scraping and entity extraction. The other capability follows a Tomita[27] methodology; it is designed for the instance where textual document follows a relatively rigorous grammar.[28] This is suitable for rapidly processing text which is expected to be well formed.

Thirdly, again at a higher level, the ECL support for Text and Structured data are BOTH built upon the same ECL record structure. Thus, once some information has been extracted from the text, it can still reside in the same record structure awaiting additional processing. Further, as keys can contain arbitrary payloads, and as joins can be made across arbitrary keys, it is even possible to combine structured and inverted index fetches within the same query.

Key-Value Pairs

Key-Value pairs are a degenerative-sub case of normal ECL batch processing. If required a Key-Value pair can be declared as:

KVPair : = RECORD

STRING KeyValue;
STRING Value;
END;

The map-reduce paradigm can then be implemented using:

(1) PROJECT/TRANSFORM statement returning a KVPair—this is equivalent to a Mapper
(2) DISTRIBUTE(KVPairFile,HASH(KeyValue))—the equivalent of the first stage of the Hadoop shuffle
(3) SORT(DistributedKVPairFile,KeyValue,LOCAL)—second stage of Hadoop shuffle
(4) ROLLUP(SortedPairFile,KeyValue,LOCAL)—the reduce.

This usage of ECL is 'degenerative' insofar as data can be distributed, sorted and rolled up using any of the fields of a record, any combination of fields of a record

---

[26]Specifically the case where linguistic rules of grammar not followed uniformly and thus the data really is 'unstructured'.

[27]A brief treatment of these is given here: http://en.wikipedia.org/wiki/GLR_parser.

[28]This case works particularly well (and easily) in the case where the text being parsed is generated against a BNF grammar.

and even any expressions involving combinations of fields of a record. Thus the 'normal' ECL case only has values[29] and any number of different keys 'appear upon demand'.

XML

XML probably represents the **biggest stretch** for ECL in terms of native support of a data model. ECL does have very strong support for XML which is well specified and gently nested. Specifically it has:

(1) XML on a dataset to allow the importing of XML records that can be reasonably represented as structured data
(2) XMLPARSE to allow the extraction of sub-sections of more free-form XML into structured data
(3) Extensive support for child datasets within record structures to allow some degree of record and file nesting on a hierarchical basis
(4) An XML import utility that will construct a good ECL record structure from an XML schema.

Unfortunately there are XML schemas out there that really do not have sane and rational structural equivalents. More specifically; they may have equivalents but they are not the best representation of the data. For this reason we have been actively researching the construction of an XPATH equivalent within ECL. This essentially extends upon our work supporting arbitrary text processing and utilizes our ability to have multiple component indexes in an inversion and to blend the results of inverted and structured index lookups. The result is the ability to store XML documents in an extremely efficient shredded form and then process XPATH queries using the index prior to performing the fetch of those parts of the document required.[30]

---

[29]Although it has any number of values; not just one per record.

[30]Fuller details of this will be published in due course; probably accompanied by a product Module offering.

RDF

Much like Key-Value pairs RDF is simply a degenerative[31] case of standard ECL processing. Pleasingly the global smart stepping technology developed for Text searching can also be utilized to produce relatively rapid RDF query processing. Unfortunately being 'relatively quick' does not avoid the fundamental issue that graph problems are extremely compute intensive and the transformation of all queries into graph queries loses a lot of academic elegance once it is appreciated that we are transforming all queries into graph queries that we cannot, in general, solve.

From an execution standpoint ECL has an immediate solution that yields many orders of magnitude performance[32] improvement. You do not restrict yourself to triples; you allow the data store to be an arbitrary n-tuple. By doing this **each** fetch returns much less data (reduced by the cardinality of the fourth and subsequent columns) and the joins required are reduced by the products of those reductions.[33] This performance boost is achieved for the same reason that multi-component keys yield staggering lift over single component keys (see the description of structural data).[34]

The downside is that the query programmer now has to adapt their coding to the more heterogeneous data layout. As an area of active research a second language KEL (Knowledge Engineering Language) has been proposed and prototyped[35] which allows the programmer to work in a homogenous environment but then translates the queries into ECL that in turn uses an execution optimal heterogeneous data model. More information will be made available as the project nears fruition.

---

[31]This term is being used technically; not in the pejorative. ECL works naturally with any combination of N-tuples. Asserting everything must be triples (or 3-tuples) is one very simple case of that.

[32]Again, this is a mathematical claim, not a marketing one.

[33]If we can find it—we could reference the Lawrence Livermore study here.

[34]Yoo and Kaplan from Lawrence Livermore have produced an excellent study and independent on the advantages of DAS for graph processing: http://dsl.cs.uchicago.edu/MTAGS09/a05-yoo-slides.pdf.

[35]A team led by LexisNexis Risk Solutions, including Sandia National Labs and Berkeley Labs has an active proposal for further development of this system.

Model Flexibility in Practice

If ECL is compared in a genuine 'apples to apples' comparison against any of the technologies here on their own data model it tends to win by somewhere between a factor of 2 and 5. There are a variety of reasons for this, and they differ from data model to data model and from competitor to competitor but the reasons usually include some of:

(1) HPCC/ECL Generates C++ which executes natively; this has a simple low-level advantage over Java based solutions
(2) The ECL compiler is heavily optimizing; it will re-arrange and modify the code to minimize operations performed and data moved
(3) The HPCC contains over a dozen patented or patent-pending algorithms that give a tangible performance boost to certain activities—sorting and keyed fetch being the two most significant
(4) Whilst ECL supports all of these data models it does not (always) support **all** of the **standards** contained in the languages that usually wrap these models. Sometimes standards written a priori in a committee room impose low-level semantics on a solution that result in significant performance degradation[36]
(5) Whilst continually evolving, the core of the HPCC has been around for over a decade and has been tweaked and tuned relentlessly since creation
(6) The ECL language has explicit support for some feature that needs to be manufactured using a combination of capabilities of the other system.[37] In this situation the extra knowledge of the intent that the HPCC system has usually gives a performance advantage.

The above is a pretty impressive list; and going 2–5× faster allows you to use less hardware, or process more data, or get the results faster or even a little bit of all three. However, a factor of two to five is not really game-changing, it just lowers the cost of playing.[38]

Those occasions when ECL is game-changing, when it delivers performance which is one or more orders of magnitude faster than the competition, usually stem from the data model. External evidence and support for this claim can be gleaned

---

[36]As a recent example: ECL provides two sets of string libraries, one which is UNICODE complaint, one of which is not. The non-compliant libraries execute five times faster than the compliant ones. That said; string library performance is not usually the dominant factor in Big Data performance.

[37]Put simply—if you can write it in one line of ECL the ECL compiler knew exactly what you were trying to do—if you write the same capability in 100 lines of Java then the system has no high level understanding of what you were doing.

[38]As an aside, the HPCC was one of the first systems available that could linearly scale work across hundreds of servers. As such it could often provide two orders of magnitude (100×) performance uplift over the extant single-server solutions; which clearly is game changing. For this paper, given the title includes 'Big Data', it is presumed that HPCC is being contrasted to other massively parallel solutions.

from the websites of many of the solutions mentioned in this paper. They all claim, and cite examples, where their solution to a data problem produces a result three to ten times faster than their competitors. Can they all claim this without deception?[39] Yes; they are citing examples where a customer with a given problem switched to them and they were providing a more suitable data model to the underlying problem. It is quite conceivable that two data solution vendors could swap customers and both customers get a performance enhancement!

Given the preceding paragraph one might naturally expect data owners to be swapping and changing vendors with very high frequency to continually benefit from the 'best model for this problem' factor. Unfortunately the reality is that many data systems have been in the same model for years (even decades) even once it is abundantly clear it is the wrong model. The reason, as described previously, is that changing model usually involves a change of language, vendor and format; all of which are high risk, high cost, and disruptive changes. The ECL/HPCC system allows for model-change without any of those factors changing. Of course, the first time ECL is used many of those changes do occur; but once within the HPCC system data can transition from one model to another on a case by case basis with zero disruption.[40]

To understand all aspects of the ECL impact one does need to accept that zero disruption is not the same thing as zero effort. Data will generally come to an ECL system in a natural or native[41] model. For reasons of time and encouragement the data will usually be made available in the original model through the ECL system. This will garner any 'ECL is quick' benefits immediately but it may not be game-changing. At that point a skilled data architect (or aspiring amateur!) is needed that can explore the options for alternative data models for the data. This might be as big as a completely different core model or it might be a change of model as the data is used for certain new or existing processes. The exploration is performed alongside the service of the data using the extant model. Then, if and when these opportunities are found and the results are shown to be as good or better, then the game can be changed.

One note of caution needs to be sounded; giving people the ability to choose also gives them the ability to choose poorly. As the data models were discussed it should have become apparent that some of them were extreme, niche models and others were more general purpose and middle of the road. ECL is flexible enough to allow

---

[39]One subtle form of deception is 'measuring something else'; the overall performance of a system needs to include latency, transaction throughput, startup time, resource requirements and system availability. There are systems that are highly tuned to one of those; this is legitimate—provided it is declared.

[40]This is a mathematical zero, not a marketing one. ECL/HPCC supports all of these models and hybrids of them within the same system and code. Many, if not all, ECL deployments run data in most of these models simultaneously and will often run the same data in different models at the same time on the same machine!.

[41]Where a native model is an extremely un-natural model imposed upon the data by an earlier system.

the most horrific extremes to be perpetrated in any direction. For this reason ECL operates best and produces its most dramatically good results when early design decisions are made by one or more experienced data personnel. If these are not available it is probably wise to **mandate** that the existing data model is adhered to; at least initially.

Conclusion

In many ways this entire chapter is a conclusion; any one of the subject headings could have been the title for a 100 page paper. I am sure that as each of the data models was discussed, the zealots of the model could have listed a dozen extra advantages and detractors could list a dozen extra flaws. Notwithstanding the purpose of the descriptions was simply to detail that there **are** many alternative models and that they each have lists of pros and cons. Further, most of them have languages, standards, and vendors wrapped around them, designed to make data processing in that particular model as proficient as possible.

Next, under the title of data abstraction, we proposed that the optimal system would allow for the problem to dictate the model to be used; rather than the model used dictating the problems which are soluble. We then detailed how the ECL/HPCC system supports each of the data models mentioned and also the large array of features and tools provided to assist in the transition between models. It was also noted that whilst ECL/HPCC generally executed against a given model more efficiently than alternative implementations it did not always support all of the standards or language features associated with a given model.

Finally the paper ended by suggesting that ECL gains performance lift from general efficiencies but that for the performance lift to reach the level of dramatic then generally a data-model shift was required. It was noted that data models can be shifted without ECL by changing vendors; the unique advantage of ECL is the ability to shift model (or even hybridize models) within the same language and system. This advantage reduces the cost and risk of model shift and therefore increases the chances of one occurring.

Chapter 10
Data Intensive Supercomputing Solutions

Anthony M. Middleton

Introduction

As a result of the continuing information explosion, many organizations are drowning in data and the resulting "data gap" or inability to process this information and use it effectively is increasing at an alarming rate. Data-intensive computing represents a new computing paradigm which can address the data gap using scalable parallel processing and allow government and commercial organizations and research environments to process massive amounts of data and implement applications previously thought to be impractical or infeasible.

The fundamental challenges of data-intensive computing are managing and processing exponentially growing data volumes, significantly reducing associated data analysis cycles to support practical, timely applications, and developing new algorithms which can scale to search and process massive amounts of data. LexisNexis believes that the answer to these challenges is a scalable, integrated computer systems hardware and software architecture designed for parallel processing of data-intensive computing applications. This paper explores the challenges of data-intensive computing and offers an in-depth comparison of commercially available system architectures including the LexisNexis Data Analytics Supercomputer (DAS) also referred to as the LexisNexis High-Performance Computing Cluster (HPCC), and Hadoop, an open source implementation of Google's MapReduce architecture.

The MapReduce architecture and programming model pioneered by Google is an example of a systems architecture specifically designed for processing and analyzing large datasets. This architecture was designed to run on large clusters of commodity machines and utilizes a distributed file system in which files are divided into blocks and stored on nodes in the cluster. In a MapReduce application, input

---

This chapter has been adopted from the white paper authored by Anthony M. Middleton, LexisNexis.

© Springer International Publishing Switzerland 2016   257
B. Furht and F. Villanustre, *Big Data Technologies and Applications*,
DOI 10.1007/978-3-319-44550-2_10

data blocks are processed in parallel by Map tasks assigned to each data block to perform specific operations and transformations on the data and Reduce tasks which aggregate results and write output data blocks. Multiple MapReduce sequences are typically required to implement more complex data processing procedures. The Hadoop architecture is functionally similar to the Google implementation but uses Java as the base programming language instead of C++. Both Google and Hadoop implemented high-level parallel dataflow languages for data analysis to improve programmer productivity. For Hadoop, this language is called Pig Latin and the associated execution environment is called Pig.

LexisNexis, an industry leader in data content, data aggregation, and information services independently developed and implemented a solution for data-intensive computing called HPCC. The LexisNexis approach also utilizes commodity clusters of hardware running the Linux operating system and includes custom system software and middleware components developed and layered on the base Linux operating system to provide the execution environment and distributed filesystem support required for data-intensive computing. Because LexisNexis recognized the need for a new computing paradigm to address its growing volumes of data, the design approach included the definition of a new high-level dataflow language for parallel data processing called ECL (Enterprise Data Control Language). The power, flexibility, advanced capabilities, speed of development, and ease of use of the ECL programming language is the primary distinguishing factor between the LexisNexis HPCC and other data-intensive computing solutions.

LexisNexis developers recognized that to meet all the requirements of data-intensive computing applications in an optimum manner required the design and implementation of two distinct processing environments, each of which could be optimized independently for its parallel data processing purpose. The first of these platforms is called a Data Refinery whose overall purpose is the general processing of massive volumes of raw data of any type for any purpose but typically used for data cleansing and hygiene, ETL processing of the raw data (extract, transform, load), record linking and entity resolution, large-scale ad hoc analysis of data, and creation of keyed data and indexes to support high-performance structured queries and data warehouse applications. The second platform is called the Data Delivery Engine. This platform is designed as an online high-performance structured query and analysis platform or data warehouse delivering the parallel data access processing requirements of online applications through Web services interfaces supporting thousands of simultaneous queries and users with sub-second response times. Both platforms can be integrated in the same processing environment, and both platforms utilize the same ECL programming language increasing continuity and programmer productivity.

This chapter presents a detailed analysis and feature comparison of the HPCC system architecture versus Hadoop, and the ECL programming language versus Pig. Results of head-to-head system performance tests based on the Terabyte sort benchmark are presented and show that HPCC is up to 4 times faster than Hadoop when using the same hardware configuration. This paper concludes that the advantages of selecting a LexisNexis HPCC architecture for data-intensive computing include: (1) an architecture which implements a highly integrated system environment with capabilities from raw data processing to high-performance queries and data analysis using a common language; (2) an architecture which provides equivalent performance at a much lower system cost based on the number of processing nodes required resulting in significantly lower total cost of ownership (TCO); (3) an architecture which has been proven to be stable and reliable on high-performance data processing production applications for varied organizations over a 10-year period; (4) an architecture that uses a mature, declarative, dataflow programming language (ECL) with extensive built-in capabilities for data-parallel processing, allows complex operations without the need for extensive user-defined functions significantly increasing programmer productivity (an important perfor-mance factor in application development), and automatically optimizes execution graphs with hundreds of processing steps into single efficient work units; (5) an architecture with a high-level of fault resilience and language capabilities which reduce the need for re-processing in case of system failures; and (6) an architecture which is available from and supported by a well-known leader in information services and risk solutions (LexisNexis) who is part of one of the world's largest publishers of information ReedElsevier.

Parallel processing approaches can be generally classified as either compute-intensive, or data-intensive [1–3]. Compute-intensive is used to describe application programs that are compute bound. Such applications devote most of their execution time to computational requirements as opposed to I/O, and typically require small volumes of data. Parallel processing of compute-intensive applica-tions typically involves parallelizing individual algorithms within an application process, and decomposing the overall application process into separate tasks, which can then be executed in parallel on an appropriate computing platform to achieve overall higher performance than serial processing. In compute-intensive applica-tions, multiple operations are performed simultaneously, with each operation addressing a particular part of the problem. This is often referred to as functional parallelism or control parallelism [4].

Data-Intensive Computing Applications

Data-intensive is used to describe applications that are I/O bound or with a need to process large volumes of data [1, 2, 5]. Such applications devote most of their processing time to I/O and movement of data. Parallel processing of data-intensive applications typically involves partitioning or subdividing the data into multiple segments which can be processed independently using the same executable appli-cation program in parallel on an appropriate computing platform, then reassembling the results to produce the completed output data [6]. The greater the aggregate distribution of the data, the more benefit there is in parallel processing of the data. Gorton et al. state that data-intensive processing requirements normally scale lin-early according to the size of the data and are very amenable to straightforward parallelization. The fundamental challenges for data-intensive computing according to Gorton et al. are managing and processing exponentially growing data volumes, significantly reducing associated data analysis cycles to support practical, timely applications, and developing new algorithms which can scale to search and process massive amounts of data.

Data-Parallelism

Computer system architectures which can support data-parallel applications are a potential solution to terabyte scale data processing requirements [6, 7]. According to Agichtein [8], parallelization is an attractive alternative for processing extremely large collections of data such as the billions of documents on the Web. Nyland et al. define data-parallelism as a computation applied independently to each data item of a set of data which allows the degree of parallelism to be scaled with the volume of data. According to Nyland et al., the most important reason for developing data-parallel applications is the potential for scalable performance, and may result in several orders of magnitude performance improvement. The key issues with developing applications using data-parallelism are the choice of the algorithm, the strategy for data decomposition, load balancing on processing nodes, message passing communications between nodes, and the overall accuracy of the results [6, 9]. Nyland et al. also note that the development of a data-parallel application can involve substantial programming complexity to define the problem in the context of available programming tools, and to address limitations of the target architecture. Information extraction from and indexing of Web documents is typical of data-intensive processing which can derive significant performance benefits from data-parallel implementations since Web and other types of document collections can typically then be processed in parallel [8].

The "Data Gap"

The rapid growth of the Internet and World Wide Web has led to vast amounts of information available online. In addition, business and government organizations create large amounts of both structured and unstructured information which needs to be processed, analyzed, and linked. Vinton Cerf of Google has described this as an "Information Avalanche" and has stated "we must harness the Internet's energy before the information it has unleashed buries us" [10]. An IDC white paper sponsored by EMC estimates the amount of information currently stored in a digital form at 281 exabytes and the overall compound growth rate at 57 % (Fig. 10.1) with information in organizations growing at even a faster rate [11]. In another study of the so-called information explosion it was estimated that 95 % of all current information exists in unstructured form with increased data processing

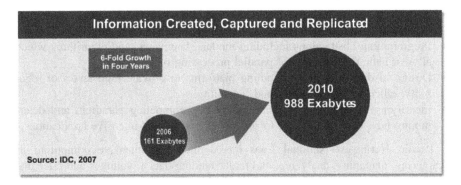

Fig. 10.1 The information explosion (*Source* IDC [11])

requirements compared to structured information [12]. The storing, managing, accessing, and processing of this vast amount of data represents a fundamental need and an immense challenge in order to satisfy needs to search, analyze, mine, and visualize this data as information [13]. LexisNexis has defined this issue as the "Data Gap": the ability to gather information is far outpacing organizational capacity to use it effectively.

Organizations build the applications to fill the storage they have available, and build the storage to fit the applications and data they have. But will organizations be able to do useful things with the information they have to gain full and innovative use of their untapped data resources? As organizational data grows, how will the "Data Gap" be addressed and bridged? LexisNexis believes that the answer is a scalable computer systems hardware and software architecture designed for data-intensive computing applications which can scale to processing billions of records per second (BORPS). What are the characteristics of data-intensive computing systems and what commercially available system architectures are available to organizations to implement data-intensive computing applications? This paper will explore those issues and offer a comparison of commercially available system architectures including the LexisNexis Data Analytics Supercomputer (DAS) also referred to as the LexisNexis High-Performance Computing Cluster (HPCC).

Characteristics of Data-Intensive Computing Systems

The National Science Foundation believes that data-intensive computing requires a "fundamentally different set of principles" than current computing approaches [14]. Through a funding program within the Computer and Information Science and Engineering area, the NSF is seeking to "increase understanding of the capabilities and limitations of data-intensive computing." The key areas of focus are:

- Approaches to parallel programming to address the parallel processing of data on data-intensive systems.
- Programming abstractions including models, languages, and algorithms which allow a natural expression of parallel processing of data.
- Design of data-intensive computing platforms to provide high levels of reliability, efficiency, availability, and scalability.
- Identifying applications that can exploit this computing paradigm and determining how it should evolve to support emerging data-intensive applications.

Pacific Northwest National Labs has defined data-intensive computing as "capturing, managing, analyzing, and understanding data at volumes and rates that push the frontiers of current technologies" [15]. They believe that to address the rapidly growing data volumes and complexity requires "epochal advances in software, hardware, and algorithm development" which can scale readily with size of the data and provide effective and timely analysis and processing results.

Processing Approach

Current data-intensive computing platforms use a "divide and conquer" parallel processing approach combining multiple processors and disks in large computing clusters connected using high-speed communications switches and networks which allows the data to be partitioned among the available computing resources and processed independently to achieve performance and scalability based on the amount of data (Fig. 10.2). This approach to parallel processing is often referred to as a "shared nothing" approach since each node consisting of processor, local memory, and disk resources shares nothing with other nodes in the cluster. In parallel computing this approach is considered suitable for data processing problems which are "embarrassingly parallel", i.e. where it is relatively easy to separate

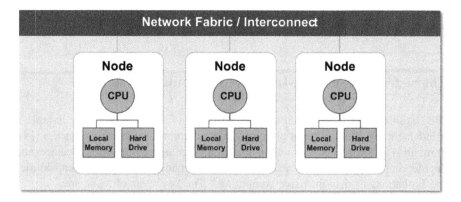

Fig. 10.2 Shared nothing computing cluster

the problem into a number of parallel tasks and there is no dependency or communication required between the tasks other than overall management of the tasks. These types of data processing problems are inherently adaptable to various forms of distributed computing including clusters and data grids.

Common Characteristics

There are several important common characteristics of data-intensive computing systems that distinguish them from other forms of computing. First is the principle of collocation of the data and programs or algorithms to perform the computation. To achieve high performance in data-intensive computing, it is important to minimize the movement of data. In direct contrast to other types of computing and supercomputing which utilize data stored in a separate repository or servers and transfer the data to the processing system for computation, data-intensive computing uses distributed data and distributed file systems in which data is located across a cluster of processing nodes, and instead of moving the data, the program or algorithm is transferred to the nodes with the data that needs to be processed. This principle—"Move the code to the data"—is extremely effective since program size is usually small in comparison to the large datasets processed by data-intensive systems and results in much less network traffic since data can be read locally instead of across the network. This characteristic allows processing algorithms to execute on the nodes where the data resides reducing system overhead and increasing performance [1].

A second important characteristic of data-intensive computing systems is the programming model utilized. Data intensive computing systems utilize a machine-independent approach in which applications are expressed in terms of high-level operations on data, and the runtime system transparently controls the scheduling, execution, load balancing, communications, and movement of programs and data across the distributed computing cluster [16]. The programming abstraction and language tools allow the processing to be expressed in terms of data flows and transformations incorporating new dataflow programming languages and shared libraries of common data manipulation algorithms such as sorting. Conventional supercomputing and distributed computing systems typically utilize machine dependent programming models which can require low-level programmer control of processing and node communications using conventional imperative programming languages and specialized software packages which adds complexity to the parallel programming task and reduces programmer productivity. A machine dependent programming model also requires significant tuning and is more susceptible to single points of failure.

A third important characteristic of data-intensive computing systems is the focus on reliability and availability. Largescale systems with hundreds or thousands of processing nodes are inherently more susceptible to hardware failures, communications errors, and software bugs. Data-intensive computing systems are designed to be fault resilient. This includes redundant copies of all data files on disk, storage of intermediate processing results on disk, automatic detection of node or processing failures, and selective re-computation of results. A processing cluster configured for data-intensive computing is typically able to continue operation with a reduced number of nodes following a node failure with automatic and transparent recovery of incomplete processing.

A final important characteristic of data-intensive computing systems is the inherent scalability of the underlying hardware and software architecture. Data-intensive computing systems can typically be scaled in a linear fashion to accommodate virtually any amount of data, or to meet time-critical performance requirements by simply adding additional processing nodes to a system configuration in order to achieve billions of records per second processing rates (BORPS). The number of nodes and processing tasks assigned for a specific application can be variable or fixed depending on the hardware, software, communications, and distributed file system architecture. This scalability allows computing problems once considered to be intractable due to the amount of data required or amount of processing time required to now be feasible and affords opportunities for new breakthroughs in data analysis and information processing.

Grid Computing

A similar computing paradigm known as grid computing has gained popularity primarily in research environments [4]. A computing grid is typically heterogeneous in nature (nodes can have different processor, memory, and disk resources), and consists of multiple disparate computers distributed across organizations and often geographically using wide-area networking communications usually with relatively low-bandwidth. Grids are typically used to solve complex computational problems which are compute-intensive requiring only small amounts of data for each processing node. A variation known as data grids allow shared repositories of data to be accessed by a grid and utilized in application processing, however the low-bandwidth of data grids limit their effectiveness for largescale data-intensive applications. In contrast, data-intensive computing systems are typically homogeneous in nature (nodes in the computing cluster have identical processor, memory, and disk resources), use high-bandwidth communications between nodes such as gigabit Ethernet switches, and are located in close proximity in a data center using high-density hardware such as rack-mounted blade servers. The logical file system typically includes all the disks available on the nodes in the cluster and data files are distributed across the nodes as opposed to a separate shared data repository such as a storage area network which would require data to be moved to nodes for processing. Geographically dispersed grid systems are more difficult to manage, less reliable, and less secure than data-intensive computing systems which are usually located in secure data center environments.

Data-Intensive System Architectures

A variety of system architectures have been implemented for data-intensive and large-scale data analysis applications including parallel and distributed relational database management systems which have been available to run on shared nothing clusters of processing nodes for more than two decades [17]. These include database systems from Teradata, Netezza, Vertica, and Exadata/Oracle and others which provide high-performance parallel database platforms. Although these systems have the ability to run parallel applications and queries expressed in the SQL language, they are typically not general-purpose processing platforms and usually run as a back-end to a separate front-end application processing system. Although this approach offers benefits when the data utilized is primarily structured in nature and fits easily into the constraints of a relational database, and often excels for transaction processing applications, most data growth is with data in unstructured form [11] and new processing paradigms with more flexible data models were needed. Internet companies such as Google, Yahoo, Microsoft, Facebook, and others required a new processing approach to effectively deal with the enormous amount of Web data for applications such as search engines and social networking. In addition, many government and business organizations were overwhelmed with data that could not be effectively processed, linked, and analyzed with traditional computing approaches.

Several solutions have emerged including the MapReduce architecture pioneered by Google and now available in an open-source implementation called Hadoop used by Yahoo, Facebook, and others. LexisNexis, an acknowledged industry leader in information services, also developed and implemented a scalable platform for data-intensive computing which is used by LexisNexis and other commercial and government organizations to process large volumes of structured and unstructured data. These approaches will be explained and contrasted in terms of their overall structure, programming model, file systems in the following sections.

Google MapReduce

The MapReduce architecture and programming model pioneered by Google is an example of a modern systems architecture designed for processing and analyzing large datasets and is being used successfully by Google in many applications to process massive amounts of raw Web data [18]. The MapReduce architecture allows programmers to use a functional programming style to create a map function that processes a key-value pair associated with the input data to generate a set of intermediate key-value pairs, and a reduce function that merges all intermediate values associated with the same intermediate key [18]. According to Dean and Ghemawat, the MapReduce programs can be used to compute derived data from documents such as inverted indexes and the processing is automatically parallelized

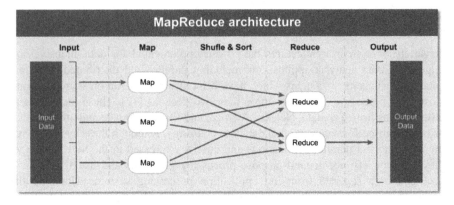

Fig. 10.3 MapReduce processing architecture [29]

by the system which executes on large clusters of commodity type machines, highly scalable to thousands of machines. Since the system automatically takes care of details like partitioning the input data, scheduling and executing tasks across a processing cluster, and managing the communications between nodes, program-mers with no experience in parallel programming can easily use a large distributed processing environment.

The programming model for MapReduce architecture is a simple abstraction where the computation takes a set of input key-value pairs associated with the input data and produces a set of output key-value pairs. The overall model for this process is shown in Fig. 10.3. In the map phase, the input data is partitioned into input splits and assigned to Map tasks associated with processing nodes in the cluster. The Map task typically executes on the same node containing its assigned partition of data in the cluster. These Map tasks perform user-specified computations on each input key-value pair from the partition of input data assigned to the task, and generates a set of intermediate results for each key. The shuffle and sort phase then takes the intermediate data generated by each Map task, sorts this data with intermediate data from other nodes, divides this data into regions to be processed by the reduce tasks, and distributes this data as needed to nodes where the Reduce tasks will execute. All Map tasks must complete prior to the shuffle and sort and reduce phases. The number of Reduce tasks does not need to be the same as the number of Map tasks. The Reduce tasks perform additional user-specified operations on the intermediate data possibly merging values associated with a key to a smaller set of values to produce the output data. For more complex data processing procedures, multiple MapReduce calls may be linked together in sequence.

Figure 10.4 shows the MapReduce architecture and key-value processing in more detail. The input data can consist of multiple input files. Each Map task will produce an intermediate output file for each key region assigned based on the number of Reduce tasks R assigned to the process (hash (key) modulus R). The reduce function then "pulls" the intermediate files, sorting and merging the files for

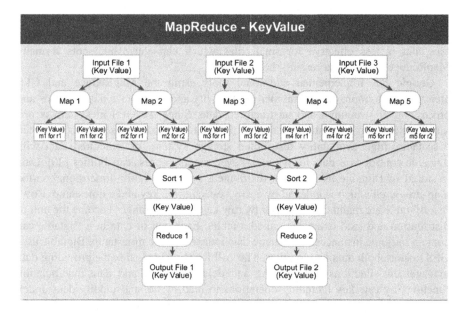

Fig. 10.4 MapReduce key-value processing [30]

a specific region from all the Map tasks. To minimize the amount of data transferred across the network, an optional Combiner function can be specified which is executed on the same node that performs a Map task. The combiner code is usually the same as the reducer function code which does partial merging and reducing of data for the local partition, then writes the intermediate files to be distributed to the Reduce tasks. The output of the Reduce function is written as the final output file. In the Google implementation of MapReduce, functions are coded in the C++ programming language.

Underlying and overlayed with the MapReduce architecture is the Google File System (GFS). GFS was designed to be a high-performance, scalable distributed file system for very large data files and data-intensive applications providing fault tolerance and running on clusters of commodity hardware [19]. GFS is oriented to very large files dividing and storing them in fixed-size chunks of 64 Mb by default which are managed by nodes in the cluster called chunkservers. Each GFS consists of a single master node acting as a nameserver and multiple nodes in the cluster acting as chunkservers using a commodity Linux-based machine (node in a cluster) running a user-level server process. Chunks are stored in plain Linux files which are extended only as needed and replicated on multiple nodes to provide high-availability and improve performance. Secondary name servers provide backup for the master node. The large chunk size reduces the need for MapReduce clients programs to interact with the master node, allows filesystem metadata to be kept in memory in the master node improving performance, and allows many operations to be performed with a single read on a chunk of data by the MapReduce client. Ideally, input splits for MapReduce operations are the size of a GFS chunk. GFS has proven to be highly effective for data-intensive computing on very large files, but is less effective for small files which can cause hot spots if many MapReduce tasks are accessing the same file.

Google has implemented additional tools using the MapReduce and GFS architecture to improve programmer productivity and to enhance data analysis and processing of structured and unstructured data. Since the GFS filesystem is primarily oriented to sequential processing of large files, Google has also implemented a scalable, high availability distributed storage system for structured data with dynamic control over data format with keyed random access capabilities [20]. Data is stored in Bigtable as a sparse, distributed, persistent multi-dimensional sorted map structured which is indexed by a row key, column, key and a timestamp. Rows in a Bigtable are maintained in order by row key, and row ranges become the unit of distribution and load balancing called a tablet. Each cell of data in a Bigtable can contain multiple instances of the same data indexed by the timestamp. Bigtable uses GFS to store both data and log files. The API for Bigtable is flexible providing data management functions like creating and deleting tables, and data manipulation functions by row key including operations to read, write, and modify data. Index information for Bigtables utilize tablet information stored in structures similar to a B + Tree. MapReduce applications can be used with Bigtable to process and transform data, and Google has implemented many large-scale applications which utilize Bigtable for storage including Google Earth.

Google has also implemented a high-level language for performing parallel data analysis and data mining using the MapReduce and GFS architecture called Sawzall and a workflow management and scheduling infrastructure for Sawzall jobs called Workqueue [21]. According to Pike et al., although C++ in standard MapReduce jobs is capable of handling data analysis tasks, it is more difficult to use and requires considerable effort by programmers. For most applications implemented using Sawzall, the code is much simpler and smaller than the equivalent C++ by a factor of 10 or more. A Sawzall program defines operations on a single record of the data, the language does not allow examining multiple input records simultaneously and one input record cannot influence the processing of another. An emit statement allows processed data to be output to an external aggregator which provides the capability for entire files of records and data to be processed using a Sawzall program. The system operates in a batch mode in which a user submits a job which executes a Sawzall program on a fixed set of files and data and collects the output at the end of a run. Sawzall jobs can be chained to support more complex procedures. Sawzall programs are compiled into an intermediate code which is interpreted during runtime execution. Pike et al. cite several reasons why a new language is beneficial for data analysis and data mining applications: (1) a programming language customized for a specific problem domain makes resulting programs "clearer, more compact, and more expressive"; (2) aggregations are specified in the Sawzall language so that the programmer does not have to provide one in the Reduce task of a standard MapReduce program; (3) a programming language oriented to data analysis provides a more natural way to think about data processing problems for large distributed datasets; and (4) Sawzall programs are significantly smaller that equivalent C++ MapReduce programs and significantly easier to program.

Hadoop

Hadoop is an open source software project sponsored by The Apache Software Foundation (http://www.apache.org). Following the publication in 2004 of the research paper describing Google MapReduce [18], an effort was begun in conjunction with the existing Nutch project to create an open source implementation of the MapReduce architecture [22]. It later became an independent subproject of Lucene, was embraced by Yahoo! after the lead developer for Hadoop became an employee, and became an official Apache top-level project in February of 2006. Hadoop now encompasses multiple subprojects in addition to the base core, MapReduce, and HDFS distributed filesystem. These additional subprojects provide enhanced application processing capabilities to the base Hadoop implementation and currently include Avro, Pig, HBase, ZooKeeper, Hive, and Chukwa. More information can be found at the Apache Web site.

The Hadoop MapReduce architecture is functionally similar to the Google implementation except that the base programming language for Hadoop is Java instead of C++. The implementation is intended to execute on clusters of commodity processors (Fig. 10.5) utilizing Linux as the operating system environment, but can also be run on a single system as a learning environment. Hadoop clusters also utilize the "shared nothing" distributed processing paradigm linking individual systems with local processor, memory, and disk resources using high-speed communications switching capabilities typically in rack-mounted configurations. The flexibility of Hadoop configurations allows small clusters to be created for testing and development using desktop systems or any system running Unix/Linux

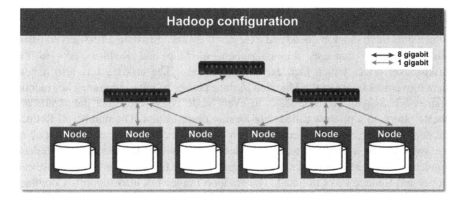

Fig. 10.5 Commodity hardware cluster [29]

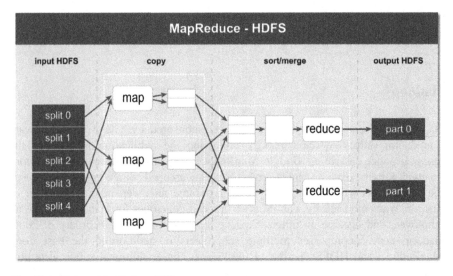

Fig. 10.6 Hadoop MapReduce [31]

providing a JVM environment, however production clusters typically use homo-geneous rack-mounted processors in a data center environment.

The Hadoop MapReduce architecture is similar to the Google implementation creating fixed-size input splits from the input data and assigning the splits to Map tasks. The local output from the Map tasks is copied to Reduce nodes where it is sorted and merged for processing by Reduce tasks which produce the final output as shown in Fig. 10.6. Hadoop implements a distributed data processing scheduling and execution environment and framework for MapReduce jobs. A MapReduce job is a unit of work that consists of the input data, the associated Map and Reduce programs, and user-specified configuration information [22]. The Hadoop frame-work utilizes a master/slave architecture with a single master server called a job-tracker and slave servers called task-trackers, one per node in the cluster. The job-tracker is the communications interface between users and the framework and coordinates the execution of MapReduce jobs. Users submit jobs to the job-tracker, which puts them in a job queue and executes them on a first-come/first-served basis. The job-tracker manages the assignment of Map and Reduce tasks to the taskt-racker nodes which then execute these tasks. The task-trackers also handle data movement between the Map and Reduce phases of job execution. The Hadoop framework assigns the Map tasks to every node where the input data splits are located through a process called data locality optimization. The number of Reduce tasks is determined independently and can be user-specified and can be zero if all of the work can be accomplished by the Map tasks. As with the Google MapReduce implementation, all Map tasks must complete before the shuffle and sort phase can occur and Reduce tasks initiated. The Hadoop framework also supports Combiner functions which can reduce the amount of data movement in a job. The Hadoop framework also provides an API called Streaming to allow Map and Reduce functions to be written in languages other than Java such as Ruby and Python and provides an interface called Pipes for C++.

Hadoop includes a distributed file system called HDFS which is analogous to GFS in the Google MapReduce implementation. A block in HDFS is equivalent to a chunk in GFS and is also very large, 64 Mb by default but 128 Mb is used in some installations. The large block size is intended to reduce the number of seeks and improve data transfer times. Each block is an independent unit stored as a dynamically allocated file in then Linux local filesystem in a datanode directory. If the node has multiple disk drives, multiple datanode directories can be specified. An additional local file per block stores metadata for the block. HDFS also follows a master/slave architecture which consists of a single master server that manages the distributed filesystem namespace and regulates access to files by clients called the Namenode. In addition, there are multiple Datanodes, one per node in the cluster, which manage the disk storage attached to the nodes and assigned to Hadoop. The Namenode determines the mapping of blocks to Datanodes. The Datanodes are responsible for serving read and write requests from filesystem clients such as MapReduce tasks, and they also perform block creation, deletion, and replication based on commands from the Namenode. An HDFS system can include additional secondary Namenodes which replicate the filesystem metadata, however there are no hot failover services. Each datanode block also has replicas on other nodes based on system configuration parameters (by default there are 3 replicas for each datanode block). In the Hadoop MapReduce execution environment it is common for a node in a physical cluster to function as both a Tasktracker and a datanode [23]. The HDFS system architecture is shown in Fig. 10.7.

The Hadoop execution environment supports additional distributed data processing capabilities which are designed to run using the Hadoop MapReduce architecture. Several of these have become official Hadoop subprojects within the Apache Software Foundation. These include HBase, a distributed column-oriented

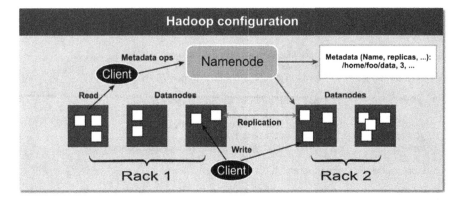

Fig. 10.7 HDFS architecture [32]

database which provides similar random access read/write capabilities as and is modeled after Bigtable implemented by Google. HBase is not relational, and does not support SQL, but provides a Java API and a command-line shell for table management. Hive is a data warehouse system built on top of Hadoop that provides SQL-like query capabilities for data summarization, ad hoc queries, and analysis of large datasets. Other Apache sanctioned projects for Hadoop include Avro—A data serialization system that provides dynamic integration with scripting languages, Chukwa—a data collection system for managing large distributed systems, ZooKeeper—a high-performance coordination service for distributed applications, and Pig—a high-level data-flow language and execution framework for parallel computation.

Pig is high-level dataflow-oriented language and execution environment origi-nally developed at Yahoo! ostensibly for the same reasons that Google developed the Sawzall language for its MapReduce implementation—to provide a specific language notation for data analysis applications and to improve programmer pro-ductivity and reduce development cycles when using the Hadoop MapReduce environment. Working out how to fit many data analysis and processing applica-tions into the MapReduce paradigm can be a challenge, and often requires multiple MapReduce jobs [22]. Pig programs are automatically translated into sequences of MapReduce programs if needed in the execution environment. In addition Pig supports a much richer data model which supports multi-valued, nested data structures with tuples, bags, and maps. Pig supports a high-level of user cus-tomization including userdefined special purpose functions and provides capabili-ties in the language for loading, storing, filtering, grouping, de-duplication, ordering, sorting, aggregation, and joining operations on the data [24]. Pig is an imperative dataflow-oriented language (language statements define a dataflow for processing). An example program is shown in Fig. 10.8. Pig runs as a client-side application which translates Pig programs into MapReduce jobs and then runs them on an Hadoop cluster. Figure 10.9 shows how the program listed in Fig. 10.8 is translated into a sequence of MapReduce jobs. Pig compilation and execution

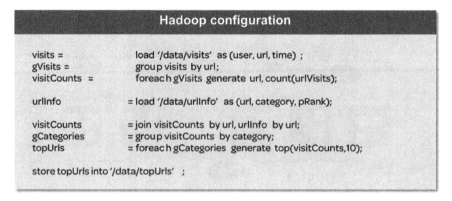

Fig. 10.8 Sample Pig Latin Program [24]

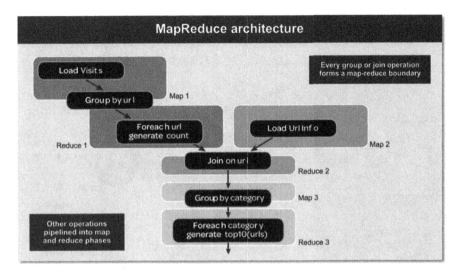

Fig. 10.9 Pig program translation to MapReduce [24]

stages include a parser, logical optimizer, MapReduce compiler, MapReduce optimizer, and the Hadoop Job Manager [25].

According to Yahoo! where more than 40 % of Hadoop production jobs and 60 % of ad hoc queries are now implemented using Pig, Pig programs are 1/20th the size of the equivalent MapReduce program and take 1/16th the time to develop [26]. Yahoo! uses 12 standard benchmarks (called the PigMix) to test Pig performance versus equivalent MapReduce performance from release to release. With the current release, Pig programs take approximately 1.5 times longer than the equivalent MapReduce (http://wiki.apache.org/pig/PigMix). Additional optimizations are being implemented that should reduce this performance gap further.

LexisNexis HPCC

LexisNexis, an industry leader in data content, data aggregation, and information services independently developed and implemented a solution for data-intensive computing called the HPCC (High-Performance Computing Cluster) which is also referred to as the Data Analytics Supercomputer (DAS). The LexisNexis vision for this computing platform is depicted in Fig. 10.10. The development of this computing platform by the Seisint subsidiary of LexisNexis began in 1999 and applications were in production by late 2000. The LexisNexis approach also utilizes commodity clusters of hardware running the Linux operating system as shown in Figs. 10.2 and 10.5. Custom system software and middleware components were developed and layered on the base Linux operating system to provide the execution

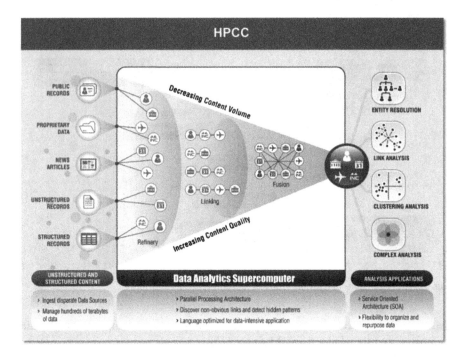

Fig. 10.10 LexisNexis vision for a data analytics supercomputer

environment and distributed filesystem support required for data-intensive computing. Because LexisNexis recognized the need for a new computing paradigm to address its growing volumes of data, the design approach included the definition of a new high-level language for parallel data processing called ECL (Enterprise Data Control Language). The power, flexibility, advanced capabilities, speed of development, and ease of use of the ECL programming language is the primary distinguishing factor between the LexisNexis HPCC and other data-intensive computing solutions. The following will provide an overview of the HPCC systems architecture and the ECL language and a general comparison to the Hadoop MapReduce architecture and platform.

LexisNexis developers recognized that to meet all the requirements of data-intensive computing applications in an optimum manner required the design and implementation of two distinct processing environments, each of which could be optimized independently for its parallel data processing purpose. The first of these platforms is called a Data Refinery whose overall purpose is the general processing of massive volumes of raw data of any type for any purpose but typically used for data cleansing and hygiene, ETL processing of the raw data (extract, transform, load), record linking and entity resolution, large-scale ad hoc analysis of data, and creation of keyed data and indexes to support high-performance structured queries and data warehouse applications. The Data Refinery is also referred to as

Thor, a reference to the mythical Norse god of thunder with the large hammer symbolic of crushing large amounts of raw data into useful information. A Thor system is similar in its function, execution environment, filesystem, and capabilities to the Hadoop MapReduce platform, but offers significantly higher performance in equivalent configurations. The second of the parallel data processing platforms designed and implemented by LexisNexis is called the Data Delivery Engine. This platform is designed as an online high-performance structured query and analysis platform or data warehouse delivering the parallel data access processing require-ments of online applications through Web services interfaces supporting thousands of simultaneous queries and users with sub-second response times. High-profile online applications developed by LexisNexis such as Accurint utilize this platform. The Data Delivery Engine is also referred to as Roxie, which is an acronym for Rapid Online XML Information Exchange. Roxie uses a special distributed indexed filesystem to provide parallel processing of queries. A Roxie system is similar in its function and capabilities to Hadoop with HBase and Hive capabilities added, but provides significantly higher throughput since it uses a more optimized execution environment and filesystem for high performance online processing. Most impor-tantly, both Thor and Roxie systems utilize the same ECL programming language for implementing applications, increasing continuity and programmer productivity.

The Thor system cluster is implemented using a master/slave approach with a single master node and multiple slave nodes for data parallel processing. Each of the slave nodes is also a data node within the distributed file system for the cluster. This is similar to the Jobtracker, Tasktracker, and Datanode concepts in an Hadoop configuration. Multiple Thor clusters can exist in an HPCC environment, and job queues can span multiple clusters in an environment if needed. Jobs executing on a Thor cluster in a multi-cluster environment can also read files from the distributed file system on foreign clusters if needed. The middleware layer provides additional server processes to support the execution environment including ECL Agents and ECL Servers. A client process submits an ECL job to the ECL Agent which coordinates the overall job execution on behalf of the client process. An ECL Job is compiled by the ECL server which interacts with an additional server called the ECL Repository which is a source code repository and contains shared ECL code. ECL programs are compiled into optimized C++ source code, which is subse-quently compiled into executable code and distributed to the slave nodes of a Thor cluster by the Thor master node. The Thor master monitors and coordinates the processing activities of the slave nodes and communicates status information monitored by the ECL Agent processes. When the job completes, the ECL Agent and client process are notified, and the output of the process is available for viewing or subsequent processing. Output can be stored in the distributed filesystem for the cluster or returned to the client process. ECL is analogous to the Pig language which can be used in the Hadoop environment.

The distributed filesystem used in a Thor cluster is record-oriented which is different from the block format used by Hadoop clusters. Records can be fixed or variable length, and support a variety of standard (fixed record size, CSV, XML) and custom formats including nested child datasets. Record I/O is buffered in large blocks to reduce latency and improve data transfer rates to and from disk Files to be loaded to a Thor cluster are typically first transferred to a landing zone from some external location, then a process called "spraying" is used to partition the file and load it to the nodes of a Thor cluster. The initial spraying process divides the file on user-specified record boundaries and distributes the data as evenly as possible in order across the available nodes in the cluster. Files can also be "desprayed" when needed to transfer output files to another system or can be directly copied between Thor clusters in the same environment. Nameservices and storage of metadata about files including record format information in the Thor DFS are maintained in a special server called the Dali server (named for the developer's pet Chinchilla), which is analogous to the Namenode in HDFS. Thor users have complete control over distribution of data in a Thor cluster, and can re-distribute the data as needed in an ECL job by specific keys, fields, or combinations of fields to facilitate the locality characteristics of parallel processing. The Dali nameserver uses a dynamic datastore for filesystem metadata organized in a hierarchical structure correspond-ing to the scope of files in the system. The Thor DFS utilizes the local Linux filesystem for physical file storage, and file scopes are created using file directory structures of the local file system. Parts of a distributed file are named according to the node number in a cluster, such that a file in a 400-node cluster will always have 400 parts regardless of the file size. The Hadoop fixed block size can end up splitting logical records between nodes which means a node may need to read some data from another node during Map task processing. With the Thor DFS, logical record integrity is maintained, and processing I/O is completely localized to the processing node for local processing operations. In addition, if the file size in Hadoop is less than some multiple of the block size times the number of nodes in the cluster, Hadoop processing will be less evenly distributed and node to node disk accesses will be needed. If input splits assigned to Map tasks in Hadoop are not allocated in whole block sizes, additional node to node I/O will result. The ability to easily redistribute the data evenly to nodes based on processing requirements and the characteristics of the data during a Thor job can provide a significant perfor-mance improvement over the Hadoop approach. The Thor DFS also supports the concept of "superfiles" which are processed as a single logical file when accessed, but consist of multiple Thor DFS files. Each file which makes up a superfile must have the same record structure. New files can be added and old files deleted from a superfile dynamically facilitating update processes without the need to rewrite a new file. Thor clusters are fault resilient and a minimum of one replica of each file part in a Thor DFS file is stored on a different node within the cluster.

Roxie clusters consist of a configurable number of peer-coupled nodes func-tioning as a high-performance, high availability parallel processing query platform. ECL source code for structured queries is pre-compiled and deployed to the cluster. The Roxie distributed filesystem is a distributed indexed-based filesystem which uses a custom B + Tree structure for data storage. Indexes and data supporting queries are pre-built on Thor clusters and deployed to the Roxie DFS with portions of the index and data stored on each node. Typically the data associated with index logical keys is embedded in the index structure as a payload. Index keys can be multi-field and multivariate, and payloads can contain any type of structured or unstructured data supported by the ECL language. Queries can use as many indexes as required for a query and contain joins and other complex transformations on the data with the full expression and processing capabilities of the ECL language. For example, the Accurint comprehensive person report which produces many pages of output is generated by a single Roxie query.

A Roxie cluster uses the concept of Servers and Agents. Each node in a Roxie cluster runs Server and Agent processes which are configurable by a System Administrator depending on the processing requirements for the cluster. A Server process waits for a query request from a Web services interface then determines the nodes and associated Agent processes that have the data locally that is needed for a query, or portion of the query. Roxie query requests can be submitted from a client application as a SOAP call, HTTP or HTTPS protocol request from a Web application, or through a direct socket connection. Each Roxie query request is associated with a specific deployed ECL query program. Roxie queries can also be executed from programs running on Thor clusters. The Roxie Server process that receives the request owns the processing of the ECL program for the query until it is completed. The Server sends portions of the query job to the nodes in the cluster and Agent processes which have data needed for the query stored locally as needed, and waits for results. When a Server receives all the results needed from all nodes, it collates them, performs any additional processing, and then returns the result set to the client requestor. The performance of query processing varies depending on factors such as machine speed, data complexity, number of nodes, and the nature of the query, but production results have shown throughput of a thousand results a second or more. Roxie clusters have flexible data storage options with indexes and data stored locally on the cluster, as well as being able to use indexes stored remotely in the same environment on a Thor cluster. Name services for Roxie clusters are also provided by the Dali server. Roxie clusters are fault-resilient and data redundancy is built-in using a peer system where replicas of data are stored on two or more nodes, all data including replicas are available to be used in the processing of queries by Agent processes. The Roxie cluster provides automatic failover in case of node failure, and the cluster will continue to perform even if one or more nodes are down. Additional redundancy can be provided by including multiple Roxie clusters in an environment.

Load balancing of query requests across Roxie clusters is typically implemented using external load balancing communications devices. Roxie clusters can be sized as needed to meet query processing throughput and response time requirements, but are typically smaller that Thor clusters. Figure 10.11 shows the various methods of accessing a Roxie cluster.

The implementation of two types of parallel data processing platforms (Thor and Roxie) in the HPCC processing environment serving different data processing needs allows these platforms to be optimized and tuned for their specific purposes to provide the highest level of system performance possible to users. This is a distinct advantage when compared to the Hadoop MapReduce platform and architecture which must be overlayed with different systems such as HBase, Hive,

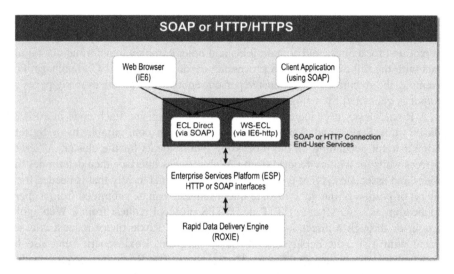

Fig. 10.11 Roxie cluster client access methods

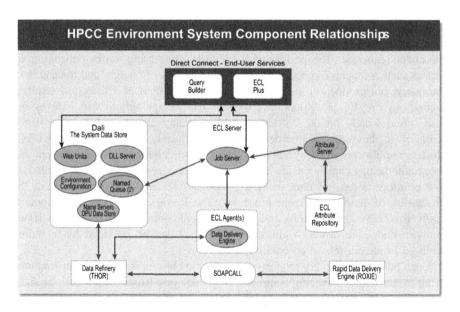

Fig. 10.12 HPCC environment system component relationships

and Pig which have different processing goals and requirements, and don't always map readily into the MapReduce paradigm. In addition, the LexisNexis HPCC approach incorporates the notion of a processing environment which can integrate Thor and Roxie clusters as needed to meet the complete processing needs of an organization. As a result, scalability can be defined not only in terms of the number of nodes in a cluster, but in terms of how many clusters and of what type are needed to meet system performance goals and user requirements. This provides a distinct advantage when compared to Hadoop clusters which tend to be independent islands of processing. The basic relationships between Thor and Roxie clusters and various middleware components of the HPCC architecture is shown in Fig. 10.12.

Programming Language ECL

The ECL programming language is a key factor in the flexibility and capabilities of the HPCC processing environment. ECL was designed to be a transparent and implicitly parallel programming language for data-intensive applications. It is a high-level, declarative, non-procedural dataflow-oriented language that allows the programmer to define what the data processing result should be and the dataflows and transformations that are necessary to achieve the result. Execution is not determined by the order of the language statements, but from the sequence of dataflows and transformations represented by the language statements. It combines data representation with algorithm implementation, and is the fusion of both a query language and a parallel data processing language. ECL uses an intuitive syntax which has taken cues from other familiar languages, supports modular code organization with a high degree of reusability and extensibility, and supports high-productivity for programmers in terms of the amount of code required for typical applications compared to traditional languages like Java and C++. Similar to the benefits Sawzall provides in the Google environment, and Pig provides to Hadoop users, a 20 times increase in programmer productivity is typical. ECL is compiled into optimized C++ code for execution on the HPCC system platforms, and can be used for complex data processing and analysis jobs on a Thor cluster or for comprehensive query and report processing on a Roxie cluster. ECL allows inline C++ functions to be incorporated into ECL programs, and external programs in other languages can be incorporated and parallelized through a PIPE facility. External services written in C++ and other languages which generate DLLs can also be incorporated in the ECL system library, and ECL programs can access external Web services through a standard SOAPCALL interface.

The basic unit of code for ECL is called an attribute. An attribute can contain a complete executable query or program, or a shareable and reusable code fragment such as a function, record definition, dataset definition, macro, filter definition, etc. Attributes can reference other attributes which in turn can reference other attributes so that ECL code can be nested and combined as needed in a reusable manner. Attributes are stored in ECL code repository which is subdivided into modules typically associated with a project or process. Each ECL attribute added to the repository effectively extends the ECL language like adding a new word to a dictionary, and attributes can be reused as part of multiple ECL queries and programs. With ECL a rich set of programming tools is provided including an interactive IDE similar to Visual C++, Eclipse and other code development environments.

The ECL language includes extensive capabilities for data definition, filtering, data management, and data transformation, and provides an extensive set of built-in functions to operate on records in datasets which can include user-defined transformation functions. Transform functions operate on a single record or a pair of records at a time depending on the operation. Built-in transform operations in the ECL language which process through entire datasets include PROJECT, ITERATE, ROLLUP, JOIN, COMBINE, FETCH, NORMALIZE, DENORMALIZE, and PROCESS. The transform function defined for a JOIN operation for example receives two records, one from each dataset being joined, and can perform any operations on the fields in the pair of records, and returns an output record which can be completely different from either of the input records. Example syntax for the JOIN operation from the ECL Language Reference Manual is shown in Fig. 10.13. Other important data

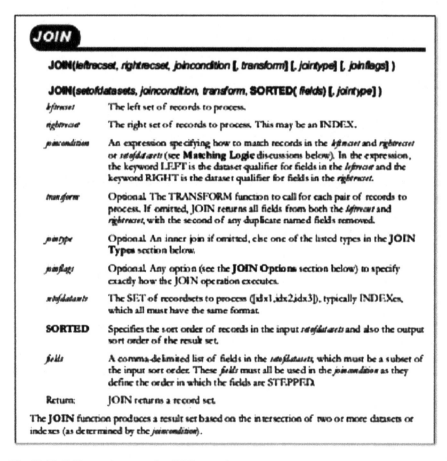

Fig. 10.13 ECL sample syntax for JOIN operation

operations included in ECL which operate across datasets and indexes include TABLE, SORT, MERGE, MERGEJOIN, DEDUP, GROUP, APPLY, ASSERT, AVE, BUILD, BUILDINDEX, CHOOSESETS, CORRELATION, COUNT, COVARIANCE, DISTRIBUTE, DISTRIBUTION, ENTH, EXISTS, GRAPH, HAVING, KEYDIFF, KEYPATCH, LIMIT, LOOP, MAX, MIN, NONEMPTY, OUTPUT, PARSE, PIPE, PRELOAD, PULL, RANGE, REGROUP, SAMPLE, SET, SOAPCALL, STEPPED, SUM, TOPN, UNGROUP, and VARIANCE.

The Thor system allows data transformation operations to be performed either locally on each node independently in the cluster, or globally across all the nodes in a cluster, which can be user-specified in the ECL language. Some operations such as PROJECT for example are inherently local operations on the part of a distributed file stored locally on a node. Others such as SORT can be performed either locally or globally if needed. This is a significant difference from the MapReduce architecture in which Map and Reduce operations are only performed locally on the input split assigned to the task. A local SORT operation in an HPCC cluster would sort the records by the specified key in the file part on the local node, resulting in the records being in sorted order on the local node, but not in full file order spanning all nodes. In contrast, a global SORT operation would result in the full distributed file being in sorted order by the specified key spanning all nodes. This requires node to node data movement during the SORT operation. Figure 10.14 shows a sample ECL program using the LOCAL mode of operation which is the equivalent of the sample PIG program for Hadoop shown in Fig. 10.8. Note the explicit programmer control over distribution of data across nodes. The colon-equals ":=" operator in an ECL program is read as "is defined as". The only action in this program is the OUTPUT statement, the other statements are definitions.

An additional important capability provided in the ECL programming language is support for natural language processing with PATTERN statements and the built-in PARSE operation. PATTERN statements allow matching patterns including regular expressions to be defined and used to parse information from unstructured data such as raw text. PATTERN statements can be combined to implement complex parsing operations or complete grammars from BNF

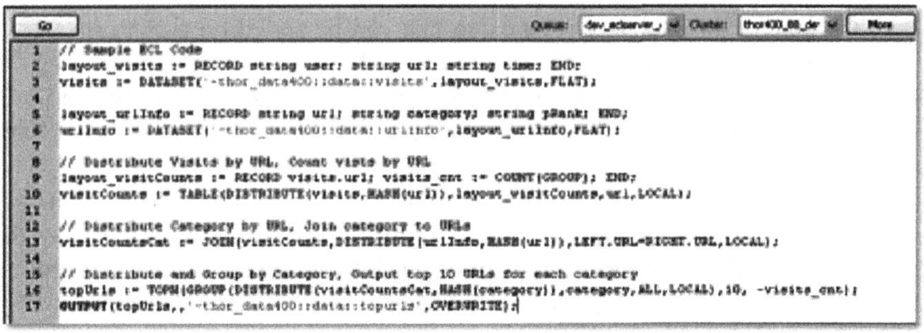

Fig. 10.14 ECL code example

definitions. The PARSE operation operates across a dataset of records on a specific field within a record, this field could be an entire line in a text file for example. Using this capability of the ECL language is possible to implement parallel processing form information extraction applications across document files including XML-based documents or Web pages. The key benefits of ECL can be summarized as follows:

- ECL incorporates transparent and implicit data parallelism regardless of the size of the computing cluster and reduces the complexity of parallel programming increasing the productivity of application developers.
- ECL enables implementation of data-intensive applications with huge volumes of data previously thought to be intractable or infeasible. ECL was specifically designed for manipulation of data and query processing. Order of magnitude performance increases over other approaches are possible.
- ECL provides a comprehensive IDE and programming tools that provide a highly-interactive environment for rapid development and implementation of ECL applications.
- ECL is a powerful, high-level, parallel programming language ideal for implementation of ETL, Information Retrieval, Information Extraction, and other data-intensive applications.
- ECL is a mature and proven language but still evolving as new advancements in parallel processing and data-intensive computing occur.

Hadoop Versus HPCC Comparison

Hadoop and HPCC can be compared directly since it is possible for both systems to be executed on identical cluster hardware configurations. This permits head-to-head system performance benchmarking using a standard workload or set of application programs designed to test the parallel data processing capabilities of each system. Currently the only standard benchmark available for data-intensive computing platforms is the Terasort benchmark managed by an industry group led by Microsoft and HP. The Terabyte sort has evolved to be the GraySort which measures the number of terabytes per minute that can be sorted on a platform which allows clusters with any number of nodes to be utilized. However, in comparing the effectiveness and equivalent cost/performance of systems, it is useful to run benchmarks on identical system hardware configurations. A head-to-head comparison of the original Terabyte sort on a 400-node cluster will be presented here. An additional method of comparing system platforms is a feature and functionality comparison, which is a subjective evaluation based on factors determined by the evaluator. Although such a comparison contains inherent bias, it is useful in determining strengths and weaknesses of systems.

Terabyte Sort Benchmark

The Terabyte sort benchmark has its roots in benchmark tests sorting conducted on computer systems since the 1980s. More recently, a Web site originally sponsored by Microsoft and one of its research scientists Jim Gray has conducted formal competitions each year with the results presented at the SIGMOD (Special Interest Group for Management of Data) conference sponsored by the ACM each year (http://sortbenchmark.org). Several categories for sorting on systems exist including the Terabyte sort which was to measure how fast a file of 1 Terabyte of data formatted in 100 byte records (10,000,000 total records) could be sorted. Two categories were allowed called Daytona (a standard commercial computer system and software with no modifications) and Indy (a custom computer system with any type of modification). No restrictions existed on the size of the system so the sorting benchmark could be conducted on as large a system as desired. The current 2009 record holder for the Daytona category is Yahoo! using a Hadoop configuration with 1460 nodes with 8 GB Ram per node, 8000 Map tasks, and 2700 Reduce tasks which sorted 1 TB in 62 s [27]. In 2008 using 910 nodes, Yahoo! performed the benchmark in 3 min 29 s. In 2008, LexisNexis using the HPCC architecture on only a 400node system performed the Terabyte sort benchmark in 3 min 6 s. In 2009, LexisNexis again using only a 400-node configuration performed the Terabyte sort benchmark in 102 s.

However, a fair and more logical comparison of the capability of data-intensive computer system and software architectures using computing clusters would be to conduct this benchmark on the same hardware configuration. Other factors should also be evaluated such as the amount of code required to perform the benchmark which is a strong indication of programmer productivity, which in itself is a significant performance factor in the implementation of data-intensive computing applications.

On August 8, 2009 a Terabyte Sort benchmark test was conducted on a development configuration located at LexisNexis Risk Solutions offices in Boca Raton, FL in conjunction with and verified by Lawrence Livermore National Labs (LLNL). The test cluster included 400 processing nodes each with two local 300 MB SCSI disk drives, Dual Intel Xeon single core processors running at 3.00 GHz, 4 GB memory per node, all connected to a single Gigabit ethernet switch with 1.4 TB/s throughput. Hadoop Release 0.19 was deployed to the cluster and the standard Terasort benchmark written in Java included with the release was used for the benchmark. Hadoop required 6 min 45 s to create the test data, and the Terasort benchmark required a total of 25 min 28 s to complete the sorting test as shown in Fig. 10.15. The HPCC system software deployed to the same platform and using standard ECL required 2 min and 35 s to create the test data, and a total

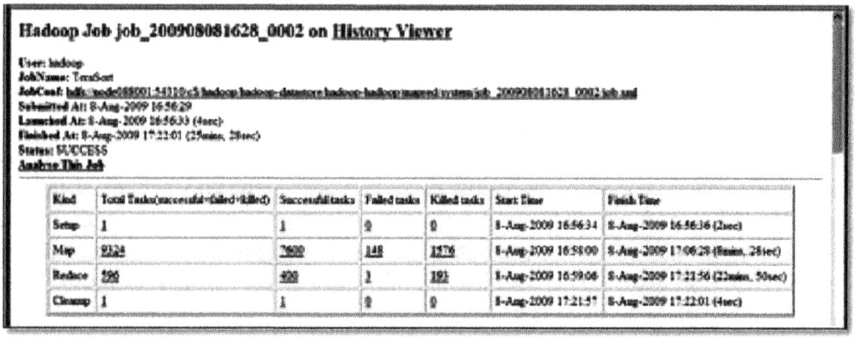

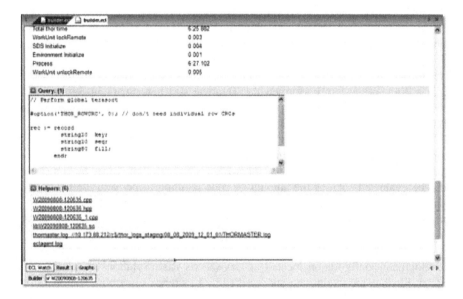

Fig. 10.15 Hadoop terabyte sort benchmark results

Fig. 10.16 HPCC terabyte sort benchmark results

of 6 min and 27 s to complete the sorting test as shown in Fig. 10.16. Thus the Hadoop implementation using Java running on the same hardware configuration took 3.95 times longer than the HPCC implementation using ECL.

The Hadoop version of the benchmark used hand-tuned Java code including custom TeraSort, TeraInputFormat and TeraOutputFormat classes with a total of 562 lines of code required for the sort. The HPCC system required only 10 lines of ECL code for the sort, a 50-times reduction in the amount of code required.

Pig Versus ECL

Although many Hadoop installations implement applications directly in Java, the Pig Latin language is now being used to increase programmer productivity and further simplify the programming of data-intensive applications at Yahoo! and other major users of Hadoop [25]. Google also added a high-level language for similar reasons called Sawzall to its implementation of MapReduce to facilitate data analysis and data mining [21]. The HPCC platform includes a high-level language discussed previously which is analogous to Pig and Sawzall called ECL. ECL is the base programming language used for applications on the HPCC platform even though it is compiled into C++ for execution. When comparing the Hadoop and HPCC platforms, it is useful to compare the features and functionality of these high-level languages.

Both Pig and ECL are intrinsically parallel, supporting transparent data-parallelism on the underlying platform. Pig and ECL are translated into programs that automatically process input data for a process in parallel with data distributed across a cluster of nodes. Programmers of both languages do not need to know the underlying cluster size or use this to accomplish data-parallel execution of jobs. Both Pig and ECL are dataflow-oriented, but Pig is an imperative programming language and ECL is a declarative programming language. A declarative language allows programmers to focus on the data transformations required to solve an application problem and hides the complexity of the underlying platform and implementation details, reduces side effects, and facilitates compiler optimization of the code and execution plan. An imperative programming language dictates the control flow of the program which may not result in an ideal execution plan in a parallel environment. Declarative programming languages allow the programmer to specify "what" a program should accomplish, instead of "how" to accomplish it. For more information, refer to the discussions of declarative (http://en.wikipedia.org/wiki/Declarative_programming) and imperative (http://en.wikipedia.org/wiki/Imperative_programming) programming languages on Wikipedia.

The source code for both Pig and ECL is compiled or translated into another language—Pig source programs are translated into Java language MapReduce jobs for execution and ECL programs are translated into C++ source code which is then compiled into a DLL for execution. Pig programs are restricted to the MapReduce architecture and HDFS of Hadoop, but ECL has no fixed framework other than the DFS (Distributed File System) used for HPCC and therefore can be more flexible in implementation of data operations. This is evident in two key areas: (1) ECL allows operations to be either global or local, where standard MapReduce is restricted to local operations only in both the Map and Reduce phases. Global operations process the records in a dataset in order across all nodes and associated file parts in sequence maintaining the records in sorted order as opposed to only the records contained in each local node which may be important to the data processing procedure; (2) ECL has the flexibility to implement operations which can process more than one record at a time such as its ITERATE operation which uses a sliding window and passes two records at a time to an associated transform function. This allows inter-record field-by-field dependencies and decisions which are not available in Pig. For example the DISTINCT operation in Pig which is used to remove duplicates does not allow this on a subset of fields. ECL provides both DEDUP and ROLLUP operations which are usually preceded by a SORT and operate on adjacent records in a sliding window mode and any condition relating to the field contents of the left and right record of adjacent records can be used to determine if the record is removed. ROLLUP allows a custom transformation to be applied to the de-duplication process.

An important consideration of any software architecture for data is the underlying data model. Pig incorporates a very flexible nested data model which allows non-atomic data types (atomic data types include numbers and strings) such as set, map, and tuple to occur as fields of a table [28]. Tuples are sequences of fields, bags are collections of tuples, and maps are a collection of data items where each data item has a key with which it can be looked up. A data record within Pig is called a relation which is an outer bag, the bag is a collection of tuples, each tuple is an ordered set of fields, and a field is a piece of data. Relations are referenced by a name assigned by a user. Types can be assigned by the user to each field, but if not assigned will default to a bytearray and conversions are applied depending on the context in which the field is used. The ECL data model also offers a nested data structure using child datasets. A user-specified RECORD definition defines the content of each record in a dataset which can contain fixed or variable length fields or child datasets which in turn contain fields or child datasets etc. With this format any type of data structure can be represented. ECL offers specific support for CSV and XML formats in addition to flat file formats. Each field in a record has a user-specified identifier and data type and an optional default value and optional field modifiers such as MAXLENGTH that enhance type and use checking during compilation. ECL will perform implicit casting and conversion depending on the context in which a field is used, and explicit user casting is also supported. ECL also allows in-line datasets allowing sample data to be easily defined and included in the code for testing rather than separately in a file.

The Pig environment offers several programmer tools for development, execution, and debugging of Pig Latin programs (Pig Latin is the formal name for the language, and the execution environment is called Pig, although both are commonly referred to as Pig). Pig provides command line execution of scripts and an interactive shell called Grunt that allows you to execute individual Pig commands or execute a Pig script. Pig programs can also be embedded in Java programs. Although Pig does not provide a specific IDE for developing and executing PIG programs, add-ins are available for several program editing environments including Eclipse, Vim, and Textmate to perform syntax checking and highlighting [22]. PigPen is an Eclipse plug-in that provides program editing, an example data generator, and the capability to run a Pig script on a Hadoop cluster. The HPCC platform provides an extensive set of tools for ECL development including a comprehensive IDE called QueryBuilder which allows program editing, execution, and interactive graph visualization for debugging and profiling ECL programs. The common code repository tree is displayed in QueryBuilder and tools are provided for source control, accessing and searching the repository. ECL jobs can be launched to an HPCC environment or specific cluster, and execution can be monitored directly from QueryBuilder. External tools are also provided including ECLWatch which provides complete access to current and historical work units (jobs executed in the HPCC environment are packaged into work units), queue management and monitoring, execution graph visualization, distributed filesystem utility functions, and system performance monitoring and analysis.

Although Pig Latin and the Pig execution environment provide a basic high-level language environment for dataintensive processing and analysis and increases the productivity of developers and users of the Hadoop MapReduce environment, ECL is a significantly more comprehensive and mature language that generates highly optimized code, offers more advanced capabilities in a robust, proven, integrated data-intensive processing architecture. Table 10.1 provides a feature to feature comparison between the Pig and ECL languages and their execution environments.

Architecture Comparison

Hadoop MapReduce and the LexisNexis HPCC platform are both scalable architectures directed towards dataintensive computing solutions. Each of these system platforms has strengths and weaknesses and their overall effectiveness for any application problem or domain is subjective in nature and can only be determined through careful evaluation of application requirements versus the capabilities of the solution. Hadoop is an open source platform which increases its flexibility and adaptability to many problem domains since new capabilities can be readily added by users adopting this technology. However, as with other open source platforms, reliability and support can become issues when many different users are contributing new code and changes to the system. Hadoop has found favor with many large Web-oriented companies including Yahoo!, Facebook, and others where data-intensive computing capabilities are critical to the success of their business. A company called Cloudera was recently formed to provide training, support and consulting services to the Hadoop user community and to provide packaged and tested releases. Although many different application tools have been built on top of the Hadoop platform like Pig, HBase, Hive, etc., these tools tend not to be well-integrated offering different command shells, languages, and operating characteristics that make it more difficult to combine capabilities in an effective manner.

However, Hadoop offers many advantages to potential users of open source software including readily available online software distributions and documentation, easy installation, flexible configurations based on commodity hardware, an execution environment based on a proven MapReduce computing paradigm, ability to schedule jobs using a configurable number of Map and Reduce tasks, availability of add-on capabilities such as Pig, HBase, and Hive to extend the capabilities of the

Table 10.1 Pig versus ECL feature comparison

Language feature or capability	Pig	ECL
Language type	Data-flow oriented, imperative, parallel language for data-intensive computing. All Pig statements perform actions in sequentially ordered steps. Pig programs define a sequence of actions on the data	Data-flow oriented, declarative, non-procedural, parallel language for data-intensive computing. Most ECL statements are definitions of the desired result which allows the execution plan to be highly optimized by the compiler. ECL actions such as OUTPUT cause execution of the data flows to produce the result defined by the ECL program
Complier	Translated into a sequence of MapReduce Java programs for execution on a Hadoop Cluster. Runs as a client application	Compiled and optimized into C++ source code which is compiled into DLL for execution on an HPCC cluster. Runs as a server application
User-defined functions	Written in Java to perform custom processing and transformations as needed in Pig language statements. REGISTER is used to register a JAR file so that UDFs can be used	Processing functions or TRANSFORM functions are written in ECL. ECL supports inline C++ in functions and external Services compiled into DLL libraries written in any language
Macros	Not supported	Extensive support for ECL macros to improve code reuse of common procedures. Additional template language for use in macros provides unique naming and conditional capabilities
Data model	Nested data model with named relations to define data records. Relations can include nested combinations of bags, tuples, and fields. Atomic data types include int, long, float, double, chararray, bytearray, tuple, bag, and map. If types not specified, default to bytearray then converted during expressions evaluation depending on the context as needed	Nested data model using child datasets. Datasets contain fields or child datasets containing fields or additional child datasets. Record definitions describe the fields in datasets and child datasets. Indexes are special datasets supporting keyed access to data. Data types can be specified for fields in record definitions and include Boolean, integer, real, decimal, string, qstring, Unicode, data, varstring, varunicode, and related operators including set of (type), typeof (expression) and recordof(dataset) and ENUM (enumeration). Explicit type casting is available and implicit type casting may occur during evaluation of expressions by ECL depending on the context. Type transfer between types is also supported. All datasets can have an associated filter express to include only records which meet the filter condition, in ECL a filtered physical dataset is called a recordset

(continued)

Table 10.1 (continued)

Language feature or capability	Pig	ECL
Distribution of data	Controlled by Hadoop MapReduce architecture and HDFS, no explicit programmer control provided. PARALLEL allows number of Reduce tasks to be specified. Local operations only are supported, global operations require custom Java MapReduce programs	Explicit programmer control over distribution of data across cluster using DISTRIBUTE function. Helps avoid data skew. ECL supports both local (operations are performed on data local to node) and global (operations performed across nodes) modes
Operators	Standard comparison operators; standard arithmetic operators and modulus division, Boolean operators AND, OR, NOT; null operators (is null, is not null); dereference operators for tuples and maps; explicit cast operator; minus and plus sign operators; matches operator	Supports arithmetic operators including normal division, integer division, and modulus division; bitwise operators for AND, OR, and XOR; standard comparison operators; Boolean operators NOT, AND, OR; explicit cast operator; minus and plus sign operators; set and record set operators; string concatenation operator; sort descending and ascending operator; special operators IN, BETWEEN, WITHIN
Conditional expression evaluation	The bincond operator is provided (condition? true_value: false_value)	ECL includes an IF statement for single expression conditional evaluation, and MAP, CASE, CHOOSE, WHICH, and REJECTED for multiple expression evaluation. The ASSERT statement can be used to test a condition across a dataset. EXISTS can be used to determine if records meeting the specified condition exist in a dataset. ISVALID determines if a field contains a valid value
Program loops	No capability exists other than the standard relation operations across a dataset. FOREACH … GENERATE provides nested capability to combine specific relation operations	In addition to built-in data transform functions, ECL provides LOOP and GRAPH statements which allow looping of dataset operations or iteration of a specified process on a dataset until a loopfilter condition is met or a loopcount is satisfied
Indexes	Not supported directly by Pig. HBase and Hive provide indexed data capability for Hadoop MapReduce which are accessible through custom user-defined functions in Pig	Indexes can be created on datasets to support keyed access to data to improve data processing performance and for use on the Roxie data delivery engine for query applications
Language statement types	Grouped into relational operators, diagnostic operators, UDF (user-defined function) statements, Eval functions, and load/store functions. The Grunt shell offers additional interactive file commands	Grouped into dataset, index and record definitions, built-in functions to define processing and dataflows, and actions which trigger execution. Functions include transform functions such as JOIN which operate on data records, and aggregation functions such as SUM. Action statements result in execution based on specified ECL definitions describing the dataflows and results for a process

(continued)

Table 10.1 (continued)

Language feature or capability	Pig	ECL
External program calls	PIG includes the STREAM statement to send data to an external script or program. The SHIP statement can be used to ship program binaries, jar files, or data to the Hadoop cluster compute nodes. The DEFINE statement, with INPUT, OUTPUT, SHIP, and CACHE clauses allow functions and commands to be associated with STREAM to access external programs	ECL includes PIPE option on DATASET and OUTPUT and a PIPE function to execute external 3rd-party programs in parallel on nodes across the cluster. Most programs which receive an input file and parameters can adapted to run in the HPCC environment
External web services access	Not supported directly by the Pig language. User-defined functions written in Java can provide this capability	Built-in ECL function SOAPCALL for SOAP calls to access external Web Services. An entire dataset can be processed by a single SOAPCALL in an ECL program
Data aggregation	Implemented in Pig using the GROUP, and FOREACH ... GENERATE statements performing EVAL functions on fields. Built-in EVAL functions include AVG, CONCAT, COUNT, DIFF, ISEMPTY, MAX, MIN, SIZE, SUM, TOKENIZE	Implemented in ECL using the TABLE statement with group by fields specified and an output record definition that includes computed fields using expressions with aggregation functions performed across the specified group. Built-in aggregation functions which work across datasets or groups include AVE, CORRELATION, COUNT, COVARIANCE, MAX, MIN, SUM, VARIANCE
Natural language processing	The TOKENIZE statement splits a string and outputs a bag of words. Otherwise no direct language support for parsing and other natural language processing. User-defined functions are required	Includes PATTERN, RULE, TOKEN, and DEFINE statements for defining parsing patterns, rules, and grammars. Patterns can include regular expression definitions and user-defined validation functions. The PARSE statement provides both regular expression type parsing or Tomita parsing capability and recursive grammars. Special parsing syntax is included specifically for XML data
Scientific function support	Not supported directly by the Pig language. Requires the definition and use of a userdefined function	ECL provides built-in functions for ABS, ACOS, ASIN, ATAN, ATAN2, COS, COSH, EXP, LN, LOG, ROUND, ROUNDUP,SIN, SINH, SQRT, TAN, TANH
Hashing functions for dataset distribution	No explicit programmer control for dataset distribution. PARALLEL option on relational operations allows the number of Reduce tasks to be specified	Hashing functions available for use with the DISTRIBUTE statement include HASH, HASH32 (32-bit FNV), HASH64 (64-bit FNV), HASHCRC, HASHMD5 (128-bit MD5)

(continued)

Table 10.1 (continued)

Language feature or capability	Pig	ECL
Creating sample datasets	The SAMPLE operation selects a random data sample with a specified sample size	ECL provides ENTH which selects every nth record of a dataset, SAMPLE which provides the capability to select non-overlapping samples on a specified interval, CHOOSEN which selects the first n records of a dataset and CHOOSESETS which allows multiple conditions to be specified and the number of records that meet the condition or optionally a number of records that meet none of the conditions specified. The base dataset for each of the ENTH, SAMPLE, CHOOSEN, and CHOOSETS can have a associated filter expression
Workflow management	No language statements in Pig directly affect Workflow. The Hadoop cluster does allow Java MapReduce programs access to specific workflow information and scheduling options to manage execution	Workflow Services in ECL include the CHECKPOINT and PERSIST statements allow the dataflow to be captured at specific points in the execution of an ECL program. If a program must be rerun because of a cluster failure, it will resume at last Checkpoint which is deleted after completion. The PERSIST files are stored permanently in the filesystem. If a job is repeated, persisted steps are only recalculated if the code has changed, or any underlying data has changed. Other workflow statements include FAILURE to trap expression evaluation failures, PRIORITY, RECOVERY, STORED, SUCCESS, WHEN for processing events, GLOBAL and INDEPENDENT
PIG relation operations		
COGROUP	The COGROUP operation is similar to the JOIN operation and groups the data in two or more relations (datasets) based on common field values. COGROUP creates a nested set of output tuples while JOIN creates a flat set of output tuples. INNER and OUTER joins are supported. Fields from each relation are specified as the join key. No support exists for conditional processing other than field equality	In ECL, this is accomplished using the DENORMALIZE function joining to each dataset and adding all records matching the join key to a new record format with a child dataset for each child file. The DENORMALIZE function is similar to a JOIN and is used to form a combined record out of a parent and any number of children
CROSS	Creates the cross product of two or more relations (datasets)	In ECL the JOIN operation can be used to create cross products using a join condition that is always true

(continued)

Table 10.1 (continued)

Language feature or capability	Pig	ECL
DISTINCT	Removes duplicate tuples in a relation. All fields in the tuple must match. The tuples are sorted prior to this operation. Cannot be used on a subset of fields. A FOREACH … GENERATE statement must be used to generate the fields prior to a DISTINCT operation in this case	The ECL DEDUP statement compares adjacent records to determine if a specified conditional expression is met, in which case the duplicate record is dropped and the remaining record is compared to the next record in a sliding window manner. This provides a much more flexible deduplication capability than the Pig DISTINCT operation. A SORT is required prior to a DEDUP unless using the ALL option. Conditions can use any expression and can reference values from the left and right adjacent records. DEDUP can use any subset of fields
DUMP	Displays the contents of a relation	ECL provides an OUTPUT statement that can either write files to the filesystem or for display. Display files can be named and are stored in the Workunit associated with the job. Workunits are archived on a management server in the HPCC platform
FILTER	Selects tuples from a relation based on a condition. Used to select the data you want or conversely to filter out remove the data you don't want	Filter expressions can be used any time a dataset or recordset is referenced in any ECL statement with the filter expression in parenthesis following the dataset name as dataset_name (filter_ expression). The ECL compiler optimizes filtering of the data during execution based on the combination of filtering expressions
ROREACH… GENERATE	Generates data transformations based on columns of data. This action can be used for projection, aggregation, and transformation, and can include other operations in the generation clause such as FILTER, DISTINCT, GROUP, etc.	Each ECL transform operation such as PROJECT, JOIN, ROLLUP, etc. include a TRANSFORM function which implicitly provides the FOREACH … GENERATE operation as records are processed by the TRANSFORM function. Depending on the function, the output record of the transform can include fields from the input and computed fields selectively as needed and does not have to be identical to the input record
GROUP	Groups together the tuples in a single relation that have the same group key fields	The GROUP operation in ECL fragments a dataset into a set of sets based on the break criteria which is a list of fields or expressions based on fields in the record which function as the group by keys. This allows aggregations and transform operations such as ITERATE, SORT, DEDUP, ROLLUP and others to occur within defined subsets of the data as it executes on each subset individually

(continued)

Table 10.1 (continued)

Language feature or capability	Pig	ECL
JOIN	Joins two or more relations based on common field values. The JOIN operator always performs an inner join. If one relation is small and can be held in memory, the "replicated" option can be used to improve performance	The ECL JOIN operation works on two datasets or a set of datasets. For two datasets INNER, FULL OUTER, LEFT OUTER, RIGHT OUTER, LEFT ONLY and RIGHT ONLY joins are permitted. For the set of datasets JOIN, INNER, LEFT OUTER, LEFT ONLY, and MOFN(min, max) joins are permitted. Any type of conditional expression referencing fields in the datasets to be joined can be used as a join condition. JOIN can be used in both a global and local modes also provides additional options for distribution including HASH which distributes the datasets by the specified join keys, and LOOKUP which copies one dataset if small to all nodes and is similar to the "replicated" join feature of Pig. Joins can also use keyed indexes to improve performance and self-joins (joining the same dataset to itself) is supported. Additional join-type operations provided by ECL include MERGEJOIN which joins and merges in a single operation, and smart stepping using STEPPED which provides a method of doing n-ary join/merge-join operations
LIMIT	Used to limit the number of output tuples in a relation. However, there is no guarantee of which tuples will be output unless preceded by an ORDER statement	The LIMIT function in ECL is to restrict the output of a recordset resulting from processing to a maximum number or records, or to fail the operation if the limit is exceeded. The CHOOSEN function can be use to select a specified number of records in a dataset
LOAD	Loads data from the filesystem	Since ECL is declarative, the equivalent of the Pig LOAD operation is a DATASET definition which also includes a RECORD definition. The examples shown in Figs. 10.8 and 10.14 demonstrate this difference
ORDER	Sorts a relation based on one or more fields. Both ascending and descending sorts are supported. Relations will be in order for a DUMP, but if the result of an ORDER is further processed by another relation operation, there is no guarantee the results will be processed in the order specified. Relations are considered to be unordered in Pig	The ECL SORT function sorts a dataset according to a list of expressions or key fields. The SORT can be global in which the dataset will be ordered across the nodes in a cluster, or local in which the dataset will be ordered on each node in the cluster individually. For grouped datasets, the SORT applies to each group individually. Sorting operations can be performed using a quicksort, insertionsort, or heapsort, and can be stable or unstable for duplicates

(continued)

Table 10.1 (continued)

Language feature or capability	Pig	ECL
SPLIT	Partitions a relation into two or more relations	Since ECL is declarative, partitions are created by simply specifying filter expressions on the base dataset. Example for dataset DS1, you could define DS2 : = DS1(filter_expression _1), DS3 : = DS1(filter_ expression _2), etc.
STORE	Stores data to the file system	The OUTPUT function in ECL is used to write a dataset to the filesystem or to store it in the workunit for display. Output files can be compressed using LZW compression. Variations of OUTPUT support flat file, CSV, and XML formats. Output can also be written to a PIPE as the standard input to the command specified for the PIPE operation. Output can write not only the filesystem on the local cluster, but to any cluster filesystem in the HPCC processing environment
UNION	The UNION operator is used to merge the contents of two or more relations into a single relation. Order of tuples is not preserved, both input and output relations are interpreted as an unordered bag of tuples. Does not eliminate duplicate tuples	The MERGE function returns a single dataset or index containing all the datasets or indexes specified in a list of datasets. Datasets must have the same record format. A SORTED option allows the merge to be ordered according to a field list that specifies the sort order. A DEDUP option causes only records with unique keys to be included. The REGROUP function allows multiple datasets which have been grouped using the same fields to be merged into a single dataset
Additional ECL transformation functions		ECL includes many additional functions providing important data transformations that are not available in Pig without implementing custom user-defined processing
COMBINE	Not available	The COMBINE function combines two datasets into a single dataset on a record-by-record basis in the order in which they appear in each. Records from each are passed to the specified transform function, and the record format of the output dataset can contain selected fields from both input datasets and additional fields as needed

(continued)

Table 10.1 (continued)

Language feature or capability	Pig	ECL
FETCH	Not available	The FETCH function processes through all the records in an index dataset in the order specified by the index fetching the corresponding record from the base dataset and passing it through a specified transform function to create a new dataset
ITERATE	Not available	The ITERATE function processes through all records in a dataset one pair of records at a time using a sliding window method performing the transform record on each pair in turn. If the dataset is grouped, the ITERATE processes each group individually. The ITERATE function is useful in propagating information and calculating new information such as running totals since it allows inter-record dependencies to be considered
NORMALIZE	Use of FOREACH … GENERATE is required	The NORMALIZE function normalizes child records out of a dataset into a separate dataset. The associated transform and output record format does not have to be the same as the input
PROCESS	Not available	The PROCESS function is similar to ITERATE and processes through all records in a dataset one pair of records at a time (left record, right record) using a sliding window method performing the associated transform function on each pair of records in turn. A second transform function is also specified that constructs the right record for the next comparison
PROJECT	Use of FOREACH … GENERATE is required	The PROJECT processes through all the records in a dataset performing the specified transform on each record in turn
ROLL UP	Not available	The ROLLUP function is similar to the DEDUP function but includes a specified transform function to process each pair of duplicate records. This allows you to retrieve and use valuable information from the duplicate record before it is thrown away. Depending on how the ROLLUP is defined, either the left or right record passed to the transform can be retained, or any mixture of data from both

(continued)

Table 10.1 (continued)

Language feature or capability	Pig	ECL
Diagnostic operators	Pig includes diagnostic operators to aid in the visualization of data structures. The DESCRIBE operator returns the schema of a relation. The EXPLAIN operator allows you to review the logical, physical, and MapReduce execution plans that are used to compute an operation in a Pig script. The ILLUSTRATE operator displays a step-by-step execution of a sequence of statements allow you to see how data is transformed through a sequence of Pig Latin statements essentially dumping the output of each statement in the script	The DISTRIBUTION action produces a crosstab report in XML format indicating how many records there are in a dataset for each value in each field in the dataset to aid in the analysis of data distribution in order to avoid skews. The QueryBuilder and ECLWatch program development environment tools provide a complete visualization tool for analyzing, debugging, and profiling execution of ECL jobs. During the execution of a job, the ECL graph can be viewed which shows the execution plan, dataflows as they occur, and the results of each processing step. Users can double click on the graph to drill down for additional information. An example of the graph corresponding to the ECL code shown in Fig. 10.14 is shown in Fig. 10.17

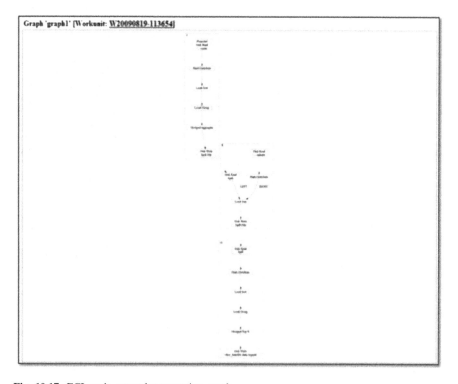

Fig. 10.17 ECL code examples execution graph

base platform and improve programmer productivity, and a rapidly expanding user community committed to open source. This has resulted in dramatic growth and acceptance of the Hadoop platform and its implementation to support data-intensive computing applications.

The LexisNexis HPCC platform is an integrated set of systems, software, and other architectural components designed to provide data-intensive computing capabilities from raw data processing and ETL applications, to high-performance query processing and data mining. The ECL language was specifically implemented to provide a high-level dataflow parallel processing language that is consistent across all system components and has extensive capabilities developed and optimized over a period of almost 10 years. The LexisNexis HPCC is a mature, reliable, well-proven, commercially supported system platform used in government installations, research labs, and commercial enterprises. The comparison of the Pig Latin language and execution system available on the Hadoop MapReduce platform to the ECL language used on the HPCC platform presented here reveals that ECL provides significantly more advanced capabilities and functionality without the need for extensive user-defined functions written in another language or resorting to a native MapReduce application coded in Java.

The following comparison of overall features provided by the Hadoop and HPCC system architectures reveals that the HPCC architecture offers a higher level of integration of system components, an execution environment not limited by a specific computing paradigm such as MapReduce, flexible configurations and optimized processing environments which can provide data-intensive applications from data analysis to data warehousing and high performance online query processing, and high programmer productivity utilizing the ECL programming language and tools. Table 10.2 provides a summary comparison of the key features of the hardware and software architectures of both system platforms based on the analysis of each architecture presented in this paper.

Table 10.2 Hadoop versus HPCC feature comparison

Architecture characteristics	Hadoop	HPCC
Hardware type	Processing clusters using commodity off-theshelf (COTS) hardware. Typically rack-mounted blade servers with Intel or AMD processors, local memory and disk connected to a high-speed communications switch (usually Gigabit Ethernet connections) or hierarchy of communications switches depending on the total size of the cluster. Clusters are usually homogenous (all processors are configured identically), but this is not a requirement	Same

(continued)

Table 10.2 (continued)

Architecture characteristics	Hadoop	HPCC
Operating system	Unix/Linux	Linux/Windows. Typically Linux is used due to the additional cost of licensing Windows
System configurations	Hadoop system software implements cluster with MapReduce processing paradigm. The cluster also functions as a distributed file system running HDFS. Other capabilities are layered on top of the Hadoop MapReduce and HDFS system software including HBase, Hive, etc.	HPCC clusters can be implemented in two configurations: Data Refinery (Thor) is analogous to the Hadoop MapReduce Cluster; Data Delivery Engine (Roxie) provides separate highperformance online query processing and data warehouse capabilities. Both configurations also function as distributed file systems but are implemented differently based on the intended use to improve performance. HPCC environments typically consist of multiple clusters of both configuration types. Although filesystems on each cluster are independent, a cluster can access files the filesystem on any other cluster in the same environment
Licensing cost	None. Hadoop is an open source platform and can be freely downloaded and used	License fees currently depend on size and type of system configurations. Does not preclude a future open source offering
Core software	Core software includes the operating system and Hadoop MapReduce cluster and HDFS software Each slave node includes a Tasktracker service and Datanode service. A master node includes a Jobtracker service which can be configured as a separate hardware node or run on one of the slave hardware nodes. Likewise, for HDFS, a master Namenode service is also required to provide name services and can be run on one of the slave nodes or a separate node	For a Thor configuration, core software includes the operating system and various services installed on each node of the cluster to provide job execution and distributed file system access. A separate server called the Dali server provides filesystem name services and manages Workunits for jobs in the HPCC environment. A Thor cluster is also configured with a master node and multiple slave nodes. A Roxie cluster is a peercoupled cluster where each node runs Server and Agent tasks for query execution and key and file processing. The filesystem on the Roxie cluster uses a distributed B + Tree to store index and data and provides keyed access to the data. Additional middleware components are required for operation of Thor and Roxie clusters

(continued)

Table 10.2 (continued)

Language feature component or capability	Pig	ECL
Language feature component or capability	Pig	ECL
Middleware components	None. Client software can submit jobs directly to the Jobtracker on the master node of the cluster. A Hadoop Workflow Scheduler (HWS) which will run as a server is currently under development to manage jobs which require multiple MapReduce sequences	Middleware components include an ECL code repository implemented on a MySQL server, and ECL server for compiling of ECL programs and queries, an ECLAgent acting on behalf of a client program to manage the execution of a job on a Thor cluster, an ESPServer (Enterpise Services Platform) providing authentication, logging, security, and other services for the job execution and Web services environment, and the Dali server which functions as the system data store for job workunit information and provides naming services for the distributed filesystems. Flexibility exists for running the middleware components on one to several nodes. Multiple copies of these servers can provide redundancy and improve performance
System tool	The dfsadmin tool provides information about the state of the filesystem; fsck is a utility for checking the health of files in HDFS; datanode block scanner periodically verifies all the blocks stored on a datanode; balancer re-distributes blocks from over-utilized datanodes to underutilized datanodes as needed. The MapReduce Web UI includes the JobTracker page which displays information about running and completed jobs, drilling down on a specific job displays detailed information about the job. There is also a Tasks page that displays info about Map and Reduce tasks	HPCC includes a suite of client and operations tools for managing, maintaining, and monitoring HPCC configurations and environments. These include QueryBuilder the program development environment, an Attribute Migration Tool, Distributed File Utility (DFU), an Environment Configuration Utility, Roxie Configuration Utility. Command line versions are also available. ECLWatch is a Web based utility program for monitoring the HPCC environment and includes queue management, distributed file system management, job monitoring, and system performance monitoring tools. Additional tools are provided through Web services interfaces

(continued)

Table 10.2 (continued)

Language feature component or capability	Pig	ECL
Ease of deployment	Assisted by online tools provided by Cloudera utilizing Wizards. Requires a manual RPM deployment	Environment configuration tool. A Genesis servier provides a central repository to distribute OS level settings, services, and binaries to all netbooted nodes in a configuration
Distributed file system	Block-oriented, uses large 64 MB or 128 MB blocks in most installations. Blocks are stored as independent units/local files in the local Unix/Linux filesystem for the node. Metadata information for blocks is stored in a separate file for each block. Master/Slave architecture with a single Namenode to provide name services and block mapping and multiple Datanodes. Files are divided into blocks and spread across nodes in the cluster. Multiple local files (1 containing the block, 1 containing metadata) for each logical block stored on a node are required to represent a distributed file	The Thor DFS is record-oriented, uses local Linux filesystem to store file parts. Files are initially loaded (Sprayed) across nodes and each node has a single file part which can be empty for each distributed file. Files are divided on even record/document boundaries specified by the user. Master/Slave architecture with name services and file mapping information stored on a separate server. Only one local file per node required to represent a distributed file. Read/write access is supported between clusters configured in the same environment. Utilizing special adaptors allows files from external databases such as MySQL to be accessed, allowing transactional data to be integrated with DFS data and incorporated into batch jobs. The Roxie DFS utilizes distributed B + Tree index files containing key information and data stored in local files on each node
Fault resilience	HDFS stores multiple replicas (userspecified) of data blocks on other nodes (configurable) to protect against disk and node failure with automatic recovery. MapReduce architecture includes speculative execution, when a slow or failed Map task is detected, additional Map tasks are started to recover from node failures	The DFS for Thor and Roxie stores replicas of file parts on other nodes (configurable) to protect against disk and node failure. Thor system offers either automatic or manual node swap and warm start following a node failure, jobs are restarted from last checkpoint or persist. Replicas are automatically used while copying data to the new node. Roxie system continues running following a node failure with a reduced number of nodes

(continued)

Table 10.2 (continued)

Language feature component or capability	Pig	ECL
Job execution environment	Uses MapReduce processing paradigm with input data in key-value pairs. Master/Slave processing architecture. A Jobtracker runs on the master node, and a TaskTracker runs on each of the slave nodes. Map tasks are assigned to input splits of the input file, usually 1 per block. The number of Reduce tasks is assigned by the user. Map processing is local to assigned node. A shuffle and sort operation is done following Map phase to distribute and sort key-value pairs to Reduce tasks based on key regions so that pairs with identical keys are processed by same Reduce tasks. Multiple MapReduce processing steps are typically required for most procedures and must be sequenced and chained separately by the user or language such as Pig	Thor utilizes a Master/Slave processing architecture. Processing steps defined in an ECL job can specify local (data processed separately on each node) or global (data is processed across all nodes) operation. Multiple processing steps for a procedure are executed automatically as part of a single job based on an optimized execution graph for a compiled ECL dataflow program. A single Thor cluster can be configured to run multiple jobs concurrently reducing latency if adequate CPU and memory resources are available on each node. Middleware components including an ECLAgent, ECLServer, and DaliServer provide the client interface and manage execution of the job which is packaged as a Workunit. Roxie utilizes a multiple Server/Agent architecture to process ECL programs accessed by queries using Server tasks acting as a manager for each query and multiple Agent tasks as needed to retrieve and process data for the query
Programming languages	Hadoop MapReduce jobs are usually written in Java. Other languages are supported through a streaming or pipe interface. Other processing environments execute on top of Hadoop MapReduce such as HBase and Hive which have their own language interface. The Pig Latin language and Pig execution environment provides a high level dataflow language which is then mapped into multiple Java MapReduce jobs	ECL is the primary programming language for the HPCC environment. ECL is compiled into optimized C++ which is then compiled into DLLs for execution on the Thor and Roxie platforms. ECL can include inline C++ code encapsulated in functions. External services can be written in any language and compiled into shared libraries of functions callable from ECL. A Pipe interface allows execution of external programs written in any language to be incorporated into jobs

(continued)

Table 10.2 (continued)

Language feature component or capability	Pig	ECL
Integrated program development environment	Hadoop MapReduce utilizes the Java programming language and there are several excellent program development environments for Java including Netbeans and Eclipse which offer plug-ins for access to Hadoop clusters. The Pig environment does not have its own IDE, but instead uses Eclipse and other editing environments for syntax checking. A PigPen add-in for Eclipse provides access to Hadoop Clusters to run Pig programs and additional development capabilities	The HPPC platform is provided with QueryBuilder, a comprehensive IDE specifically for the ECL language. QueryBuilder provides access to shared source code repositories and provides a complete development and testing environment for developing ECL dataflow programs. Access to the ECLWatch tool is built-in, allowing developers to watch job graphs as they are executing. Access to current and historical job Workunits is provided allowing developers to easily compare results from one job to the next during development cycles
Database capabilities	The basic Hadoop MapReduce system does not provide any keyed access indexed database capabilities. An add-on system for Hadoop called HBase provides a columnoriented database capability with keyed access. A custom script language and Java interface is provided. Access to HBase is not directly supported by the Pig environment and requires user-defined functions or separate MapReduce procedures	The HPCC platform includes the capability to build multi-key, multivariate indexes on DFS files. These indexes can be used to improve performance and provide keyed access for batch jobs on a Thor system, or be used to support development of queries deployed to Roxie systems. Keyed access to data is supported directly in the ECL language
Online query and data warehouse capabilities	The basic Hadoop MapReduce system does not provide any data warehouse capabilities. An add-on system for Hadoop called Hive provides data warehouse capabilities and allows HDFS data to be loaded into tables and accessed with an SQL-like language. Access to Hive is not directly supported by the Pig environment and requires userdefined functions or separate MapReduce procedures	The Roxie system configuration in the HPCC platform is specifically designed to provide data warehouse capabilities for structured queries and data analysis applications. Roxie is a highperformance platform capable of supporting thousands of users and providing sub-second response time depending on the application

(continued)

Table 10.2 (continued)

Language feature component or capability	Pig	ECL
Scalability	1 to thousands of nodes. Yahoo! has production clusters as large as 4000 nodes	1 to several thousand nodes. In practice, HPCC configurations require significantly fewer nodes to provide the same processing performance as a Hadoop cluster. Sizing of clusters may depend however on the overall storage requirements for the distributed file system
Performance	Currently the only available standard performance benchmarks are the sort benchmarks sponsored by http://sortbenchmark.org. Yahoo! has demonstrated sorting 1 TB on 1460 nodes in 62 s, 100 TB using 3452 nodes in 173 min, and 1 PB using 3658 nodes in 975 min	The HPPC platform has demonstrated sorting 1 TB on a high-performance 400-node system in 102 s. In a recent head-to-head benchmark versus Hadoop on a another 400-node system conducted with LLNL, The HPPC performance was 6 min 27 s and the Hadoop performance was 25 min 28 s. This result on the same hardware configuration showed that HPCC was 3.95 times faster than Hadoop for this benchmark
Training	Hadoop training is offered through Cloudera. Both beginning and advanced classes are provided. The advanced class includes Hadoop add-ons including HBase and Pig. Cloudera also provides a VMWare based learning environment which can be used on a standard laptop or PC. Online tutorials are also available	Basic and advanced training classes on ECL programming are offered monthly in several locations or can be conducted on customer premises. A system administration class is also offered and scheduled as needed. A CD with a complete HPCC and ECL learning environment which can be used on a single PC or laptop is also available

Conclusion

As a result of the continuing information explosion, many organizations are drowning in data and the data gap or inability to process this information and use it effectively is increasing at an alarming rate. Data-intensive computing represents a new computing paradigm which can address the data gap and allow government and commercial organizations and research environments to process massive amounts of data and implement applications previously thought to be impractical or infeasible. Some organizations with foresight recognized early that new parallel-processing architectures were needed including Google who initially developed the MapReduce architecture and LexisNexis who developed the HPCC architecture. More recently the Hadoop platform has emerged as an open source alternative for the MapReduce approach. Hadoop has gained momentum quickly, and additional add-on capabilities to enhance the platform have been developed including a dataflow programming language and execution environment called Pig. These architectures and their relative strengths and weaknesses are described in this paper, and a direct comparison of the Pig language of Hadoop to the ECL language used with the LexisNexis HPCC platform was presented.

The suitability of a processing platform and architecture for an organization and its application requirements can only be determined after careful evaluation of available alternatives. Many organizations have embraced open source platforms while others prefer a commercially developed and supported platform by an established industry leader. The Hadoop MapReduce platform is now being used successfully at many so-called Web companies whose data encompasses massive amounts of Web information as its data source. The LexisNexis HPCC platform is at the heart of a premier information services provider and industry leader, and has been adopted by government agencies, commercial organizations, and research laboratories because of its high-performance cost-effective implementation. Existing HPCC applications include raw data processing, ETL, and linking of enormous amounts of data to support online information services such as LexisNexis and industry-leading information search applications such as Accurint; entity extraction and entity resolution of unstructured and semi-structured data such as Web documents to support information extraction; statistical analysis of Web logs for security applications such as intrusion detection; online analytical processing to support business intelligence systems (BIS); and data analysis of massive datasets in educational and research environments and by state and federal government agencies. There are many tradeoffs in making the right decision in choosing a new computer systems architecture, and often the best approach is to conduct a specific benchmark test with a customer application to determine the overall system effectiveness and performance. The relative cost-performance characteristics of the system in additional to suitability, flexibility, scalability, footprint, and power consumption factors which impact the total cost of ownership (TCO) must be considered.

A comparison of the Hadoop MapReduce architecture to the HPCC architecture in this paper reveals many similarities between the platforms including the use of a high-level dataflow-oriented programming language to implement transparent data-parallel processing. The advantages of choosing a LexisNexis HPCC platform include: (1) an architecture which implements a highly integrated system environment with capabilities from raw data processing to high-performance queries and data analysis using a common language; (2) an architecture which provides equivalent performance at a much lower system cost based on the number of processing nodes required as demonstrated with the Terabyte Sort benchmark where the HPCC platform was almost 4 times faster than Hadoop running on the same hardware resulting in significantly lower total cost of ownership (TCO); (3) an architecture which has been proven to be stable and reliable on high-performance data processing production applications for varied organizations over a 10-year period; (4) an architecture that uses a dataflow programming language (ECL) with extensive built-in capabilities for data-parallel processing which allows complex operations without the need for extensive user-defined functions and automatically optimizes execution graphs with hundreds of processing steps into single efficient workunits; (5) an architecture with a high-level of fault resilience and language capabilities which reduce the need for re-processing in case of system failures; and (6) an architecture which is available from and supported by a well-known leader in information services and risk solutions (LexisNexis) who is part of one of the world's largest publishers of information ReedElsevier.

References

1. Gorton I, Greenfield P, Szalay A, Williams R. Data-intensive computing in the 21st century. IEEE Comput. 2008;41(4):30–2.
2. Johnston WE. High-speed, wide area, data intensive computing: a ten year retrospective. In: Proceedings of the 7th IEEE international symposium on high performance distributed computing: IEEE Computer Society; 1998.
3. Skillicorn DB, Talia D. Models and languages for parallel computation. ACM Comput Surv. 1998;30(2):123–169.
4. Abbas A. Grid computing: a practical guide to technology and applications. Hingham: Charles River Media Inc; 2004.
5. Gokhale M, Cohen J, Yoo A, Miller WM. Hardware technologies for high-performance data-intensive computing. IEEE Comput. 2008;41(4):60–8.
6. Nyland LS, Prins JF, Goldberg A, Mills PH. A design methodology for data-parallel applications. IEEE Trans Softw Eng. 2000;26(4):293–314.
7. Ravichandran D, Pantel P, Hovy E. The terascale challenge. In: Proceedings of the KDD workshop on mining for and from the semantic web; 2004.
8. Agichtein E. Scaling information extraction to large document collections: Microsoft Research. 2004.
9. Rencuzogullari U, Dwarkadas S. Dynamic adaptation to available resources for parallel computing in an autonomous network of workstations. In: Proceedings of the eighth ACM SIGPLAN symposium on principles and practices of parallel programming, Snowbird, UT; 2001. p. 72–81.
10. Cerf VG. An information avalanche. IEEE Comput. 2007;40(1):104–5.
11. Gantz JF, Reinsel D, Chute C, Schlichting W, McArthur J, Minton S, et al. The expanding digital universe (White Paper): IDC. 2007.
12. Lyman P, Varian HR. How much information? 2003 (Research Report): School of Information Management and Systems, University of California at Berkeley; 2003.
13. Berman F. Got data? A guide to data preservation in the information age. Commun ACM. 2008;51(12):50–6.
14. NSF. Data-intensive computing. National Science Foundation. 2009. http://www.nsf.gov/funding/pgm_summ.jsp?pims_id=503324&org=IIS. Retrieved 10 Aug 2009.
15. PNNL. Data intensive computing. Pacific Northwest National Laboratory. 2008. http://www.cs.cmu.edu/~bryant/presentations/DISC-concept.ppt. Retrieved 10 Aug 2009.
16. Bryant RE. Data intensive scalable computing. Carnegie Mellon University. 2008. http://www.cs.cmu.edu/~bryant/presentations/DISC-concept.ppt. Retrieved 10 Aug 2009.

17. Pavlo A, Paulson E, Rasin A, Abadi DJ, Dewitt DJ, Madden S, et al. A comparison of approaches to large-scale data analysis. In: Proceedings of the 35th SIGMOD international conference on management of data, Providence, RI; 2009. p. 165–8.
18. Dean J, Ghemawat S. MapReduce: simplified data processing on large clusters. In: Proceedings of the sixth symposium on operating system design and implementation (OSDI); 2004.
19. Ghemawat S, Gobioff H, Leung S-T. The Google file system. In: Proceedings of the 19th ACM symposium on operating systems principles, Bolton Landing, NY; 2003. p. 29–43.
20. Chang F, Dean J, Ghemawat S, Hsieh WC, Wallach DA, Burrows M, et al. Bigtable: a distributed storage system for structured data. In: Proceedings of the 7th symposium on operating systems design and implementation (OSDI '06), Seattle, WA; 2006. p. 205–18.
21. Pike R, Dorward S, Griesemer R, Quinlan S. Interpreting the data: parallel analysis with Sawzall. Sci Program J. 2004;13(4):227–98.
22. White T. Hadoop: the definitive guide. 1st ed. Sebastopol: O'Reilly Media Inc; 2009.
23. Venner J. Pro Hadoop. Berkeley: Apress; 2009.
24. Olston C, Reed B, Srivastava U, Kumar R, Tomkins A. Pig latin: a not-so-foreign language for data processing (presentation at sigmod 2008). 2008a. http://i.stanford.edu/~usriv/talks/sigmod08-pig-latin.ppt#283,18,User-CodeasaFirst-ClassCitizen. Retrieved 10 Aug 2009.
25. Gates AF, Natkovich O, Chopra S, Kamath P, Narayanamurthy SM, Olston C, et al. Building a high-level dataflow system on top of Map-Reduce: the Pig experience. In: Proceedings of the 35th international conference on very large databases (VLDB 2009), Lyon, France; 2009.
26. Olston C. Pig overview presentation—Hadoop summit. 2009. http://infolab.stanford.edu/~olston/pig.pdf. Retrieved 10 Aug 2009.
27. O'Malley O, Murthy AC. Winning a 60 second dash with a yellow elephant. 2009. http://sortbenchmark.org/Yahoo2009.pdf. Retrieved 10 Aug 2009.
28. Olston C, Reed B, Srivastava U, Kumar R, Tomkins A. Pig Latin: a not-so_foreign language for data processing. In: Proceedings of the 28th ACM SIGMOD/PODS international conference on management of data/principles of database systems, Vancouver, BC, Canada; 2008b. p. 1099–110.
29. O'Malley O. Introduction to Hadoop. 2008. http://wiki.apache.org/hadoop-data/attachments/HadoopPresentations/attachments/YahooHadoopIntro-apachecon-us-2008.pdf. Retrieved 10 Aug 2009.
30. Nicosia M. Hadoop cluster management. 2009. http://wiki.apache.org/hadoopdata/attachments/HadoopPresentations/attachments/Hadoop-USENIX09.pdf. Retrieved 10 Aug 2009.
31. White T. Understanding MapReduce with Hadoop. 2008. http://wiki.apache.org/hadoop-data/attachments/HadoopPresentations/attachments/MapReduce-SPA2008.pdf. Retrieved 10 Aug 2009.
32. Borthakur D. Hadoop distributed file system. 2008. http://wiki.apache.org/hadoopdata/attachments/HadoopPresentations/attachments/hdfs_dhruba.pdf. Retrieved 10 Aug 2009.

Chapter 11
Graph Processing with Massive Datasets: A Kel Primer

David Bayliss and Flavio Villanustre

Introduction

Graph theory and the study of networks can be traced back to Leonhard Euler's original paper on the Seven Bridges of Konigsberg, in 1736 [1]. Although the mathematical foundations to understanding graphs have been laid out over the last few centuries [2–4], it wasn't until recently, with the advent of modern computers, that parsing and analysis of large-scale graphs became tractable [5]. In the last decade, graph theory gained mainstream popularity following the adoption of graph models for new applications domains, including social networks and the web of data, both generating extremely large and dynamic graphs that cannot be adequately handled by legacy graph management applications [6].

Two data models and methodologies are prevalent to store and process graphs: graph databases using a property graph data model, storing graphs in index-free adjacency representations and using real-time path traversal as a querying methodology; and more traditional relational databases that use relational data models, store representations of graphs in rows and use real-time joins to perform queries [7].

Unfortunately, these two approaches are not exempt from drawbacks, particularly when graph scales exceed from millions of nodes and billions of edges, which is not uncommon in many problem domains. Exploring hundreds of thousands of edges and nodes in real-time in graph database can be very time consuming, unless the entire graph can be stored in RAM in a unified memory architecture. Relational databases, with the potential for the combinatorial explosion of intermediate candidate sets resulting from joins are in no better position to handle large graphs; to cope with this issue, alternative approaches using summarization have been proposed, with varying degrees of success [8].

© Springer International Publishing Switzerland 2016   307
B. Furht and F. Villanustre, *Big Data Technologies and Applications*,
DOI 10.1007/978-3-319-44550-2_11

More recently, other computational models for graph processing have arisen, including the storage of graphs as adjacency matrices and modeling graph algorithms as a composition of linear algebra operations, and even BSP models such as Map/Reduce and Pregel [9].

In this article, we'll cover a novel methodology to store and process very large-scale graphs, using a distributed architecture and a high level denotational programming model that can efficiently handle hybrid storage (hard-disk/flash/RAM), partitioning and parallel and distributed execution, and still exhibit adequate real-time performance at query processing.

Motivation

Social Network Analysis (SNA) is the use of Network Theory to analyze Social Networks [10]. Its relevance has increased significantly partly due to the abundance of data that can be used to identify networks of individuals and the indication of correlation between individual behavior and the surrounding network of relationships [11, 12]. Moreover, Social Network Analysis has demonstrated usefulness across a broad range of disciplines, which has driven numerous research initiatives and practical applications, from law enforcement to modelling of consumer behavior [13]. Unfortunately, the size of the generated graph in many of these networks can be substantially larger than what it can fit in RAM in a single computer, creating challenges to the traditional approaches that graph databases, using index-free adjacency representations, utilize. Partitioning these graphs to scale out graph databases is complex and can be quite inefficient due to the significant increase of internode communication and consequent processing latencies and/or the need to create redundant sections of the graph to decrease the amount of network traffic [14, 15]. The use of relational models to process un-partitioned and partitioned graphs poses other challenges, mainly due to the exponential growth in intermediate results as a consequence of the very large self-join operations required to parse the table-based representation of these networks.

Besides the tractability aspect of the processing of very large graphs, there is the complexity behind expressing the algorithms needed to perform the actual parsing, matching, segmentation or metric calculation. Efficient graph algorithms are non-trivial, and require deep knowledge of graph theory, computer architecture, algorithms and data structures, which severely limits in practice the number of people that can perform this analysis. In this article, we will describe both, a systems architecture to process large-scale graphs and a novel declarative programming model that can be used for effective Social Network Analysis in very large graphs.

Background

The Open Source HPCC Systems Platform Architecture

The Open Source HPCC Systems platform is a distributed data intensive computing platform originally developed by LexisNexis circa 2000 [16]. The platform utilizes a cluster of commodity nodes, each consisting of one or more x86-64 processors, local RAM, local persistent storage (Hard Drives, SSD) and an IP based interconnect, and runs as a user-space application on the GNU Linux Operating System. Data is usually partitioned and distributed across all the nodes and the platform parallelizes the execution of data parallel algorithms across the entire cluster. There are two distinct components in this platform: a batch oriented data processing system called Thor and a real-time data retrieval and analytics system called Roxie; both of these components are programmed using a declarative dataflow language called ECL. Data transformations and data queries are declaratively expressed using ECL, which is compiled into C++ and natively executed by the platform. Parallelism, partitioning and hardware architecture are all abstracted out from the data analyst perspective, who can express algorithms in terms of operations over data records. Different from imperative programming paradigms, control-flow is implicitly determined by the data dependencies, rather than by direct indication by the programmer, increasing the overall programming efficiency and code succinctness, and minimizing the opportunity for bugs.

Before ECL is compiled, a compiler parses the declarations, and an optimizer identifies opportunities for improved execution time efficiencies leveraging the fact that the declarative nature of ECL gives the optimizer the ability to more easily comprehend the ultimate goal of the programmer.

KEL—Knowledge Engineering Language for Graph Problems

Even though ECL is a high level data oriented language, it can still be considered low-level from the perspective of an analyst operating at a Social Graph level. While manipulating graphs, the programmer is concerned with entities and associations rather than with specific data records and even simple operations on graphs can require multiple steps when approached at a low data level. To increase programmers' efficiencies, one of the authors designed a new graph oriented declarative programming language called KEL (Knowledge Engineering Language). In a way similar to how ECL abstracts out underlying details of system architecture, data partitioning, parallel execution and low-level data structures, KEL detaches the analyst from the details of data layouts and data operations, focusing on the graph itself. In ECL, data records are first class citizens of the language; in KEL, entities and associations are the first class citizens over which algorithms are built.

KEL, like ECL, is a compiled language, but KEL compiles into ECL, which can then be optimized and compiled into C++ utilizing the standard machinery available in the HPCC Systems platform. By stacking abstractions, code reuse is maximized and the benefits already built into the underlying layers can be leveraged by the upper ones. More important, improvements to any of the bottom layers instantly benefit every upper layer relying on it.

KEL—A Primer

The basic premise underlying KEL is that there is an important distinction between a "Graph Language" and a "language for solving graph problems". The purpose of this section is to detail this distinction and then outline the proposed features of KEL that would enable it to claim the title of "Language to Solve Graph Problems".

The first assertion that underlies KEL is that most 'graph problems' are not themselves graph problems but are problems which are considered to be isomorphic to a graph problem. Typically one has real world data about real world entities and perhaps information regarding potential associations between those entities. The real problem of concern is defining further information based upon this real world information. Therefore before one really has a 'graph' problem to solve one is actually performing two or three separate mappings a priori. Firstly the world view that is of interest is mapped onto a data model (or ontology). Then all of the real world data is mapped into that data structure. Finally the 'important question' that is of interest is mapped into a query against that data structure.

Even once the real world problem has been mapped into a 'pure graph' problem there are still important decisions to be made prior to encoding in the 'graph language'. Even the simplest of graph queries, such as "find the shortest path between two nodes" yield themselves to a number of different algorithms and the best algorithm based upon the data model and the data being queried has to be selected. Finally that algorithm can be encoded in the "Graph Language" and then executed. A good graph language will have isolated the graph programmer from the vagaries of the underlying hardware; a poorer one may not have.

The real picture is therefore shown in Fig. 11.1.

Some of the arrows have been given letter labels because exactly what occurs in there is open to discussion:

Arrow A: In a language such as the "Semantic Graph Query Language" proposed by Ian Kaplan of LLNL the language is composed of low level primitives; therefore if a "depth first search" is required then someone has to code that search up in terms of the underlying data model. If the Graph Language in question has the facility to encapsulate query segments then that code may be re-usable; provided the data model doesn't change. Of course the query writer still has to know which libraries to use in which situation. Further it is the quality of the encoding of these libraries that will drive performance.

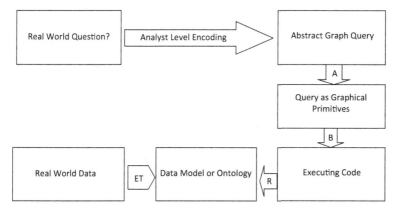

Fig. 11.1 The proposed KEL system

Arrow B: The performance of the query is going to hinge largely upon the mapping from the graphical primitives to the underlying machine. At a simple level this is driven by whether or not the primitives have easily executed machine equivalents. Rather more subtly it also hinges upon whether or not the primitives allow the algorithms of arrow 'A' to be readily expressed. If the graph algorithm writer had to use long sequences of graph primitives to express a function that could have been coded easily outside of a graph language then a large amount of inefficiency has been introduced. In a worst case scenario a larger number of inefficient graph primitives will have replaced a single efficient primitive in some other language.

Arrow R: Of course the performance of the query will also be driven by the efficiency of the run time systems that are available on the target machine. The machine architecture including the availability of memory, processors and threading will also drive the performance of the query.

Whilst fully acknowledging, and offering innovative solutions to each of the three arrows above, the driving distinctive behind KEL is the belief that none of them are quite as critical to the ultimate efficiency of programmer and processor as the selection of data model and the encoding of the real world problem as a graphical problem. The reason is simple: many graph algorithms are combinatory or worse in depth of the tree being traversed. Thus any data model or algorithm that increases the amount of tree walking being done catastrophically degrades the performance of the system. Similarly any modeling method which increases the branching factor of each node will swamp any linear performance gain from the underlying machinery. Conversely any data model or algorithm that manages to decrease the walk depth or branching factor will achieve performance gains well beyond a linear speed up.

The second driving reason for believing that having the real world mapping close to the heart of the language is that it is an area in which real world development will otherwise do the opposite of that which is needful. The data model has to exist before any significant work on a data driven system can occur; therefore it is typically developed early. Once the data model exists then large amounts of ETL[1] code are written to map the actual data into the required data model. Further if productivity efficiencies are to be gleaned from arrow 'A' then libraries of graph algorithms will be developed to the underlying data model. Both of these activities essentially 'lock' a very early and possibly damaging decision into the architecture. Even if the data model is evolved a little it is quite common for the graph libraries to not be re-coded to take advantage of the altered performance profile.

The third driving reason for an insistence upon focusing on the real world mapping is that alterations to the data model are **required** if one wishes to allow the system to benefit from information that has already been gleaned from it. Suppose one analyst has produced a sub-graph pattern to identify a person exhibiting a particular terrorist threat. A second analyst has produced a different sub-graph pattern to reveal a weakness to a particular threat. The detection of either pattern is an NP-Complete problem in its own right. A third analyst now wishes to run a query looking for two 'suspects' that are linked to any potential weakness by up to three degrees of separation. We now have a combinatory problem executing against the graph where every one of the individual nodes is itself an NP-Complete problem. Even leaving aside the pain inflicted on the machinery the third analyst has to find and coral three different algorithms into the same piece of code. A key principle that we wish to build into our solution is that accepted and known 'facts' or 'models' can be burned (or learned) back into the data model to allow the system to be faster and more efficient to use as time progresses. *** **This is one of their cases of needing a dynamic data model from their example—I don't know if we should call that out that a little** ***

The fourth and final driving reason for insisting that real world mapping is essential is that the nature of a new query might fundamentally alter the requirements on the underlying data model. Worse many crucial real-world queries may not fit into a purely graphical data model. In the example above the third analyst was happy to look for links by two degrees of separation. Suppose she instead wanted to look for those within a twenty mile radius of each other? Unless the 'twenty mile radius' relation had been burned in from day one then that query simply could not have been answered. Even if 'twenty mile radius' had been burned in; what happens when she wants to expand to twenty five? In fact geo-spatial clustering is one of those tasks that best yields itself to a non-graphical data model as the 'graphical' model requires either a fixed and pre-defined radius *or* a relationship from every entity to every entity. *** **This is a sub-case of their second 'dynamic' data model requirement*****

---

[1]Extract Transform and Load; a generic term in the data industry for data manipulation that occurs prior to the exercise of real interest.

Proposed Solution

The proposed solution thus embodies designing a language which allows the real world problem to be expressed accurately yet succinctly and that allows a rich data model to be produced over the life of the project. The 'architecture' will then map that problem onto a library of graph algorithms which will in turn be mapped onto a series of graph and data primitives which will in turn be mapped onto a machine executable to run against the current data model. The succession of mappings may have given cause for concern except that LexisNexis already has a decade of experience in the latter two mappings. Further LexisNexis believes it has demonstrated that a well written expert system is capable of performing those mappings at a level consistently ahead of a human programmer (Fig. 11.2).

In the diagram above everything to the right of the vertical line constitutes the proposed KEL system; the lower right-hand corner represents Lexis' existing area of expertise in the field of very large scale data manipulation on clustered commodity hardware. In the following each box and arrow will be discussed in some depth calling out what is believed and what is yet to be researched. Finally some tentative ideas will be shared regarding the KEL language itself.

Data Primitives with Graph Primitive Extensions

Lexis' experience of large scale data manipulation suggests that there is a fairly small set of fundamental operations that can be performed upon data. Recent research with Lawrence Livermore has shown that Kaplan's graph primitives can all be built upon the ECL data operations. More recently the PIG language from Yahoo has emerged which implements a sub-set of the ECL operations. The recent

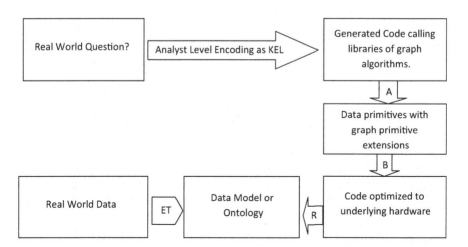

Fig. 11.2 Main functional areas

Table 11.1 Comparison between ECL and Pig

Operation	ECL	PIG
SORT—render a relation (or set of tuples) in a particular sequence	SORT—includes a patented algorithm for dealing with highly skewed data	ORDER
JOIN—allows all nodes with certain common attributes to be compared	JOIN—allows for hard (equivalence) and fuzzy comparisons between nodes	JOIN—only hard comparison allowed
AGGREGATE—allows information regarding node trends to be computed	TABLE—many different optimized forms allowed	GROUP/FOREACH —requires redistribution of data
DENORMALIZE—allows a set of records to form one record	DENORMALIZE or ROLLUP	GROUP
FILTER	Any record set can be filtered by following the set name with a filter condition	FILTER
ITERATE—process records in sequence allowing data to be copied from one record to the next	ITERATE—both node local and global forms	Not supported by MAP-REDUCE paradigm
DEDUP—allows multiple copies of records to be removed	DEDUP—both local and global allowing both hard and fuzzy conditions on any number of fields	DISTINCT—only allows for removal of fully identical records
LOOP—allows a process to iterate	LOOP—allows for a fixe d number of iterations or iterating until a condition is met. Also allows for iterations happening sequentially or in parallel	Not supported by MAP-REDUCE paradigm

Hadoop conference revealed that many are already trying to perform Graph work based upon these Hadoop primitives. We do **not** assert that the existing data operations will efficiently achieve all that is required for graph algorithms but believe them to be a strong starting point from which to begin. Our proposal therefore includes time to identify and efficiently implement new data operations as required to support all graph operations and also to develop any specific graph traversal primitives that are required to efficiently support graph algorithms.

For the purposes of this proposal the ECL run-time also suffers from two significant draw-backs; it is not open source and it is not optimized towards highly threaded shared memory architectures. The proposal therefore calls for the team to collaborate to develop a highly optimized open-source run-time to implement these data operations on shared memory architectures. Preliminary work in such a run-time exists in the MAESTRO and qthreads runtime systems from RENCI and Sandia. This run-time will then be wrapped in an Open Source 'Bootstrap-ECL' language. This Bootstrap language will be a small sub-set of ECL that supports KEL. It is envisaged that this bootstrap language can then be ported to any back-end that wishes to support KEL. Using this technique a very high level language (KEL) becomes portable with a minimum of effort (Table 11.1).

Generated Code and Graph Libraries

One key research question of the KEL proposal is the construction of the graph algorithm libraries. This is deliberate, new graph algorithms are being developed yearly and the intent is to allow KEL to become a repository in which the very best of shared knowledge can be stored. The proposal therefore allows for a number of these algorithms to be coded but is relying upon the open source nature of the project to encourage others to supplement an ever growing library of these algorithms that will become part of the KEL system. Sandia's experience developing the Multithreaded Graph Library (MTGL) and working in collaboration on the Parallel Boost Graph Library (PBGL) will be utilized in the creation of KEL's new runtime system.

Finding the correct way to specify, implement and select these algorithms is the major new field of research that the KEL project aims to extend. The proposition is as follows:

(a) There are fewer graph **problems**[2] than there are **algorithms**. Therefore each different graph problem will be given a logical name with defined logical parameters at the KEL level.

(b) For each logical name at the KEL level one or more algorithmic interfaces will exist. The different interfaces will exist to allow a given problem to be solved for differing physical data structures. The KEL compiler will be responsible for mapping the actual data model to the algorithm interface prior to calling.

(c) Each graph algorithm will then implement one or more algorithmic interfaces for each problem that it is believed to solve.

(d) Each interface to each algorithm will need a description or parameter file which describes to the KEL compiler the circumstances under which it performs best. Number of nodes, branching factors, ability to fit into memory etc. The compiler will then take this information, along with the availability of an algorithmic interface suitable for the underlying data structure, into account which selecting the algorithm for a particular problem. It will also be possible to override the selection from within KEL.

Another related field of research is discovering the correct language for these algorithms to be written in. It might at first appear than any good graph language should be the correct language to write graph algorithms in; however this may not be true. The fundamental data structure for a "best first" search is a prioritorized queue; not a graph. The best data structure for finding global shortest paths in a densely connected graph is an array; not a graph. It is currently expected that many of these algorithms will be coded in a sub-set of ECL implemented on top of the data operation libraries discussed above. For those that need more control it is

---

[2]Examples of 'problem' are minimal closure, shortest path, subgraph-matching, graph-isomorphism, etc.

proposed that a mechanism exist to allow the graph algorithms to be integrated into the data-operation layer itself.

KEL Compiler

The KEL compiler is not a compiler in the traditional sense; it will not be tied to any particular processor or executable format. In fact the KEL compiler is really a language translator as it will produce output in another high-level language which will then be converted to a machine executable. That said, it would also be misleading to call the KEL compiler a translator as it is translating code from a language operating at a very high level to a rather lower level language; typically it produces[3] 5–10 lines of ECL for every line of KEL. Some may refer to a program of this form as a source-to-source compiler. For brevity, from here forward, the program which takes in KEL and outputs ECL will be described as a compiler.

As the diagram above shows the principle responsibilities of the KEL compiler are:

(a) Syntax checking and providing useful error reporting for the KEL language itself
(b) Selecting suitable graph algorithms to solve the required graphical problems.
(c) Mapping from the presented data format to the required data model.
(d) Marshalling the data model in order to apply the graphical algorithms.
(e) Generating code to handle those graphical and other primitives that are not explicitly coded within the graph libraries.

The compiler still represents a significant undertaking; however it is believed that the abstraction out of the actual data operation piece has rendered the problem tractable.

KEL Language—Principles

Of course the design of the KEL language is one of the major areas of proposed research and thus anything discussed below should be considered extremely tentative. That said our experience of large scale informatics (both in terms of data, time and programmers) combined with some earlier research has given us some strong yet flexible ideas regarding how we want the language to look and behave. Whilst the exact syntax may well change some language principles are primary drivers and so are called out here:

---

[3]Based upon a very early prototype of a small sub-set of the proposed KEL language.

Productivity is driven by the speed the data can be mastered; not by the brevity of the tutorial exercises: many scripting languages such as PIG aim to garner productivity by skimping on detail as the data is declared to the system.[4] We believe that the collection of large datasets is a time consuming process and that the extra few seconds taken to declare the precise data structure to the system is time well spent. It allows the optimizer to work and for the syntax checking to be more thorough. Further the productivity can be reclaimed by allow the data declaration phase to be reused (leveraged) from one query to the next.

It is much more productive to catch a semantic error in the compiler than in the run-time: In a real world graph the nodes will not all have the same type; neither will the edges. Ensuring that the correct type of node/edge is being considered at any given point can be done one of three ways:

1. Simply ignore the typing issues and hope the programmer got it right. Optimal in terms of machine cycles and *possibly* even in terms of programmer cycles until the first time the programmer gets it wrong. Once the programmer gets it wrong debugging a huge complex graph algorithm could take days or even weeks.
2. Move the type checking into run-time. This still allows the programmer to ignore the typing issues and allows the run-time to trap any miss-matches. The problem here is that it slows the run-time considerably and doesn't allow the programmer to compare apples and oranges **if** he wants to.
3. Enforce strong, clear and overrideable type checking in the compiler. This is efficient at run-time (0 overhead) and catches errors early whilst allowing the programmer full and explicit control.

For this reason KEL will be strongly typed.

No query is genuinely and completely one off: This may seem like a bold and even ludicrous assertion. However, we would argue that if a query provided a valid answer then it will almost certainly spawn a derivative query (to discover more along the same line) or it will spawn a query with certain key components changed (to try a different approach). Of course if the query provided an invalid answer then another similar query will almost certainly be issued to try to get the correct answer. Even if a particular line of query is now dead one would expect the analyst to be going forward to write queries along a different line of questioning. All of the foregoing is there to argue that easy and natural code reuse combined with ease of code maintainability are the keys to long-term *radical* productivity gains—not the terseness of trivial queries.

Be nice to your compiler and it will be nice to you: any detailed study of compiler optimization theory will quickly show that the nemesis of strong optimization is sloppiness of expression in the compiled language. If the compiler cannot accurately predict the expected behavior of a given primitive then it cannot

---

[4]In PIG you do not need to provide data types or even declare the schema for data although the PIG manual warns "it may not work very well if you don't".

optimize it. To quote Martin Richards[5]: "once they assign to a pointer—all bets are off". Conversely by producing a tightly specified and predictable language the optimizer is able to produce game-changing optimizations.

The best encoding of a particular problem is the one in which it is thought: it is possible to produce an extremely elegant language using some academic abstraction such as a lambda calculus or regular expression.[6] The problem is that unless the coder naturally thinks at the same level of abstraction as the language designer then the elegance of the language will be lost in the clumsiness of the encoding. Therefore we aim to encapsulate the thoughts of the encoder using the terminology and at the level of granularity in which the encoder is thinking. Put differently; KEL is designed to allow the encoders to develop their own abstractions.

KEL Language—Syntax

A KEL program consists of four distinct phases or sections although it is envisaged that the first two and most of the third will be shared between many of the same programs or queries running upon the same system.

Section One—Model Description
The job of the first section is to declare the logical data model for all of the following graph work. The two principle data types are Entity (node or vertex) and Association (link). Both Entity and Association can have an arbitrary number of properties. Explicit support will exist for unique identifiers for each entity existing in data or being created during data load. Support will exist at run time for those identifiers being replaced by memory pointers when that is possible and appropriate.

The entity declaration principally declares the properties of an entity which come directly from the underlying data. It also defines the *default* mapping expected from that data if it is coming from FLAT, XML or PARSE[7] (natural language) files.

Here are some examples generated during some earlier research:

```
Address := ENTITY( UID⁸( prim_range,prim_name,zip ) );
```

---

[5]The man that created BCPL which eventually led to C; and a great 'Compiler Theory' lecturer too!.

[6]In fact some of the elements and benefits of both appear within KEL. However, at all points we believe that the thought process of the encoder should be paramount rather than the academic purity of a particular abstraction.

[7]The PARSE format allows entities and facts which have been extracted from text files to appear within the knowledge base.

[8]Here the UID does not exist and so is generated based upon those fields

```
Person := ENTITY( FLAT(UID⁹=DID,First_Name=fname,string30
Last_Name=lname, Address At) )
Company := ENTITY( XML(UID='\ORG\BDID',Company_Name='\ORG
\company_name') )
Vehicle := ENTITY( UID(vin), INT value );
WorksFor := ASSOCIATION(Person¹⁰ Who, Company What, TEXT
CompanyType,CALLED(works_for,employs));
Relatives := ASSOCIATION(Person Who, Person WhoElse,CALLED
(Related_To));
LivesAt := ASSOCIATION(Person Who, Address Where, CALLED
(Lives_At,Has_Living_At));
Owns := ASSOCIATION(Person Who, Vehicle What, CALLED¹¹(Owns,
Owned_By));
```

In addition to specifying the fields of an entity it is possible to express the MODEL in which the entity is represented. The essentially comes down to specifying which *unique combinations* of fields are interesting. By default every unique combination of every field is stored. The MODEL attribute allows you to specify those patterns which are interesting.

Consider, for example, an entity with the fields FNAME, MNAME, LNAME, DOB, PRIM_RANGE, PRIM_NAME then by default there will be one record stored for every different combination of those fields that exist in the underlying data. This corresponds to

MODEL({FNAME, MNAME, LNAME, DOB, PRIM_RANGE, PRIM_NAME})

Suppose however that you were only interested in the different names and different addresses and were not concerned about whether they co-occurred. Then a model of:

MODEL(NAME{FNAME, MNAME, LNAME}, DOB,{PRIM_RANGE, PRIM_NAME})

Would be more appropriate. It is also possible to specify a logical or conceptual field name for these combinations: thus—

MODEL(NAME{FNAME,   MNAME,   LNAME},   DOB,ADDRESS{PRIM_RANGE, PRIM_NAME})

Section 2—Model Mapping
Provided the model is declared with all the default mappings in place for the various file types and provided the data to be ingested actually conforms to the default

---

⁹Here the UID is existing and called UID in the underlying data

¹⁰Once an entity with a UID has been declared then the type can be used to implicitly declare a foreign key existing within another part of the data.

¹¹Called allows for both bi-directional and unidirectional links to be used in text

mappings then the second section can be very thin indeed. It simply declares those files from which the data is going to be taken and the entities and associations that are to be taken from each file. Thus given the model above one may have:

```
USE  header.File_Headers(FLAT,Person,Address),  Business_
Header.File_Header(XML,Company,Address),           header.
Relatives(FLAT,Relatives)
```

Which states that for this query the file header.File_Headers will provide Person and Address entities, Business.File_Header will provide Company and Address entities and that the Relatives associate will come from header.Relatives. The USE clause will also allow for explicit field-by-field overriding of the default field mappings if the data is mal-formed.

Section 3—Logic Phase

Perhaps the heart of a KEL program is the third phase; this is where new information is declared to exist based upon logic that exists already. Every 'line' of the logic phase (called an 'assertion') consists of three parts: the scope, the predicate and the production. Again this is best understood by example:

Here we are asserting that 'for all COMPANY' (the scope) if the CompanyType is within a particular list then the following property is also true. Essentially this allows for property derivation within a single entity.

COMPANY: CompanyType   IN   ['CHEMICAL','FERTILIZER', 'AGRICULTURAL'] => SELF.ExplosiveMaterial = TRUE;

It is important to note that the predicate and production refer to a single entity; it is the scope that applies them to every entity within the scope.

All the immediate associations of an entity are also available within the scope of that entity; thus the below would be a way to define a person as being orphaned if they have no recorded relatives.

PERSON: ~EXISTS(Relatives(Who=SELF)) => SELF.Orphaned = TRUE;

In fact KEL would make SELF the default scope of an assertion and the first matching field of any associate would be deemed to refer to self unless overridden. Thus the line above would more normally be coded as:

PERSON: ~EXISTS(Relatives) => Orphaned = TRUE;

A large part of the point and power of KEL comes from the ability to cascade assertions; thus the below defines a risky person as an orphan working for a company that provides explosive materials.

PERSON: Orphaned AND EXISTS(WorksFor(Company(ExplosiveMaterial))) =>Risky = TRUE;

Of course KEL is designed to support mathematics as well as logic. It is also possible to view an aggregation of neighboring nodes as a property of a local node. Thus the value of someone's vehicles could be garnered using:

PERSON: EXISTS(Owns) => VehicleWorth = SUM(Owns.What.Value);

Of course some productions are valid upon the entire scope; in this case the predicate can be omitted.

PERSON:=> VehicleWorth = SUM(Owns.What.Value);

The aggregations can be performed based upon computed as well as original field values. Thus the vehicle value of all of the vehicles of all the people in a building could be:

ADDRESS: EXISTS(Has_Living_At) => TotalWorth = SUM(Has_Living_At. Who.VehicleWorth);

*** Should I mention the need to have 'iterative' computation to solve cyclic dependencies here ??? ***

It is important to realize that whilst no explicit graph primitives have yet been considered or discussed it is already possible to express complex and derivative facts about graph areas and nodes. The syntax covers the same logic as the union, intersect, diff, filter and pattern primitives in Kaplan's Semantic Graph Query Language (***** should I give a detailed proof of this?*****). In particular the syntax already detailed covers the sub-graph matching problem.

If one is seeking to develop a pragmatic language which still has the power to express the full range of graphical algorithms then the next important task to tackle is the fuzzy sub-graph matching problem. An alternative way of viewing this is that one is replacing a particular combination of constrained relationships with a single concrete relationship. As a gentle example lets express that two people are 'linked' if they have lived at two or more of the same addresses; for good measure we will store the strength of the link too.

Linked := ASSOCIATION(Person Who, Person WhoElse, INT Weight);
GLOBAL: LivesAt(#1,#2),LivesAt(#3,#2) ,#1<>#3,COUNT(#2)>1 => Linked(#1, #3,COUNT(#2));

This introduces the pattern matching elements; #1, #2 etc. They function similarly to Prolog. Starting left to right the system finds all unique solutions to the primitives that are expressed for the entire scope although the predicate and production are still referring to one entity (or really solution) at a time.

A special and important case of this allows for far more flexible paths to be found between two entities. These are problems usually characterized as search or path-finding algorithms and more colloquially as the 'Kevin Bacon Problem'. We believe this problem to be fundamental enough to be the first of the graph algorithms baked directly into the language: as the LINK predicate.

At its simplest this allows for a path up to a certain length to be specified:

GLOBAL: LINK(#1,Linked*3,#3) => Networked(#1,#3)

It is also possible to use heterogeneous associations as the links and also to constrain aggregate properties of those associations. Thus—

Colleague := ASSOCIATION(Person Who, Person WhoElse, INT Weight);
GLOBAL: WorksAt(#1,#2),WorksAt(#3,#2) => Colleague(#1,#3,COUNT(#2));
GLOBAL: LINK(#1,#2=[Linked,Colleague]*3,#3),SUM(#2,Weight)>1   =>
Networked(#1,#3);

***** There are a bunch of other important cases of LINK—should I call
them out ??? ******

What is not handled in the foregoing are algorithms that can only be solved
viewing the graph (or sub-graph) as a whole; classical examples are the travelling
salesman problem, graph coloring and graph-partitioning. It is quite possible that
some of these algorithms will prove sufficiently fundamental that some more
sophisticated syntax is useful to handle them. Notwithstanding KEL also aims to
have a generic 'graph function calling' syntax to allow the system to be fully
extensible. This syntax is based upon the observation that graph functions, with
graph results, either partition, sequence or sequence and partition the nodes and
edges of a graph.[12] Thus the scope is the graph the algorithm acts upon, the
predicate lists the algorithm and any parameters required. The predicate should
succeed for all of those entities for which a data change is required and the output of
the algorithm is then recorded in the production. Thus:

PERSON: GRAPHCOLORING([Colleague,Linked],4) => MyColor = #1[13]

Section 4—Query Phase

It cannot be stressed enough that a primary indicator of the success of KEL will be
the extent to which logic exists within the Logic Phase and not Query Phase. That
said, with the exception of 'scope' much of the syntax is common between the two
phases. The object of the query phase is to define one or more parameterized
questions to which answers are required. Thus:

//Finds all occurrences of a person with a particular name and brings back some
information for them aggregated from moderately complex local graph operations

QUERY: MyQueryName(string fname1,string lname1) <= Person(fname=fname1,
lname=lname1){fname,lname,VehicleWorth,Count(Linked)};

//Does a much fuzzier match using derived linkage information and also the
output of a global graph algorithm.

//Notice that the => (production) syntax has been reversed <=. This is to express
the conceptual difference between A&B => C and if you want to know C then look
for A&B.

---

[12]There is a separate category of graph function which returns scalar results; these are covered by
the syntax discussed already.

[13]At the moment it is envisaged that the outputs of an algorithm are only recorded in the algorithm
declaration and implicitly appear as #1, #2 etc. in the production. This may prove too sloppy if
algorithms with many, many outputs are invented.

QUERY: MyQueryName(string fname1,string fname2) <= Person(fname=fname1, MyColor='Blue',Networked(Person(fname=fname2),MyColor='Pink'));

It is possible for a query to have multiple outputs and for second and subsequent outputs to reference earlier ones.

It is important to understand that the fact that two queries are defined at the same time and the fact that they are parameterized is one of the most important, and painful, aspects of the KEL implementation. The job of the KEL optimizer is to load and execute as much of the logic phase as is necessary to service the two queries with as low latency as possible. In the instance given here it would probably decide that VehicleWorth was worthy of pre-computation and storage; it would also probably compute the Linked association and perform the global graph coloring algorithm putting the result into temporary storage. A rather more controversial and interesting question is whether or not it should also construct an index of the nodes by first name and even a Pink/Blue index of networked pairs.

It is possible that the QUERY statement will need to grow some 'hint' syntax to allow the expected frequency and variety of query calls to be evaluated to ascertain the value of precomputation.

KEL—The Summary

It cannot be stressed enough that the foregoing is the result of detailed experienced in data and data operations but only exploratory research into the best syntax and semantics for the KEL language itself. That said it is believed that the basic structure and framework described above is sound and will prove useful for expressing knowledge engineering problems. An extremely simple prototype of KEL has been produced and was deployed against the IMDB database which essentially consists of pairs of Actor/Movie name. The KEL program was:

```
Actor := ENTITY( FLAT(UID(ActorName),Actor=ActorName) )
Movie := ENTITY( FLAT(UID(MovieName),Title=MovieName) )
Appearance := ASSOCIATION( FLAT(Actor Who,Movie What) )
USE IMDB.File_Actors(FLAT,Actor,Movie,Appearance)
CoStar := ASSOCIATION( FLAT(Actor Who,Actor WhoElse) )
GLOBAL:  Appearance(#1,#2)  Appearance(#3,#2)  =>  CoStar
(#1,#3)
QUERY:FindActors(_Actor) <= Actor(_Actor)
QUERY:FindMovies(_Actor) <= Movie(UID IN Appearance(Who IN
Actor(_Actor){UID}){What})
QUERY:FindCostars(_Actor) <= Actor(UID IN CoStar(Who IN
Actor(_Actor){UID}){WhoElse})
```

```
QUERY:FindAll(_Actor)    <=   Actor(_Actor),Movie(UID    IN
Appearance(Who IN $1{UID}){What}),Actor(UID IN CoStar(Who
IN $1{UID}){WhoElse})
```

This (working) ten line program produced just over 150 lines of ECL:

```
//KEL V0.1alpha generated ECL

//Layout for ENTITY type actor
actor_layout := RECORD
  UNSIGNED8 uid; // Usually comes from  a list of values
  STRING actor {MAXLENGTH(2048)}; // Usually comes from
actorname
  KEL_Instances := 1;
END;

MAC_actor_Into_Default(i,o) := MACRO
#uniquename(into)
actor_layout %into%(i le) := TRANSFORM
  SELF.uid := HASH(le.actorname); // Produces HASH of values in
'unique' field combo
  SELF.actor := le.actorname;
END;
o := PROJECT(i,%into%(LEFT));
ENDMACRO;


MAC_actor_Into_Default(imdb.file_actors,imdb_file_actors_as_acto
r) // Cast imdb.file_actors into type actor

//Layout for ENTITY type movie

movie_layout := RECORD
  UNSIGNED8 uid; // Usually comes from  a list of values
  STRING title {MAXLENGTH(2048)}; // Usually comes from
moviename
  KEL_Instances := 1;
END;

MAC_movie_Into_Default(i,o) := MACRO
#uniquename(into)
movie_layout %into%(i le) := TRANSFORM
  SELF.uid := HASH(le.moviename); // Produces HASH of values in
'unique' field combo
  SELF.title := le.moviename;
END;
o := PROJECT(i,%into%(LEFT));
ENDMACRO;


MAC_movie_Into_Default(imdb.file_actors,imdb_file_actors_as_movi
e) // Cast imdb.file_actors into type movie
```

```
//Layout for ENTITY type appearance
appearance_layout := RECORD
  UNSIGNED8 who; // Usually comes from  a list of values
  UNSIGNED8 what; // Usually comes from  a list of values
  KEL_Instances := 1;
END;

MAC_appearance_Into_Default(i,o) := MACRO
#uniquename(into)
appearance_layout %into%(i le) := TRANSFORM
  SELF.who := HASH(le.actorname); // Produces HASH of values in
'unique' field combo
  SELF.what := HASH(le.moviename); // Produces HASH of values in
'unique' field combo
END;
o := PROJECT(i,%into%(LEFT));
ENDMACRO;


MAC_appearance_Into_Default(imdb.file_actors,imdb_file_actors_as
_appearance) // Cast imdb.file_actors into type appearance

actor_before_dedup := imdb_file_actors_as_actor;
actor_layout actor_roll_transform(actor_layout le,actor_layout
ri) := transform
  self.kel_instances := le.kel_instances+ri.kel_instances;

  self := le;
end;
actor_value := rollup( sort( distribute( actor_before_dedup,
hash(uid) ),   RECORD, local), actor_roll_transform(left,right),
RECORD, EXCEPT KEL_Instances,local);

findactors(STRING _actor) := MODULE
  EXPORT Q1_value := actor_value(_actor = actor);
  EXPORT D1 := output(Q1_value,NAMED('findactors1'));
  EXPORT DoAll := PARALLEL(D1);
END;

movie_before_dedup := imdb_file_actors_as_movie;
movie_layout movie_roll_transform(movie_layout le,movie_layout
ri) := transform
  self.kel_instances := le.kel_instances+ri.kel_instances;
  self := le;
end;
movie_value := rollup( sort( distribute( movie_before_dedup,
hash(uid) ),   RECORD, local), movie_roll_transform(left,right),
RECORD, EXCEPT KEL_Instances,local);
```

```
appearance_before_dedup := imdb_file_actors_as_appearance;
appearance_layout appearance_roll_transform(appearance_layout
le,appearance_layout ri) := transform
  self.kel_instances := le.kel_instances+ri.kel_instances;
  self := le;
end;
appearance_value := rollup(
sort(appearance_before_dedup,RECORD),
appearance_roll_transform(left,right), RECORD, EXCEPT
KEL_Instances,local);

findmovies(STRING _actor) := MODULE
  EXPORT Q1_value := movie_value(uid IN
SET(TABLE(appearance_value(who IN SET(TABLE(actor_value(_actor =
actor),{uid}),uid)),{what}),what));
  EXPORT D1 := output(Q1_value,NAMED('findmovies1'));
  EXPORT DoAll := PARALLEL(D1);
END;

//Layout for ENTITY type costar
costar_layout := RECORD
  UNSIGNED8 who; // Usually comes from  a list of values
  UNSIGNED8 whoelse; // Usually comes from  a list of values
  KEL_Instances := 1;
END;

MAC_costar_Into_Default(i,o) := MACRO
#uniquename(into)
costar_layout %into%(i le) := TRANSFORM
  SELF.who := HASH(le.actorname); // Produces HASH of values in
'unique' field combo
  SELF.whoelse := HASH(le.actorname); // Produces HASH of values
in 'unique' field combo
END;
o := PROJECT(i,%into%(LEFT));
ENDMACRO;

// Make a value for Production_3
// Prepare those entity expressions used in the assertions
//Now the encapsulated function itself ....
Production_3 := FUNCTION
  // First move all of the assertion data into a format that
matches the #templates used
  Production_3_j0 :=
TABLE(appearance_value(true,true),{UNSIGNED8 Value_1 :=
who,UNSIGNED8 Value_2 := what});
  Production_3_assert_1 :=
TABLE(appearance_value(true,true),{UNSIGNED8 Value_3 :=
who,UNSIGNED8 Value_2 := what});
  //Now build up the production; one assertion at a time left to
right
```

```
  Production_3_after_2_asserts := RECORD
    typeof(Production_3_j0.Value_1) Value_1;
    typeof(Production_3_assert_1.Value_3) Value_3;
  END;
  Production_3_after_2_asserts Production_3_jt2(Production_3_j0
le,Production_3_assert_1 ri) := TRANSFORM
    SELF := le;
    SELF := ri;
  END;
  Production_3_j1 :=
JOIN(Production_3_j0,Production_3_assert_1,left.Value_2 =
right.Value_2,Production_3_jt2(left,right));
  //Finally the intermediate format of the assertions needs to
be mapped to the production
  RETURN PROJECT(Production_3_j1,TRANSFORM(costar_layout,
SELF.who := LEFT.Value_1; SELF.whoelse := LEFT.Value_3));
END;
Production_3_as_costar := Production_3;

costar_before_dedup := Production_3_as_costar;
costar_layout costar_roll_transform(costar_layout
le,costar_layout ri) := transform
  self.kel_instances := le.kel_instances+ri.kel_instances;
  self := le;
end;
costar_value := rollup( sort(costar_before_dedup,RECORD),
costar_roll_transform(left,right), RECORD, EXCEPT
KEL_Instances,local);

findcostars(STRING _actor) := MODULE
  EXPORT Q1_value := actor_value(uid IN
SET(TABLE(costar_value(who IN SET(TABLE(actor_value(_actor =
actor),{uid}),uid)),{whoelse}),whoelse));
  EXPORT D1 := output(Q1_value,NAMED('findcostars1'));
  EXPORT DoAll := PARALLEL(D1);
END;

findall(STRING _actor) := MODULE
  EXPORT Q1_value := actor_value(_actor = actor);
  EXPORT D1 := output(Q1_value,NAMED('findall1'));
  EXPORT Q2_value := movie_value(uid IN
SET(TABLE(appearance_value(who IN
SET(TABLE(Q1_value,{uid}),uid)),{what}),what));
  EXPORT D2 := output(Q2_value,NAMED('findall2'));
  EXPORT Q3_value := actor_value(uid IN
SET(TABLE(costar_value(who IN
SET(TABLE(Q1_value,{uid}),uid)),{whoelse}),whoelse));
  EXPORT D3 := output(Q3_value,NAMED('findall3'));
  EXPORT DoAll := PARALLEL(D1,D2,D3);
END;
```

KEL Present and Future

KEL version 0.4 has been released, and version 0.5 is underway. The first LexisNexis production systems relying on KEL for graph processing are being developed and the interest on KEL from the HPCC Systems Open Source community is slowly growing. An interface to support RDFS and SPARQL has been proposed and the work on this is about to start. While the language is still in development, most general aspects have been defined and additional optimizations are being constantly introduced into the KEL and ECL compilers to introduce further efficiencies. For those readers that are interested in additional information about KEL, there is a very good blog series by one of the authors, which includes a general tutorial on the basic functionality of the language [17].

References

1. http://en.wikipedia.org/wiki/Seven_Bridges_of_K%C3%B6nigsberg.
2. Euler L. Solutio Problematis ad Geometriam Situs Pertinentis. Novi Commentarii Academiae Scientarium Imperialis Petropolitanque 7(1758–59), 9–28.
3. Hierholzer C. Uber die Moglichkeit, einen Linienzug ohne Wiederholung und ohne Unterbrechnung zu umfahren. Math Ann. 1873;6:30–2.
4. Biggs NL, et al. Graph theory 1736–1936. Oxford: Clarendon Press; 1986.
5. Agnarsson G. Graph theory: modeling, applications, and algorithms. Upper Saddle River: Prentice Hall; 2006.
6. Cudre-Mauroux P et al Graph data management systems for new application domains. In: Proceedings of the VLDB Endowment, vol 4, No 12; 2011.
7. Vicknair C et al. A comparison of a graph database and a relational database. ACMSE '10, April 15–17, Oxford, MS, USA; 2010
8. Yang X et al. Summary graphs for relational database schemas. In: Proceedings of the VLDB Endowment, vol 4, No 12; 2011.
9. Shao B et al. Managing and mining large graphs: systems and implementations. SIGMOD '12, May 20–24, Scottsdale, Arizona, USA; 2012.
10. http://en.wikipedia.org/wiki/Social_network_analysis.
11. Singla P et al. Yes, there is a correlation—from social networks to personal behavior on the web. WWW 2008, April 21–25, Beijing, China; 2008.
12. Malm A, et al. Social network and distance correlates of criminal associates involved in Illicit Drug Production. Secur J. 2008;21:77–94. doi:10.1057/palgrave.sj.8350069.
13. Latour J. Understanding consumer behavior through data analysis and simulation: Are Social Networks changing the World economy? Master Thesis. http://essay.utwente.nl/58146/.
14. Averbuch A et al. Partitioning graph databases—a quantitative evaluation. Master of Science Thesis Stockholm, Sweden; 2010. arXiv:1301.5121.
15. Plantikow S et al. Latency-optimal walks in replicated and partitioned graphs. In: DASFAA Workshops 2011, LNCS 6637, pp 14–27; 2011.
16. Middleton A. Data-intensive technologies for cloud computing. In: Handbook of cloud computing. Berlin: Springer; 2010
17. http://hpccsystems.com/blog/adventures-graphland-v-graphland-gets-reality-check.

Part III
Big Data Applications

Chapter 12
HPCC Systems for Cyber Security Analytics

Flavio Villanustre and Mauricio Renzi

Many of the most daunting challenges in today's cyber security world stem from a constant and overwhelming flow of raw network data. The volume, variety, and velocity at which this raw data is created and transmitted across networks is staggering; so staggering in fact, that the vast majority of data is typically regarded as background noise, often discarded or ignored, and thus stripped of the immense potential value that could be realized through proper analysis. When an organization is capable of comprehending this data in its totality—whether it originates from firewall logs, IDS alerts, server event logs, or other sources—then it can begin to identify and trace the markers, clues, and clusters of activity that represent threatening behavior (Fig. 12.1).

Today's biggest cyber challenges, which include the emergence of the advanced persistent threat, take advantage of the data deluge described above to establish long-term footholds, exploit multiple vulnerabilities, and deliver malicious payloads, all while avoiding detection. This white paper will focus on the big data processing platform from LexisNexis called HPCC Systems, (High Performance Computing Cluster) as a technology platform to ingest and analyze massive data that can offer meaningful indicators and warnings of malicious intent.

In contrast to current approaches, the effectiveness of the HPCC Systems solution increases as data volumes grow into the hundreds of terabytes to petabyte range. Not only does this solution provide the ability to fuse an organization's own network data (e.g. firewall logs, access logs, IDS alert logs, etc.), but also it delivers enrichment routines that can automatically incorporate and fuse any relevant 3rd party data set, including blacklists, known bad domains, geo-location data, etc. Finally, HPCC Systems delivers this capability at speeds that cannot be achieved by

This chapter has been adopted from the LexisNexis white paper on "HPCC Systems for Cyber Security Analytics: See through patterns, hidden relationships and networks to find threats in Big Data", LexisNexis.

© Springer International Publishing Switzerland 2016
B. Furht and F. Villanustre, *Big Data Technologies and Applications*,
DOI 10.1007/978-3-319-44550-2_12

331

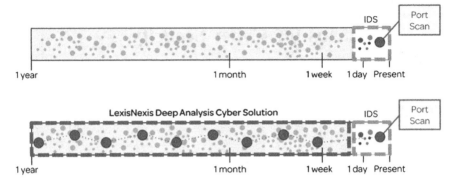

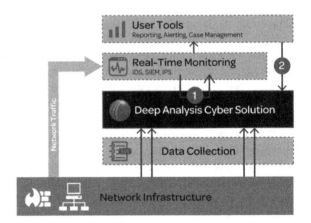

Fig. 12.1 Continuous monitoring technologies alone cannot perform long-term contextual analysis. When deployed in conjunction with the Deep Analysis Cyber Solution, new alerts from these technologies can be linked to historical behavior and attack patterns to identify persistent, "low and slow" attacks

Fig. 12.2 While real-time monitoring tools continuously analyze network traffic, the Deep Analysis Cyber Solution collects raw data, and routinely applies a series of long-term analytics. The results of these queries are made available to (1) continuous monitoring tools to enrich and validate alerts, and (2) end users via web-based searches or other applications

typical database-oriented approaches. The results of HPCC Systems large-scale analytics can provide an administrator with a significant forensic advantage and a tremendous head start in quickly verifying the significance of a potential incident (Fig. 12.2).

The Advanced Persistent Threat

Consider the following situation: the young man emerged from his supervisor's office with a job to do. This man—a systems administrator at a large laboratory in on the west coast—was asked to investigate a minuscule accounting error in his lab's computer usage accounts. He likely had more interesting things to do on a sunny August day than research a $0.75 discrepancy, but the more he pored over access logs and system files, the more curious—and suspicious—he became. As he followed the faint trail of electronic breadcrumbs, he realized that this seemingly benign "accounting error" was the result of intentionally malicious activity. It was, in fact, the first of many subtle clues left behind by a hacker who had gained access to the administrator's network by exploiting a vulnerability in the lab's email system.

Was this simply a prank? Could it have been a bored college student with too much free time and a mischievous streak? This seemingly minor incident turned out to be the start of a 10-month journey that would take the systems administrator from his network's data center all the way to the heart of several U.S. military networks. He would employ all his accumulated knowledge to monitor, analyze, and deploy various kinds of electronic bait for the hacker. Over time, he was able to discover not only the root of these intrusions, but the motivation behind them as well. He traced the attacks to a hacker living in central Europe. Using a persistent, methodical approach—over a period of months—this hacker had not only gained access into the laboratory's network, but was able to exploit a number of inter-connections between national labs, government agencies, and government contractors to gain root level access to military computers around the United States. Furthermore, the hacker used this access to download hundreds of sensitive documents related to nuclear weapons and defense programs. This hacker, as it turned out, had been systematically exploiting network vulnerabilities to access these facilities and was selling the information he had illegally obtained to agents of a foreign government.

While this tale reads like a cyber attack that might have been pulled from today's headlines, these events actually unfolded in a year remembered more for Chernobyl, the Iran-Contra affair, and the Space Shuttle Challenger disaster—1986. This was an early form of a network attack now referred to as "Advanced Persistent Threat" (APT). Like this example, a typical modern day APT is characterized by:

- Attackers who are typically funded and directed by external entities, organizations and governments.
- Attackers who utilize the full spectrum of intelligence collection methods, which may include computer intrusion technologies as well as coordinated human involvement and social engineering techniques to reach and compromise their target.
- An attack that is conducted through continuous monitoring and interaction in order to achieve the defined objectives.
- An attack that, rather than relying on a barrage of continuous intrusions and malware, employs a "low-and-slow" approach.

As our world grows increasingly more connected, LexisNexis recognizes that Advanced Persistent Threats have the potential to cause increasingly significant damage to critical infrastructure, financial systems, and to sensitive military operations. Examples are frighteningly numerous, and include well-publicized incidents such as Stuxnet,[1] Titan Rain,[2] and Operation Aurora.[3] While the basic features of the threat are not much different than they were in 1986, the potential for damage has been greatly magnified.

Try as we might, these attacks are difficult to detect. Why? The attackers are careful, patient, well funded, and highly motivated. They tend to apply methodical techniques that keep their activities under the radar. Whereas an attack such as a distributed denial of service is generally hard to miss; the probes, connections, and malicious downloads performed by sophisticated actors over the course of months or years are easily obscured by huge volumes of routine network activity.

Therein lies the crux of the problem. With the rise of mobile computing, distributed data storage, cloud infrastructures, and Internet-enabled telecommuting, we are witnessing three phenomena which provide attackers tremendous opportunities to do harm:

1. Multimedia, networked collaboration, and mass participation have resulted in a constant deluge of heterogeneous network activity, both within private networks and across the public Internet. This provides constant "cover" to the activities of malicious users.
2. Through technologies like connected mobile devices, virtual private networks, and cloud computing, the notion of a corporate network has expanded well beyond the traditional firewall-based perimeter. This provides significantly more vulnerabilities and access points through which malicious users can gain entry to protected resources.
3. The combination of increased activity and expanded network architectures has complicated network security. Maintaining a secure posture is a result of constant vigilance and mitigating one's risk through vulnerability management and continuous monitoring. However, security technologies are unable to keep pace with the evolving threat landscape, and as a result traditional approaches have shown severe limitations. These limitations, and the ability to overcome them, are the focus of this chapter.

---

[1]Stuxnet is a Microsoft Windows computer worm discovered in July 2010 that targets industrial software and equipment. Source: http://en.wikipedia.org/wiki/Stuxnet.

[2]Titan Rain was the U.S. government's designation given to a series of coordinated attacks on American computer systems since 2003. Source: http://en.wikipedia.org/wiki/Titan_Rain.

[3]Operation Aurora is a cyber-attack that originated in China, and occurred from mid-2009 through December 2009. The attack targeted dozens of major corporations, including Google. Source: http://en.wikipedia.org/wiki/Operation_Aurora.

LexisNexis HPPS Systems for Deep Forensic Analysis

HPCC Systems is a massively parallel analytics platform that delivers two large-scale, long-term data fusion capabilities. When applied to the cyber security domain, HPCC Systems provides network security teams the ability to transform massive data to intelligence in a manner that would be impossible for traditional data mining and analysis technologies. The HPCC Systems technology is optimized for aggregating, fusing, and analyzing massive, multi-source, multi-format data sets. It delivers an analytics capability that bridges the data gap between today's short-term operational data analysis and the deep situational understanding that only comes with large-scale data analytics.

The two core capabilities of the HPCC Systems include:

1. **Pre-computed Analytics**—combined with continuous monitoring tools such as IDS and SIEM, pre-computed analytics improve the quality and accuracy of alerts by instantly comparing alert metadata against a comprehensive repository of historical network patterns and computed behaviors. This adds relevance and context to real-time alerts to not only identify and tag false positives, but to also associate seemingly benign activity with longer term, more serious threats, thereby addressing the "false negative" dilemma.

2. **Deep Forensics Analysis**—using an advanced query engine, security analysts with deep domain expertise and technical skills can routinely perform highly customized, sophisticated analysis as needed—against massively complex data sets. For instance, an analyst might want to execute a complex correlation across months' worth of log files originating from dozens of device types. For a typically large network, not only would this analysis need to fuse data of multiple formats, but it would also be required to correlate potentially hundreds of Terabytes (or more) of raw data. HPCC Systems delivers this capability at speeds that cannot be achieved by typical database oriented approaches.

Combined, these capabilities are meant to tackle the challenge of "big data" in order to eliminate the cover attackers rely on and to provide better visibility into the increasing number of network entry points and vulnerabilities.

Pre-computed Analytics for Cyber Security

There is a compelling need for new types of analytics, focused on massive, long-term data sets. The federal government is pressing its agencies for "continuous monitoring" of government networks. However, traditional approaches to continuous monitoring are limited by the amount of data they can analyze. As a result, systems such as Intrusion Detection Systems are restricted to performing "selective analysis"—either looking at some metadata subset (selected fields from packet headers or netflow messages, for instance), or sampling network data to seek statistically significant patterns (analyze one of every thousand packets). Additionally, signature-based detection algorithms typically analyze data over relatively small time windows (hours or days), and so tend only to be able to detect short-term activities. In contrast to these technologies, LexisNexis approach to automated information analysis:

- Processes all data, regardless of the overall volume.
- Merges data from many sources, whether they are structured data sets, unstructured text, or feeds from external sources.
- Performs full-text analysis on all fields in the data.
- Executes large scale analysis in a timely fashion, thanks to a massively parallelized data processing technology.

HPCC Systems delivers a default library of configurable input adapters and pre-computed cyber security analytics. The computed results of these analytics can be called by end users through a web-based search interface, or automatically queried by 3rd party solutions such as Intrusion Detection Systems (IDS) and Security Information Event Managers (SIEM). They are "pre-computed" in that analytic results are routinely calculated over all available data, and prepared for consumption by end users or third party systems.

The solution works by routinely analyzing all collected data from relevant sources, and then computing a series of analytical results. While real-time network data is fed to continuous monitoring systems via techniques such as packet capture; the HPCC Systems can routinely operate on aggregated log output from network devices (firewalls, routers, servers, etc.) and security systems (vulnerability management, NIDS, HIDS, antivirus, etc.). When this aggregated output is ingested, HPCC Systems can fuse all data, and re-computes a series of cyber analytics against either the entire set of data, or just incremental portions. Once the analytics have been applied to the target data, computed results are persisted, indexed and prepared for delivery via standard web services or web-based interfaces.

From a forensics perspective, this query library comprises a comprehensive set of long term patterns that a cyber security operator would want to identify as part of an investigation into any potential exploit or alert.

Unlike typical analytical approaches, these queries are asked at scale, across a period of months and against tremendous volumes of data. The combined results of these queries can provide an administrator with a significant forensic advantage and a tremendous head start in concluding the significance of a potential incident.

Some of these default queries include:

Wavelet transforms. This mathematically derived set of functions was developed to analyze data across both frequency and temporal scales. This type of algorithm is especially useful for recovering a "true signal" from very noisy data sets, without requiring prior knowledge of anomalous patterns or needing explicit signatures to be defined. Standard wavelet algorithms can be effectively distributed across the Deep Analysis Cyber Solution's distributed computing cluster, and therefore can be applied to uncover a "true signal" within tens or hundreds of TBs of network "noise".

Information theory. This is a technique for distinguishing malicious network scan traffic from normal activity using general data compression tools. This method is particularly effective when a malicious user employs techniques that cause repetitive communication patterns, such as port scans occurring over large time periods utilizing multiple IP addresses to obfuscate malicious activities.

Slow oscillators. This query identifies groups of suspicious hosts by detecting communications patterns characterized by a small number of packets being sent between two hosts, or infrequent and irregular communications intervals.

Low hitters. This query identifies groups of suspicious hosts by detecting small numbers of systems in a large network that are suddenly making large quantities of DNS queries for new/previously unseen DNS entries. This can be representative of compromised machines trying to communicate back to command and control nodes that utilize "fast flux" techniques to constantly change the IP address.

Network activity clustering. Clustering can be used to find groups of features in the data. These might be IP addresses that demonstrate similar traffic patterns such as beacons, botnets and Trojans or groups of email servers. Various clustering techniques are employed, including K-means and leader clustering for processing IP packet data and NetFlow data.

Data exfiltration. This query identifies hosts, for any given time period, that are transmitting above or below a bytes per packet threshold for traffic on a given port.

Session hijacking detection. This query analyzes activity over every network connection to identify anomalies, such as a changing user-agent string, that are indicative of session hijacking activity.

Rare targets. This query seeks out long-term activities that are limited to a small number of hosts. For instance, it may look for external domains that are targeted by fewer than 10 different IP addresses, or oscillators that have targeted only rare domains.

Given a combination of search parameters (e.g. source IP, destination IP, source port, etc.), these queries can be executed and the results merged with the output of a traditional alerting system, such as IDS. This enhanced intelligence layer is used to enrich the IDS alert and to more effectively validate and prioritize it.

The Benefits of Pre-computed Analytics

This capability improves an organization's continuous monitoring capability by allowing security administrators to relax tuning constraints on their alerting systems. Today, there is a tendency to over-tune continuous monitoring tools, like IDS, to reduce alert volumes to a manageable amount. The consequence of this, however, is that the monitoring system is then only able to detect major violations and blatant exploits, allowing nearly any action that is part of an APT to go unnoticed. By relaxing IDS rules to trigger on more events, and relying on deep, historical context to prioritize the resulting alerts, this solution delivers a twofold benefit: first, deep historical context helps eliminate false positives so security administrators are left with a manageable workload, and more importantly, it reduces the number of false negatives, or malicious activity which would otherwise have gone unnoticed by an over-tuned continuous monitoring system.

Deep Forensics Analysis

While the pre-computed analytical routines of HPCC Systems represent an important capability, the key to effectively detecting and responding to Advanced Persistent Threats is in persistent vigilance and continuous human analysis. In fact, there are many cases when certain indicators and warnings push security experts into action, and when those experts are required to perform unique, "one off" analyses to make sense of malicious behavior. In these cases, a security analyst is expected to behave more like a traditional intelligence analyst, performing very deep, multi-dimensional analysis of their data.

Today's data mining technologies do not support the kind of improvisational analysis that's required to fully comprehend the nature and extent of an attack. The reasons for this vary; from the rigid relational database models that restrict the kinds of queries an investigator can perform to the inability to execute such queries against massive amounts of data within a desired timeframe (e.g. being able to perform calculations in minutes or hours instead of days).

This is where HPCC Systems shines. The technology is exceptional at this type of ad hoc usage—it is currently leveraged in multiple Federal programs specifically to fuse and link disparate structured and unstructured datasets and discover non-obvious relationships and anomalous patterns across truly massive content sets. It allows subject matter experts, possessed of deep technical skills, to write any kind of query against multi-terabyte, multi-source data sets.

The solution's non-relational storage architecture frees analysts from the shackles of traditional database-oriented approaches, where the kinds of analysis they can perform are restricted by how the data was modeled, and which individual fields the data modeler determined needed to be searchable.

In addition to fully customizing and re-purposing any of the included analytics, cyber analysts can develop, share, and reuse highly complex, custom analyses such as:

- Graph analytics to identify "hubs" of activity within massive data sets.
- Multi-watch list analysis and fusion against all-source network data.
- N2 analysis where, for instance, every internal IP address might be compared against every other internal IP address that connected to the same external host on the same day. This kind of analysis can be performed by the Deep Analysis Cyber Solution in a fraction of the time it could be performed on a traditional relational database system. In fact, this kind of query at this scale will typically fail on most relational database servers.

As new threats emerge, analysts leverage the data-oriented, declarative programming language of HPCC Systems to model new attack patterns. These routines are executed against any size data set across the system's distributed computing cluster—so the analyst can run a custom query against 1 weeks' worth of data, or 6 months' worth of data without having to worry about how to optimize the query for the target data set.

Conclusion

As more and more critical systems and mission operations become interconnected, the nature of national security and corporate threats is quickly shifting. It seems increasingly likely that the next major terrorist attack will be launched through an infected network host or a compromised industrial process rather than a suicide bomber or explosives-laden truck. Likewise the nature of conflict itself is evolving, as armies of hackers continually attempt to penetrate our most sensitive networks to steal classified information, trade secrets, intellectual property; or worse. Attacks of this nature require time, resources, and a motivated party to successfully execute. HPCC Systems has been developed to provide cyber security experts one of the tools they need to uncover the markers of an Advanced Persistent Threat before it has the opportunity to achieve its intended objectives. The HPCC Systems platform provides the capability to perform sophisticated analysis against all data over a long period of time. This comprehensive analysis—where no potential clue is ignored—is what allows the detection of subtle activities and "low and slow" attack patterns. The combination of pre-computed analytics and an ad hoc deep forensics analysis capability allows cyber security teams to both improve the quality of their ongoing automated monitoring and to quickly react to and understand new threats in an improvisational fashion.

Chapter 13
Social Network Analytics: Hidden and Complex Fraud Schemes

Flavio Villanustre and Borko Furht

Introduction

In this chapter we briefly describe several case studies of using HPCCC systems in social network analytics.

Case Study: Insurance Fraud

LexisNexis HPCC system has been applied to detect fraud insurance claims and additional linked claim. In this example, the insurance company was able to find a connection between two of the seven claims, and identified only one other claim as being weekly connected, as illustrated in Fig. 13.1.

However, LexisNexis HPCC system explored two additional degrees of relative separation, and the results showed two family groups interconnected on all of these seven claims, as shown in Fig. 13.2. The links were much stronger than the carrier data previously supported.

Case Study: Fraud in Prescription Drugs

Healthcare insurers need better analytics to identify drug seeking behavior and schemes that recruit members to use their membership fraudulently. Groups of people collude to source schedule drugs through multiple members to avoid being detected by rules based systems. Providers recruit members to provide and escalate services that are not rendered.

This chapter has been developed by Flavio Villanustre and Borko Furht.

© Springer International Publishing Switzerland 2016   341
B. Furht and F. Villanustre, *Big Data Technologies and Applications*,
DOI 10.1007/978-3-319-44550-2_13

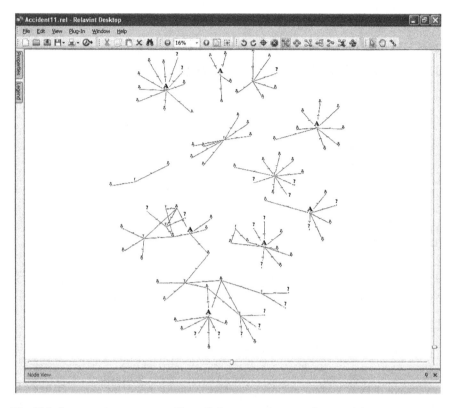

Fig. 13.1 Insurance company was able to find a connection two of seven claims

We used LexisNexis HPCC technology to resolve the following task. For a given a large set of prescriptions, we calculated normal social distributions of each brand to detect where there is an unusual socialization of prescriptions and services.

The analysis detected social groups that are sourcing Vicodin and other schedule drugs, as illustrated in Fig. 13.3. The system identified prescribers and pharmacies involved to help the insurer focus investigations and intervene strategically to mitigate risk.

Case Study: Fraud in Medicaid

The HPCC system was applied to proof of concept for Office of the Medicaid Inspector Generation (OMIG) of large Northeastern state.

The task was set as follows: for given large list of names and addresses, identify social clusters of Medicaid recipients living in expensive houses and driving expensive cars.

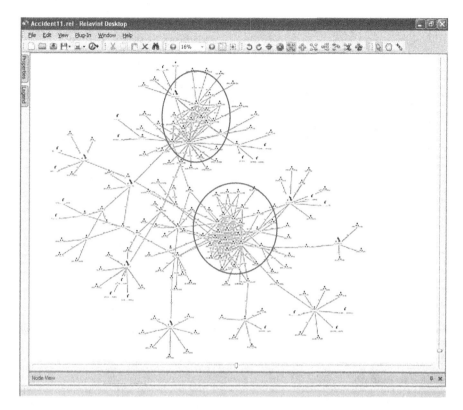

Fig. 13.2 HPCC system identified two family groups interconnected on all of the seven claims

Interesting recipients were identified using asset variables, revealing hundreds of high-end automobiles and properties. Leveraging the Public Data Social Graph, shown in Fig. 13.4, large social groups of interesting recipients were identified along with links to provider networks. Table 13.1 illustrates the number of Medicaid recipients identified by the HPCC system, who were driving expensive cars.

The analysis identified key individuals not in the data supplied along with connections to suspicious volumes of "property flipping" potentially indicative of mortgage fraud and money laundering.

Case Study: Network Traffic Analysis

Conventional network sensor and monitoring solutions are constrained by inability to quickly ingest massive data volumes for analysis. Typically 15 min of network traffic can generate 4 Terabytes of data, which can take 6 h to process. Similarly, 90 days of network traffic can add up to 300+ Terabytes.

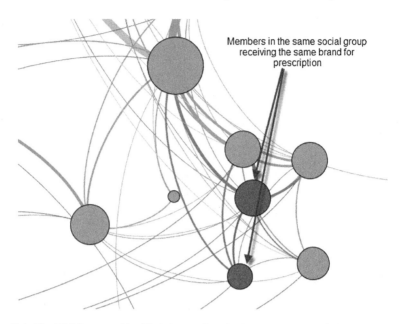

Fig. 13.3 The HPCC system identified the members in the same group receiving the same brand for prescription

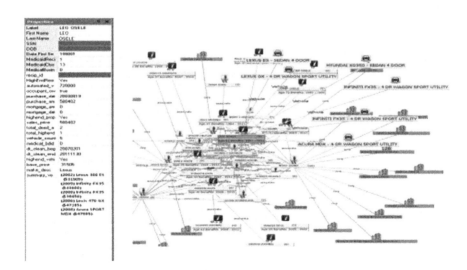

Fig. 13.4 Public Data Social Graph used in the HPCC system to identify medicaid recipients with expensive house and driving expensive cars

Make description	#	Make description	#
Table 13.1 The HPCC system identified the Medicaid recipients who were driving expensive cars			
Mercedes-Benz	46	Chevrolet	2
Lexus	41	Hummer	2
BMW	27	Jeep	2
Infiniti	13	Nissan	2
Acura	9	Toyota	2
Lincoln	8	Aston Martin	1
Audi	7	Bentley	1
Land Rover	7	Cadillac	1
Porsche	6	GMC	1
Jaguar	5	Honda	1
Mercedes Benz	3	Volkswagen	1
Saab	3	Volvo	1

In this project we analyzed all the data to see if any US government systems have communicated with any suspect systems of foreign organizations in the last 6 months. In this scenario, we look specifically for traffic occurring at unusual hours of the day.

In seconds, HPCC Systems sorted through months of network traffic to identify patterns and suspicious behavior, as illustrated in Fig. 13.5.

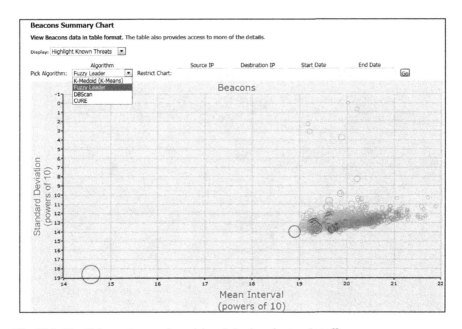

Fig. 13.5 Identifying patterns and suspicious behavior of network traffic

Case Study: Property Transaction Risk

The objective of this project was to perform large scale measurement of influences to identify suspicious transactions. The HPCC system used big data measuring over a decade of property transfers nationwide. Large Scale Graph Analytics was applied to identify suspicious equity stripping clusters. Three core transaction variables that were measured included velocity, profit, and buyer to seller relationship distance.

Based on the derived public data relationships from about 50 Terabytes database, the Collusion Graph Analytics has been performed providing chronological analysis of all property sales. Based on this analysis, large scale suspicious cluster ranking has been done, and persons and network indicators were counted.

The system was able to identify known ringleaders in flipping and equity stripping schemes; they were typically not connected directly to suspicious transactions. The system also identified clusters offloading property and generating defaults and clusters with high levels of potential collusion.

Chapter 14
Modeling Ebola Spread and Using HPCC/KEL System

Jesse Shaw, Flavio Villanustre, Borko Furht, Ankur Agarwal and Abhishek Jain

Introduction

Epidemics have disturbed human lives for centuries causing massive numbers of deaths and illness among people and animals. Due to increase in urbanization, the possibility of worldwide epidemic is growing too. Infectious diseases like Ebola remain among the world's leading causes of mortality and years of life lost. Addressing the significant disease burdens, which mostly impact the world's poorest regions, is a huge challenge which requires new solutions and new technologies. This paper describes some of the models and mobile applications that can be used in determining the transmission, predicting the outbreak and preventing from an Ebola epidemic.

Infections can be caused by various infectious agents like viruses, viroids, prions, bacteria, nematodes and many other macro parasites. Infectious disease if are easily transmitted to others by contact with an ill person or their secretions it is referred to as contagious disease. There are many contagious diseases like HIV/AIDS, Tuberculosis, Pneumonia, Malaria, Ebola. Table 14.1 shows the death tolls of some of the contagious diseases.

As can be seen from the above table, contagious diseases are one of the major cause of deaths in many regions. An outbreak of a communicable disease associated with a high fatality rate in the rural forest communities of Guinea has spiraled into an epidemic that is ravaging West Africa and evoking fear around the globe. Ebola virus (EBOV, formerly Zaire ebolavirus), one of the five species of genus Ebolavirus in the family Filoviridae, has been identified as the causative agent of this unprecedented epidemic, in terms of initial geographic occurrence, magnitude, complexity and persistence. EBOV was involved in previous outbreaks in remote regions of Central Africa, with the largest, in the Democratic Republic of Congo in

This chapter has been developed by Jesse Shaw and Flavio Villanustre, LexisNexis Risk Solutions, and Borko Furht, Ankur Agarwal, and Abhishek Jain, Florida Atlantic University.

Table 14.1 Death tolls of some of the infectious disease till date

Disease	Death tolls
HIV/AIDS	34 million [21]
Tuberculosis	1.5 million [22]
Malaria	438,000 [23]
Ebola	11,315

1976, accounting merely 318 cases, including 280 deaths [1]. The first case of current West African Epidemic of Ebola virus disease was reported on March 22, 2014 with the report of 49 cases in Guinea [2]. By February 28, 2016, the World Health Organization had reported, 28,639 probable, confirmed and suspected cases out of which 15,484 in Liberia, 18,080 in Sierra Leone and 6340 in Guinea.

Five different Ebola virus strains have been identified, namely Zaire Ebola virus (EBOV), Sudan ebolavirus (SUDV), Tai Forest Ebola virus (TAFV), Bundibugyo Ebola virus (BDBV) and Reston Ebola virus (RESTV), with fruits bats considered as the most likely reservoir host. The great majority of past Ebola outbreaks in humans have been linked to three Ebola strains: EBOV, SUDV and BDBV [3]. EBOV is identified as the deadliest of the five Ebola virus strains [3] and its name was derived from the Ebola River located near the source of the first outbreak. Ebola is characterized by a high case fatality ratio which was nearly 90 % in a past outbreak.

After an incubation period mostly ranging from 2 to 21 days, influenza like symptoms appear, including sudden onset of fever, weakness, vomiting, diarrhea, decreased appetite, muscular pain, joint pain, headache and a sore throat. Fever is usually higher than 101 °F [4]. In some cases, skin may even develop a maculopapular rash within 5–7 days. A fraction of patients may later develop severe internal and external hemorrhagic manifestations and experience multiple organ failures.

Human epidemics subsequently take off by direct human-to-human contact via bodily fluids or indirect contact with contaminated surfaces. Body fluids that may contain Ebola viruses include saliva, mucus, vomit, feces, sweat, tears, breast milk, urine and semen [4]. Unsafe burials that involve direct contact with Ebola-infected bodies also pose a major infection risk [CH NI 14]. Entry point for the virus include nose, mouth, eyes, open wounds, cuts and abrasions. Healthcare workers treating people with Ebola infection are at greatest risk, which is increased if the workers do not have proper protective clothing such as masks, gowns, gloves and eye protection. Human-to-human transmission of EBOV through air has not been reported to occur during the Ebola outbreaks. However, it has been demonstrated in very strict laboratory conditions [4].

After the outbreak various models have been developed to predict and prevent the spread of the disease. Since social media has become one of the primary means by which people learn about worldwide developments, characterization of both news and rumors on Twitter about Ebola has been done in order to understand the popularity of misinformation. Tweets about Ebola peaked in late September through mid-October 2014, when there was extensive reporting on the disease in the US and Europe. On September 30, 2014, the Centers for Disease Control and Prevention (CDC) confirmed first importation of Ebola in US when Thomas Eric

Duncan exposed to virus in Monrovia travelled to Dallas. On October 8, Duncan died at Texas Health Presbyterian Hospital in Dallas and few days later a healthcare worker who was attending Duncan tested positive for the disease. On October 14, another healthcare worker at the hospital reported low grade fever and was isolated.

In the last decade, digital health tools like mobile apps have become very useful tool in stopping these infectious diseases from spreading. As said by Dr. Hussain to New York Times "Physicians are becoming more knowledgeable in this area. Apps are going to be most effective when used in conjunction with your physician" [5]. During the Ebola outbreak in 2014, many different mobile apps were developed in order to inform and educate people to prevent people from future outbreak occurrences. When Ebola was first discovered in Nigeria, Google trends was the first app used to stop the spread the deadly virus. It was also used by journalist to pin point the most frequently asked questions about Ebola and provided answers to them in order to educate the public [6].

Survey of Ebola Modeling Techniques

In the following section, we provide a detailed analysis of various contributions done towards the Ebola transmission, prediction, and prevention.

Basic Reproduction Number (R_0)

The basic reproduction number also known as basic reproductive ratio is the number of cases one case generates on average over the course of its infectious period. This metric is very useful in determining whether an infectious disease can spread through the population [7]. If R_0 is less than 1, the transmission chain is not self-sustaining and is unable to generate any major epidemic. But if R_0 is greater than 1, an epidemic is likely to occur. When measured over time t, the effective reproduction number R_t, can be helpful to quantify the time-dependent transmission potential and evaluate the effect of control interventions in almost 'real time' [3].

The incidence can be modelled as:

$$i(t) = k \exp(rt)$$

where k is constant. This equation is integrated from starting time t_0 to the latest time t,

$$I(t) = k/r[\exp(rt) - \exp(rt_0)]$$

The growth rate of Sierra Leone is divided into two phases i.e. early phase and late phase for which R_0 is 3.07 and 1.30 respectively and for Liberia it is 1.96 [3].

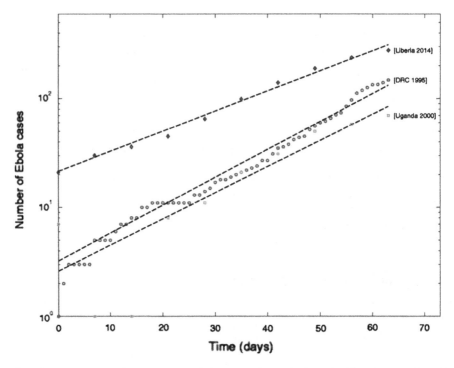

Fig. 14.1 Comparison of the growth trends for past outbreaks in Central Africa (Congo 1995 and Uganda 2000) with the ongoing Ebola epidemic in Liberia [3]

The comparison growth trends for past outbreaks in Central Africa with the current outbreak in Liberia is shown in Fig. 14.1.

There are some limitations with R_0, especially when calculated from mathematical models, particularly ordinary differential equations do not provide a true value of R_0 as they cannot compare different diseases [7]. Therefore, these values should be used with caution, especially if the values are calculated from mathematical models.

Case Fatality Rate (CFR)

Case Fatality rate or case fatality risk is calculated by dividing the total number of deaths that have occurred due to a certain condition by total number of cases. For infectious disease, this is very important measure to estimate because it tells us the probability of dying after infection [8]. The so-called Zaire strain is considered to be slightly more fatal than the Sudan strain. While the CFR for the Sudan strain ranges from 41 to 65 %, the CFR for the Zaire strain ranges from 61 to 89 %. The CFR of the ongoing epidemic among cases with definitive recorded clinical outcomes for

Guinea, Liberia and Sierra Leone has been consistently estimated at 70.8 %, which is in good agreement with estimates from prior outbreaks. It is important to follow up the reasons why the estimated 53 % in real-time has been much lower than the published estimate of 70.8 % among a portion of cases [3].

The case fatality rate is not a reliable calculation for an ongoing epidemic as firstly, it doesn't take into account the infections that have yet to run their course and if many few cases are being reported then it is under estimate of the CFR. Secondly, there can be bias in reporting and diagnosing towards severe cases of the disease which will give the overestimate of CFR. With Ebola virus, bias can occur if patients are being looked at home or are hospitalized only if severe or if patient dies [9].

SIR Model

Individuals in these model are labelled in three compartments: Susceptible (S), Infected (I) and Recovered (R). The model is based on the following assumptions: (1) the networks are homogenous which means that all nodes have same linkage and the probability that there is a link between any two nodes are equal, (2) susceptible individuals can get infected from infected individuals via contacts, and (3) an infected individual becomes immune after recovering from the disease [10].

When there is a significant number of infected individuals in a community, the effected contacts become susceptible and infected individuals do not grow quickly. This phenomenon is called "crowding" or "protection effect". The particular SIR model ignores the crowding effect and hence have some unrealistic assumptions. Figure 14.2 shows the schematic description of the model [10].

P_{SI} and P_{RI} are the probabilities with which an individual can transmit from Susceptible to Infected and from Infected to Recovered respectively. The basic SIR model can be represented mathematically by the following nonlinear Ordinary Differential Equations (ODE) [ZH WA 15]:

$$dS/dt = -P_{SI}S$$
$$dI/dt = P_{SI}S - P_{IR}I$$
$$dR/dt = P_{IR}I$$

where $S(t) + I(t) + R(t) = N$, N is the total number of individuals in the community.

Fig. 14.2 Schematic representation of the model [10]

Improved SIR (ISIR) Model

In the standard SIR model, it is assumed that community is fully fixed with each individual having equal probability to come in contact with others. However, many recent studies have shown that social contact network is heterogeneous in nature and not homogenous [10]. In social contact networks, the contact number per unit time can be reduced by the "crowding effect" and hence the force of infection should include the adaptation of individuals to the infection risk. In this improved SIR model (ISIR) infection rate is not fixed but is a function of the number of infected individuals $\lambda = P_{SI}(I)$. Hence the ISIR model can be represented mathematically as [10]:

$$
\begin{aligned}
dS/dt &= -\lambda(I)S \\
dI/dt &= \lambda(I)S - P_{IR}I \\
dR/dt &= P_{IR}I
\end{aligned}
$$

The force of infection $\lambda(I)$ can be represented as f(I), [ZH WA 15]:

$$
f(I) = P_{SI}/(1 + \alpha I)
$$

where α is the parameter defining the level of "crowding effect". Figure 14.3 shows an example solution of the above mentioned ODE for ISIR model.

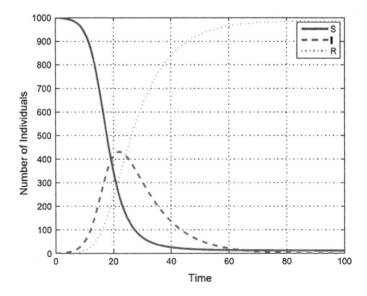

Fig. 14.3 Example solution of the ODE in ISIR model [10]

Fig. 14.4 SIS Model framework [11]

SIS Model

This model divides individuals in only two compartments: susceptible and infected. The infected individuals can return to susceptible class on recovery as the disease confers no immunity against reinfection [JI DO]. In order to adapt this model to the social networking, these terms are given new meaning: an individual is identified as infected (I) if he/she post about the topic of interest, and susceptible (S) if he/she has not posted regarding the topic. Figure 14.4 shows the schematic representation of the model.

This model can be represented in mathematical form by the following ODE's [11]:

$$dS/dt = -P_{SI}SI + P_{IS}I$$
$$dI/dt = P_{SI}SI - P_{IS}I$$

There is a consequence in this interpretation, that if an individual post about a topic he/she is remained in the infected compartment i.e. he/she cannot propagate back to susceptible class which is possible in epidemiological application.

SEIZ Model

One drawback with SIS model once a susceptible user gets exposed to the disease, he/she is directly transitioned to the infected state. Once a user is exposed to a news/rumor he/she may take time to adopt an idea or may be skeptical to some facts. It is even possible that the user is exposed to the news or rumor but never posted about it. Based on this reasoning a more robust and applicable model is introduced known as SEIZ model which compartmentalizes the users into the following classes: Susceptible (S), where user have not heard about the news yet; Infected (I), are the users who have posted about the news; Skeptic (Z), represent the users who have heard about the news but have not posted about it and Exposed (E), are the users who have heard about the news and have taken some time (exposure delay) prior to posting it [JI DO]. Figure 14.5 shows the SEIZ model framework [11].

The different parameter definitions used in representing the SEIZ model mathematically as ODE are shown in Table 14.2.

Fig. 14.5 SEIZ model
Framework [11]

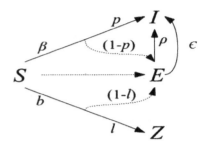

Table 14.2 Parameter
definitions in the SEIZ model

Parameter	Definition
N	Total population
β	S–I contact rate
b	S–Z contact rate
ρ	E–I contact rate
ε	Incubation rate
1/ε	Average incubation rate
bl	Effective rate of S–Z
βρ	Effective rate S–I
b(1-l)	Effective rate of S–E via contact with Z
β(1-p)	Effective rate S–E via contact with I
l	S–Z probability given contact with skeptics
1-l	S–E probability given contact with skeptics
p	S–I probability given contact with adopters
1-p	S–E probability given contact with adopters

This model can be represented mathematically by the following ODE's [11]:

$$dS/dt = -\beta S(I/N) - bS(Z/N)$$
$$dE/dt = (1-p)\beta S(I/N) + (1-I)bS(Z/N) - \rho E(I/N) - \varepsilon E$$
$$dI/dt = p\beta S(I/N) + \rho E(I/N) + \varepsilon E$$
$$dZ/dt = lbS(Z/N)$$

There were many constraints during the adoption of SIS and SEIZ model like: transition rate between different compartments and also the initial value of these compartments was unknown. Another limitation was the inability to quantify the total population size. The total population size appears to be simply the total number of social accounts however, the value truly needed is the number of users who could be exposed to the news or rumor. For example, if we take a total of 175 million registered Twitter accounts, out of these 90 million have no followers and 56 million follow no one. Also, there are many fake accounts which are used just to enhance the popularity. Taking all these factors into consideration, estimating the users receiving the tweet is quite difficult.

Agent-Based Model

Mathematical models are very useful in providing future projections of the ongoing health crises and also in assessing the impact interventions might have towards transmission control. The model was developed with N individuals that interact through a small network having variable edge density which can be defined as the number of links divided by the total possible links [1].

Individuals can be in one of the following five discrete states: Susceptible (S), Exposed (E), Infected (I); Dead of disease but not yet buried (D_I) and Dead of the disease and safely buried (D_b). The D_I infectious state includes agents who die but whose burial entails risk for onward virus transmission [1], while virus transmission stops with dead individuals that have been buried safely (D_b).

Figure 14.6 depicts the model scheme, where:

P_{SE} is the probability of an individual being transmitted from Susceptible to Exposed compartment.

P_{EI} is the probability of an individual being transmitted from Exposed to Infected compartment.

P_{ID} is the probability of an individual being transmitted from Infected to Dead compartment.

P_{IR} is the probability of an individual being transmitted from Infected to Recovered compartment.

The inverse of the probability P_{EI} that determines the rate by which an exposed individual becomes infectious that is incubation period which is set to a constant value of 1/9 as reported by WHO Ebola Response Team. There are many rules governing the system dynamics from time to time [1], like:

$$p(Y_{vk}(t+1)) = D_b|Y_{vk}(t-1) = D_I = 1$$

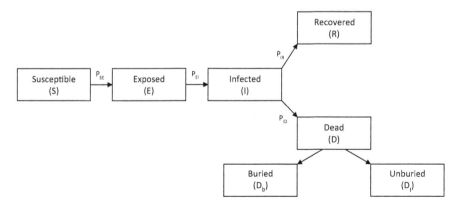

Fig. 14.6 Schematic representation of the model [1]

This rule sets the time period from death to burial to two days during which family members can get infected due to physical contact with the dead.

$$p(Y_{vk}(t+1)) = E|Y_{v1}(t) = I, Y_{v1}(t) = D_I = P_{SE}, V_1 \in R_{vk}$$

This rules tells that a susceptible individual can get exposed to a disease with a rate determined by the probability P_{SE}.

$$p(Y_{vk}(t+1)) = I|Y_{vk}(t) = E = P_{EI}$$

The third rule implies that an exposed individual can become infectious with a rate determined by the probability P_{EI}, inverse of this gives the incubation period.

$$p(Y_{vk}(t+1)) = D_I|Y_{vk}(t) = I = P_{ID}$$

The fourth rule implies that an individual die with a rate determined by the probability P_{ID}

$$p(Y_{vk}(t+1)) = R|Y_{vk}(t) = I = P_{IR}$$

The final rule tells that an individual can recover from infection at a rate determined by the probability P_{IR}.

Figures 14.7 and 14.8 show the cumulative number of infected and dead predicted by the model which are compared with the cases reported in Liberia and Sierra Leone, respectively.

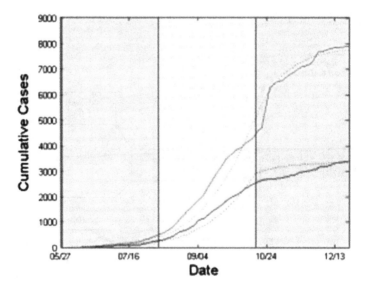

Fig. 14.7 Results for Liberia from May 27 to December 21, 2014 [1]

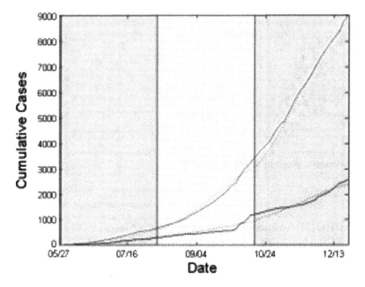

Fig. 14.8 Results for Sierra Leone from May 27 to December 21, 2014 [1]

A Contact Tracing Model

Along with features of the standard model, this model includes some additional features like rate of transmission to susceptible from infectious and improperly handled deceased cases, rates of reporting and isolating these cases and rates of recovery and mortality of these cases [BR HU 14].

Individuals are compartmentalized into six compartments at any given time t, like Susceptible S(t), capable of being infected; Exposed E(t), individuals who are exposed to infection and can grow the infection; I(t), who are infected with the disease; Contaminated deceased C(t), improperly handled corpses of infected; Isolated infectious II(t), exposed and infectious infected who have been identified and isolated from susceptible population; and Removed R(t), infectious individuals who have been recovered or are dead. Classes II(t) and R(t) can be separated from other classes and can be obtained from S(t), C(t), I(t) and E(t) [BR HU 14]. Figure 14.9 shows the schematic diagram of the model.

The model can be explained with the differential equations as follows [12]:

$$S(t) = -(\beta S(t)(I(t)/N)) - (\varepsilon S(t)(C(t)/N))$$

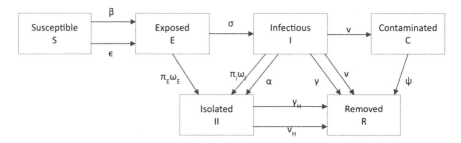

Fig. 14.9 Schematic diagram of the model [12]

where:

$\beta S(t)(I(t)/N)$ is infection rate of susceptible
$\varepsilon S(t)(C(t)/N)$ is infection rate due to improper handling of the deceased

$$E(t) = (\beta S(t)(I(t)/N)) + (\varepsilon S(t)(C(t)/N)) - \sigma E(t) - \kappa(\alpha I(t) + \psi C(t))\Pi_E \omega_E(E(t)/N)$$

where,

$\sigma E(t)$ is rate of progression of infectiousness
$\kappa(\alpha I(t) + \psi C(t))\Pi_E \omega_E(E(t)/N)$ is removal of exposed individual due to contact tracing

$$I(t) = \sigma E(t) - \alpha I(t) - \gamma I(t) - vI(t) - \kappa(\alpha I(t) + \psi C(t))\Pi_I \omega_I(I(t)/N)$$

$\alpha I(t)$ is the general rate of identifying and isolating infection
$\gamma I(t)$ is rate of recovery
$vI(t)$ is the rate of mortality outside hospital
$\kappa(\alpha I(t) + \psi C(t))\Pi_I \omega_I(I(t)/N)$ is the rate of identifying and isolating infectious individual due to contact tracing

$$C(t) = vI(t) - \psi C(t)$$

As the data is available in cumulative manner as per the definitions given by WHO, the cumulative reported cases at time t and $(t + \Delta t)$ can be given as [BR HU 14]:

$$CUM(t + \Delta t) = CUM(t) + \int_t^{(t+\Delta t)} \alpha I(s)ds + \int_t^{(t+\Delta t)} \psi C(s)ds$$

Table 14.3 Model parameters [12]

Parameter	Definition
N	Total population (assumed to be constant)
S(t)	Number of susceptible individuals at a given time t
E(t)	Number of exposed individuals at a given time t
I(t)	Number of infectious individuals at given time t
C(t)	Number of individuals deceased but improperly handled at time t
Q(t)	Number of susceptible individuals under quarantine at time t
II(t)	Number if infectious individuals under isolation at time t
R(t)	Number of individuals recovered or are deceased and properly handled at t
β	Transmission rate excluding the improper handling of the deceased
ϵ	Transmission rate due to improper handling of the deceased
κ	Average number of contacts traced per identified/isolated infectious individuals
$1/\alpha$	Average time for symptoms onset to isolation of individuals independent of contact tracing
Π_I	Probability of contact traced infectious individuals isolated without causing new case
ω_I	Ratio of the probability of contact traced infectious individuals to probability of random infectious individual
Π_E	Probability of contact traced exposed isolated individual without causing a new case
ω_E	Ratio of probability of contact traced exposed isolated individual to probability of random exposed individual
$1/\gamma$	Average time from symptoms onset to recovery
$1/\nu$	Average time from symptoms onset to death
$1/\sigma$	Average incubation period
$1/\psi$	Average time until improperly handled deceased is handled properly

where:

$\int_{t}^{t+\Delta t} \alpha I(s)ds$ is the number of identified/isolated infectious individuals in the time interval of t to (t + Δt)

$\int_{t}^{t+\Delta t} \psi C(s)ds$ is the number of deceased identified and properly handled in the time interval of t to (t + Δt)

The different parameter definitions used in representing this model are shown in Table 14.3.

Figures 14.10, 14.11 and 14.12 show the simulation of cumulative cases in Sierra Leone, Liberia and Guinea, respectively from May 27 to September 23, 2014 [12].

The entire tracing process is dependent on the public health resources and changes with different location and epidemic stages.

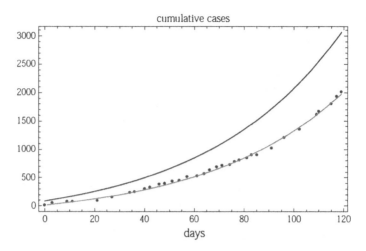

Fig. 14.10 Simulation of cumulative cases in Sierra Leone without contact tracing. The parameter values are N = 6,000,000; β = 0.32; ε = 0.0078; ψ = 0.2; α = 0.1; σ = 1/9; γ = 1/30; ν = 1/8. The initial conditions are S(0) = N; E(0) = 47; I(0) = 26; C(0) = 12 and R_0 = 1.26 [12]

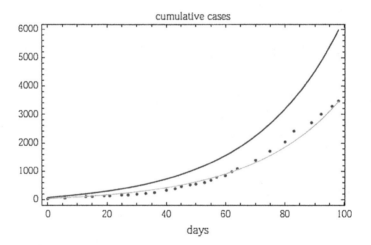

Fig. 14.11 Simulation of cumulative cases in Liberia without contact tracing. The parameter values are N = 4,000,000; β = 0.3; ε = 0.316; ψ = 0.18; α = 0.18; σ = 1/9; γ = 1/30; ν = 1/8. The initial conditions are S(0) = N; E(0) = 40; I(0) = 22; C(0) = 12 and R_0 = 1.54 [12]

Spatiotemporal Spread of 2014 Outbreak of Ebola Virus Disease

This model is generated to overcome the limitations of the standard models like homogenous mixing assumption and lack of spatial structure. In this model, individuals are in susceptible state if they acquire infection from an infectious

Fig. 14.12 Simulation of cumulative cases in Guinea without contact tracing. The parameter values are N = 12,000,000; β = 0.24; ε = 0.224; ψ = 0.18; α = 0.2; σ = 1/7; γ = 1/32; ν = 1/8. The initial conditions are S (0) = N; E(0) = 6; I(0) = 3; C (0) = 15 and R_0 = 1.12 [12]

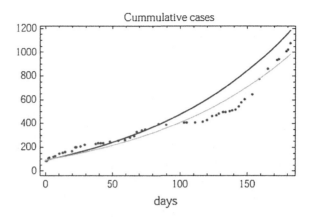

individual and become exposed without symptoms; after some amount of time this infectious individual can transmit infection to home. This individual then might be admitted to hospital, or die or may even recover. Also the individuals admitted to hospitals may die or recover. Deceased individuals may transmit the infection during their funeral and are then removed from the model [2].

For the spatial spread of the infection, movements of individuals which included patients not infected with Ebola virus, individuals seeking assistance in the health care facilities, health care workers taking care of these individuals and also the attendance of funerals was modelled.

It is assumed that infectious individuals can transmit infection to general community on a daily basis which is restricted to individuals living within a 10 km radius of the infected individual. Although, local population movement cannot be ruled out and can be used as a possible factor in Ebola virus disease dynamics in the future [2].

Quarantine Model

If no cure is found for Ebola, it will become out of control and hence a way has to be defined to prevent this spread and this is quarantine. In the standard SIER model another compartment is added Quarantine (Q) which indicates the number of individuals being hospitalized by government and other medical organizations along with Susceptible, Infected, Exposed and Removed compartments and a variable α to denote the rate of infectious individuals being hospitalized. Now the new system can be represented as [13]:

$$dS/dt = -(\beta/\gamma)S(1-\alpha)I$$
$$dE/dt = (\beta/\gamma)S(1-\alpha)I - (\delta/\gamma)E$$
$$dI/dt = (\delta/\gamma)E - I$$
$$dR/dt = I$$
$$dQ/dt = \alpha I$$

It is assumed in this model, that the individuals hospitalized have the same probability of death as the other infectious individuals but they do not infect other exposed or susceptible individuals. It is not possible to have as many quarantine places as there are infected individuals, but in order to prevent the disease spread it is necessary to have satisfactory amount of these spots. This sufficient amount is reached when the growth rate of infectious individuals is not greater than that of removed population [13].

Global Epidemic and Mobility Model

Global Epidemic and Mobility Model (GLEAM) produces a realistic simulations of global spread of infectious disease and is integrated in three layers namely: real world data on global population, real world data on mobility of this population and individual based stochastic mathematical model of infection dynamics [14].

The real-world population and mobility data is used to determine when and where people will interact and potentially transmit the infection. This data divides the world into a grid of small square cells and assigned an estimated population value. GLEAM uses cells that are approximately 25 × 25 km, dividing the globe into over 250,000 populated cells.

GLEAM uses set of twelve different flight networks which contains more than 3800 commercial airports in about 230 countries [14]. There are some airports with lots of connections and large volumes and many airports with few connections and low volumes. This characteristic is sometimes called the "long tail", and has a significant impact on how infections spread around the globe.

The GLEAM engine simulates the infection dynamics according to the characteristics of the disease coupled with any prevention and intervention measures. The characteristics of the infection are defined in the compartmental model and each individual fits into these compartmental models at any given time. These compartments are connected to each other with paths that define how an individual can travel from one compartment to another and associated parameters tells the probability of such a transition. GLEAM uses stochastic algorithms mathematically defined through individual based stochastic chain binomial and multinomial processes to calculate the proportion of the population within each compartment for each subpopulation, and how these proportions change over time as individuals transition from one compartment to the next [14].

Table 14.4 below depicts the models that are used in predicting and preventing the transmission of Ebola virus disease along with some of the characteristics of these models.

Table 14.4 Different models used to predict and stop the Ebola spread

Models	Characteristics
Basic reproduction number	This is the number of cases one case generates on average over the course of its infectious period. This is useful in determining whether the infectious disease can spread throughout the population
Case fatality ratio	This is the ratio of total number of deaths to the number of cases and helps in estimating the [probability of dying after the infection]
SIR model	There are three compartments: Susceptible, Infected and Recovered. This model works on some assumptions like, all networks are homogenous, susceptible individuals can get infected from infectious individuals via contact and infected individuals become immune after recovering. This model also ignores the crowding effect
Improved SIR (ISIR) model	This is an improved version of SIR model and takes crowding effect into consideration. Therefore, force of infection includes the adaptation to individuals to infection risk and is a function of number if infectious individuals
SIS model	In this model there are only two compartments: Susceptible and Infected. According to this an infectious individual can return to susceptible class on recovery as the virus confers no immunity against reinfection
SEIZ model	This model overcomes the drawback of SIS model that once a user is exposed to the disease, he/she is directly transitioned to infected state. This model has four compartments: Susceptible, Exposed, Infected and Skeptic. There are some constraints in this model like transition rate between different compartments and also the initial rate of these compartments are unknown. There is also an inability to quantify the total population size
Agent based model	This model is developed with N individuals who interact through a small network having variable edge density. This model has five discrete stages: Susceptible, Exposed, Infected, Dead of disease but not yet buried and Dead of disease and safely buried
Contact tracing model	Some of the features of this model are rate of transmission to susceptible from infectious and improperly handled deceased cases, rate of reporting and isolating these cases and rates of recovery and mortality of these cases. There are six compartments in this model: Susceptible, Exposed, Infected, Contaminated deceased, Isolated infectious and Removed
Spatiotemporal model	This model overcomes the limitation of some models like homogenous assumption of networks and also the lack of spatial structure. Infection can be transmitted to general community from an infected individual living within 10 km radius of that individual and hence this model can be used as a possible factor in Ebola virus disease dynamics

(continued)

Table 14.4 (continued)

Models	Characteristics
Quarantine model	This model along with standard SIER model has another compartment namely Quarantine. It is important to have maximum number of quarantine spots in order to stop the disease spread and this number can be reached when the growth rate of infectious individuals is not greater than that of removed population
GLEAM	This model is integrated in three layers: real world data on global population, real world data on mobility of this population and individual based stochastic mathematical model of infection dynamics. The real world population and mobility data determines when and where people will interact and potentially transmit the infection and the model determines the rate at which an individual transition from one compartment to another

Other Critical Issues in Ebola Study

Delays in Outbreak Detection

There are several factors which hampers the identification of Ebola outbreaks in Africa. Firstly, only a small number of Ebola outbreaks have occurred in East and Central Africa since the first outbreak in 1976 compared to other infectious disease like Malaria. There is also a limitation to community level knowledge of the diseases some areas at risk of Ebola have yet to experience Ebola outbreak. Secondly, early symptoms of Ebola are not specific which tends to misdiagnosing Ebola with malaria or other locally infectious epidemic disease. Thirdly, lack of epidemiological surveillance systems and diagnostic testing in poor countries increases the delay in detecting outbreaks. Longer the delay in detecting the outbreak and implementing control interventions, more are the chances that the virus spreads from remote and sparsely populated areas into higher populated areas [CH NI 14]. Ebola outbreak is directly dependent on the initiation of the control interventions which is depicted in Fig. 14.13.

As seen from above figure, a 5-day delay is highly unlikely to result in a major outbreak but if this delay exceeds to 2 weeks it may lead to a major Ebola outbreak [3].

The timely detection of an outbreak will minimize transmission in healthcare facility and in community by reducing the case fatality due to epidemic, strengthening coordination for the response, and building capacity for ongoing surveillance and control [15].

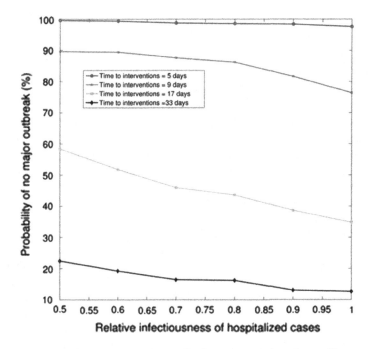

Fig. 14.13 Effects of size of baseline isolation effectiveness and timing of control interventions on likelihood of observing an outbreak [3]

Lack of Public Health Infrastructure

Basic infection control measures in health care settings are essential to avoid further spread of disease. However, under-resourced African regions from a low ratio of health-care worker's total population, but also lack in personal protective equipment and local capacity to effectively trace contacts and isolate infectious individuals. Therefore, Ebola outbreak has been amplified in health care settings. In Guinea, it took nearly 3 months for the health officials to identify Ebola [16] and by that time there were already few scattered cases which were imported to Liberia and Sierra Leone from Guinea. Most of the aid allocated is given to combat human immunodeficiency virus, malaria and tuberculosis and the rest going to maternal and child health services [17], which leaves very less or almost zero to support development of health systems. This lack of balanced investment is a continuous challenge in controlling the current Ebola outbreak. Also, individuals need to be screened for Ebola and the ones tested negative still needs to be treated. National government along with the external partners need to implement strategies to make health systems stronger and meet the essential health needs of the population in order to address the outbreak.

Health Worker Infections

The first health worker infected by Ebola was from Gueckedou, Guinea in January 2014. It is seen that based on health workers position they were 21–32 times at greater risk of Ebola infection than the general adult population out of which the most affected were the nursing workers with 52 % of the cases [18]. There were some gaps in implementing the Infection Prevention and Control (IPC) standards in the area where transmission took place. It is difficult to pin point the area where the health worker got infected. It may be in the health facility or the health worker would have been infected while providing care for Ebola-infected patient unknowingly.

From January 2014 to March 31, 2015, 815 confirmed and probable health worker cases were recorded in VHF (Viral Haemorrhagic Fever) database with 328 in Sierra Leone, 288 in Liberia and 199 in Guinea.

61 % of health workers infected with the disease were males, nearly 50 % of the infections occurred between the age of 30–44 years and 22 % of health workers infected were aged between 15–29 years old. 77 % of the health workers were hospitalized compare to 62 % non-health workers which reflects a greater awareness and access to care among health workers.

Table 14.5 shows the total number of health worker cases updated on November 4, 2015 in various countries. This number includes all the confirmed, probable and suspected cases. It is evident from the table that out of the cases that are reported many of them are from the heath workers hence it is necessary first to prevent the heath workers from getting infected from this virus so that pubic and patient safety can be improved.

WHO and partners have worked actively with managers and health workers to put IPC and Occupational Health and Safety (OHS) strategies and supplies in place to prevent health worker infection and improve patient safety.

Table 14.5 Total health worker cases from different countries as on November 4, 2015 [24]

Country	Total cases
Guinea	196
Liberia	378
Mali	2
Nigeria	11
Sierra Leone	307
Spain	1
United Kingdom	1
USA	3

Misinformation Propagation in Social Media

In conjunction with news reports about Ebola, rumors about the disease began to propagate wildly on Twitter. Some tweets were gathered from late September through late October 2014 which were filtered by the keyword "Ebola" or relevant hashtags such as #ebola, #EbolaVirus, #EbolaOutbreak, #EbolaWatch, #EbolaEthics, #EbolaChat, #nursesfightebola, #ebolafacts, #StopEbola, #FightingEbola and #UHCRevolution. Several widespread rumors were circulating on twitter, top 10 of which are described in Table 14.6.

Since social media has become one of the primary means by which people learn about worldwide developments, characterization of both news and rumors on social media about Ebola has been done in order to understand the popularity of misinformation. Since the first diagnosed case of Ebola in US, public has been curious to gain more knowledge about Ebola and hence has been leaned towards social media. Social media is a platform to reach millions of users, hence public health officials and medical experts are using it to educate and inform. But, some users share the same platform to share half-truths and rumors which increases the number of irrational fears about Ebola. Also because of the poor Internet and lack of roads for communication, the outbreak is believed to be three times worse than all the previous outbreaks.

Hence many models have been developed in order to study the rumor propagation and how users respond to the ideas, whether they adopt it readily or skeptical initially about it.

Table 14.6 Top 10 Ebola related rumors by tweet volume from September 28 to October 18, 2014

Rumor No.	Content	Label
1.	Ebola vaccine only works on white people	White
2.	Ebola patients have risen from dead	Zombie
3.	Ebola could be airborne in some cases	Airborne
4.	Health officials might inject Ebola patients with lethal substances	Inject
5.	There will be no 2016 elections and complete anarchy	Vote
6.	The US government owns a patent on the Ebola virus	Patent
7.	Terrorists will purposely contract Ebola and spread it around	Terrorists
8.	The new iPhone 6is infecting people with Ebola	iPhone
9.	There is a suspected Ebola case in Kansas City	Kansas
10.	Ebola has been detected in hair extensions	Hair

Risk Score Approach in Modeling and Predicting Ebola Spread

Modern disease compartmental models are developed to the point where the most significant factors controlling propagation make up components in the name. For example: Susceptible-exposed-infectious-recovered/removed (SEIR). Since propagation varies from disease to disease, this model naming convention can loosely serve as a classification for disease type which represents simple diseases: from the common cold or influenza (SIS) to pathogens more complex in nature such as Ebola (SEIR). Compartmental models produce efficient estimates for pathogen prevalence and duration, and this insight is vital in stopping highly contagious diseases like Ebola. This infections period would also be marked with an asymptomatic characteristic meaning: a host is infected but no symptoms are presenting (SEIaR).

Because of its protracted[1] asymptomatic period and virulence, Ebola can spread quickly unless strategic precautions are taken, including re-examining the compartmental model to account for newly observed spread characteristics. During the 2014 West African outbreak, it was observed surviving males carried live Ebola in seminal fluid for more than 30 days. While the US Center for Disease Control[2] and the World Health Organization[3] have not published definitive proof Ebola may be contracted as an STI, pathogen screening guidance has been provided for survivors.

Beyond Compartmental Modeling

The basis for compartmental models are making assumptions about social networks or graphs. Common assumptions can include: number of individuals, infection probability, incubation period, infected recovery time, etc. These phenomenological assumptions limit the scope of the model while preserving the most realistic aspects of it, but some model dimension assumptions are necessary because actual social network data does not exist. In the era of "big data" this is quickly changing.

Corporations across the globe are becoming experts at the collection of transactional data. While some of the data captured is specifically to enhance automated decision-making systems, the majority of data collected is still in a raw, unleveraged form making knowledge extraction the next field to experience an explosion of growth. On the forefront of knowledge extraction, LexisNexis Risk has produced the RELX Social Graph consisting of over 4 billion relationships built from applied identity analytics on a 4 petabyte core of content.

Physical and Social Graphs

Unlike user curated social graphs such as Facebook, the RELX graph coalesces as people experience life events. Sharing employers, addresses, insurance policies, and vehicle or property ownership are examples of the life events linking two people together. Applied graph analytics appends useful measures to help describe the quality of clusters. For the purposes of measuring the risk of a cluster contracting/propagating a disease, physical proximity of nodes (regardless of social connection) also plays a critical role. The physical proximity calculation between nodes is a simple distance calculation for each of the subject's most current address. A traditional social network does not imply a physical network, but a physical network may imply the sub-set of a social network. A physical network is constructed by proximity resulting in a 'nearest neighbor' linking, as illustrated in Fig. 14.14. Proximity, however, does not guarantee contact, therefore, a combination of proximity and social linking should be considered.

Graph Knowledge Extraction

Tools such as Gephi, NodeXL, or SVAT offer intuitive visual searches and a basic set of network measures, but to move beyond superficial graph descriptors to real-world application a different approach must be taken. Similar in nature to Neo4j, the RELX Knowledge Engineering Language (KEL) provides the ability to blend massive graph databases (billions of records) and derive dimensions beyond simple relational properties. As mentioned earlier, performing a distance calculation between nodes creates an additional edge weight distance. KEL can not only calculate the most recent difference in addresses, but also a chronology of addresses providing metrics such as cluster mobility, average move distance, physical cluster expansion/contraction, address density, occupant density, and many others.

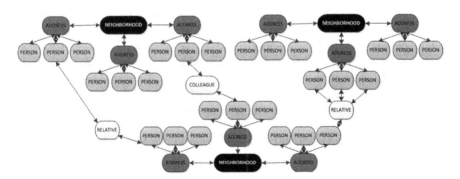

Fig. 14.14 A physical network is constructed by proximity resulting in a 'nearest neighbor' linking

Since the SEIR model places emphasis on physical proximity and social cohe-sivity, can the spread of information about the disease outpace the spread of the disease, thereby slowing its progression? Is disease transmission highest when a cluster is highly proximal, but non-cohesive socially? Do friendly people keep us safe from diseases like Ebola by serving as connectors helping to propagate awareness between disparate social groups faster than the disease can spread? Research conducted by Damon Centola from the University of Pennsylvania titled *"The Social Origins of Networks and Diffusion"* suggests the diffusion of ideas is as sensitive to the homogeneity of the network as was Little Red Riding Hood to porridge [29]. Idea diffusion requires a network to be "just right": moderately homogenous and moderately connected. A highly homogenous or under-connected graph population results in poor idea propagation.

Applying this idea on the national scoped RELX Graph: *"Which clusters have: the largest first degree count, the lowest average degree (cohesivity measure), the highest neighbor count, the highest colleague count, haven't moved in 5 years, and live in an area where there are few single family dwellings and car ownership is 1:10?"* These people are highly connected, live in a metro area, rely on public transportation, commute to work, and know their neighbors.

The nodes identified by this filter are key influencers and could be leveraged to proactively slow the propagation of a physically communicable disease like Ebola, potentially limiting the exposure to health care workers and their networks. Future refinements could include the incorporation of a health care worker flag or prox-imity to a health care facility; homogeneity dimensions such as: political affiliation, economic trajectory, or migration velocity; or proximity to public transportation hubs: bus and train stations or airports.

Graph Propagation

Points of intervention can also be identified by simulating the propagation of a disease based on SEIR model dimensions as edge characteristics. The node selected for intervention would be the first non-exposed node found on the most infectious, shortest path. The most infectious, shortest path is defined as: the shortest path in a sub-graph through which the number of first degree nodes is maximized. KEL does not have native graph traversal rules distinguishing between a *walk* and a *path*; however, KEL does allow for the creation of such rules. To control backtracking, or double counting nodes as n^{th} degree relatives, the GLOBAL primitive is used.

A first degree relatives with a distance edge weight:

$$\textbf{GLOBAL} : \textbf{Relations}(\#1, \#2, \#dist1) => \textbf{D1}(\#1, \#2);$$

A second degree relationship excluding backtracking:

GLOBAL : Relations(#1, #2, #dist1), Relations(#2, #3, #dist2),
1 <> #3 = > D2(#1, #2, #dist1, COUNT(#3));

A second degree path based on the second degree pattern including the calculation of total traversed distance and first degree node count:

GLOBAL : D2(#1, #2, #dist1, #cnt1), D2(#2, #3, #dist2, #cnt2),
#1 <> #3, NOT D2(#1, #3)
= > D2Paths(#1, #3, #dist1 + #dist2, #cnt1 + #cnt2, #2);

KEL will then apply the *D2Paths* pattern rule to each node in the entire graph aggregating: root, intermediate node, sink, edge distance, and first degree nodes encountered, as illustrated in Fig. 14.15.

The resulting rules produce an interim output as follows:

Node 1	Node 2	Node 3	Distance	Count
A	B	D	5	4
A	C	D	4	4

Expanding the rules out to eight degrees exceeds the largest inter-cluster diameter found in the RELX graph. Applying these rules to a sample data set produced the desired results. Graph traversal rules identifies the root, sink, inter-mediate nodes, total distance traveled, the number of unique first degree nodes encountered along the path, the total path length, and percent of nodes encountered during traversal, as shown in Fig. 14.16. Shortest path does not guarantee most infectious.

With the most virulent path discovered, proactive steps can be taken to interrupt the transmission of the disease by contacting the best connected person on the

Fig. 14.15 Using the GLOBAL primitive to define graph traversal rules

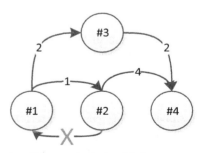

When calculating D2 Count, do not include {121}.

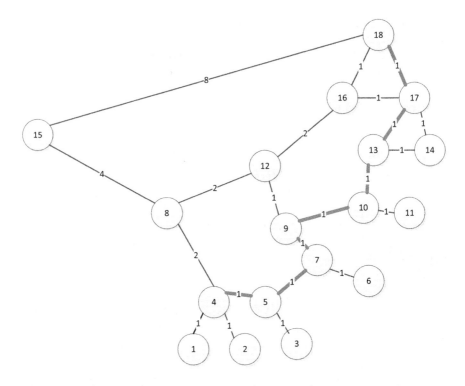

Root	Sink	Path	Distance	D1 Count	Path Length	PCT of Graph
4	18	4, 5, 7, 9, 10, 13, 17, 18	7	10	8	100%
4	18	4, 8, 12, 16, 18	7	6	5	61%
4	18	4, 8, 15, 18	15	6	4	56%

Fig. 14.16 Graph traversal rules identifies the root, sink, intermediate nodes, total distance traveled, the number of unique first degree nodes encountered along the path, the total path length, and percent of nodes encountered during traversal

shortest path, most infectious path. In this case, node 9 would be an effective intervention point.

With the most virulent path discovered, proactive steps can be taken to interrupt the transmission of the disease by contacting the best connected person on the shortest path, most infectious path. In this case, node 9 would be an effective intervention point.

With each outbreak, disease compartmental models become more sophisticated as new dimensions are added, but even the most sophisticated models make many assumptions impacting their real-world application. Due to the lack of actual social network data, modeling has been restricted to hypothetical populations, but two emerging RELX technologies could accelerate the accumulation of knowledge around disease propagation in the United States. As the RELX Social Graph and

KEL continue to evolve, disease spread simulations may move past hypothetical scenarios to identifying individuals key to hindering disease propagation based on an array of dimensions from blending: demographical, geographical, and social.

Mobile Applications Related to Ebola Virus Disease

As the mobile technology is taking a prominent role in the healthcare systems, it is necessary to have some mobile apps which can spread knowledge on the Ebola virus disease. In this section, we present a list of apps on which one can rely in preparing and responding to Ebola.

ITU Ebola—Info—Sharing

The International Telecommunication Unit (ITU) has made a freely available "Ebola—Info—Sharing" smartphone application which has the inputs of the organization who were involved in the fight against Ebola, organizations who were directly working on the ground in Ebola-affected countries. This app focuses on sharing precise information on Ebola with public and organizations, official news and key points on the map is also one of the feature in the app. Also, organizations have specific feature of sharing useful information with the staff on the ground [19].

Ebola Prevention App

This is a free mobile app which help users from the prevention of the disease anywhere he/she goes around the affected or unaffected regions. Some of the features of the app include affected area mapping where app shows the user location with respect to the affected areas in a map, Ebola hot zone detection where the app notifies the user if his/her region has been affected by Ebola virus disease, Preventive Measures in which app gives the user preventive measures based on his/her proximity of current location to affected areas, Latest info keeps up to date information on Ebola from all regions.

Ebola Guidelines

This app was created by Medical eGuides using the contents from WHO and CDC. It is a free of charge app which helps in improving the patient outcome by updating the healthcare workers with correct information at any point in an easy to use format. The healthcare workers are always up to date as the updates are pushed automatically to the app.

About Ebola

This app provides the knowledge and simple precautions one can take to protect his/her family and friends from this disease. This app mainly focuses on educating people about the information, prevention and risk regarding Ebola disease.

Stop Ebola WHO Official

This is official mobile application World Health Organization Africa to fight against Ebola. This app host features like preventive measures against Ebola, Testimonies from Ebola survivors, FAQS about Ebola, Symptoms of Ebola, reporting of a suspected case of Ebola to trigger an investigation, Awareness feedback survey, updates on Ebola, WHO contacts in countries and Local Ebola treatment centers detailed and updated information. User will also receive WHO alerts for important information in his/her area.

HealthMap

This app tracks disease outbreak information in real time and let him see all the current outbreaks around him. This app has been credited with spotting Ebola outbreak 9 days before the WHO.

#ISurvivedEbola

It is a global campaign which places the stories of Ebola survivors in order to inform, protect and spread hope among others. The #ISurvivedEbola website houses the stories of a growing community of Ebola survivors from Liberia, Sierra Leone, and Guinea and is part of the #TackleEbolainitiative.

Ebola Report Center

This app was created as an emergency response mechanism to prevent Ebola from spreading in West African region. This app takes relevant data about Ebola from growing mobile density which acts as information center for health and health care workers and also builds history of Ebola occurrence for future reference.

What is Ebola

This app is created to educate everyone about Ebola so that it can be stopped. This app tells everyone about the risks of the disease and also the preventive measures that are needed to be taken to protect oneself.

Ebola

This app presents the information pertaining to Ebola like how it is transmitted, what are the signs and symptoms of this disease, the risk of exposure, diagnosis, various treatments options available and also tells the user the preventive methods of the disease.

Stop Ebola

This app gives all information about Ebola, history of the disease, transmission modes, statistics of all the affected countries, conditions under which one can be infected from this disease, treatments, various signs and symptoms of the disease, preventive measures of the disease. This app also gives real time notifications to protect the users from this disease in everyday activities.

Virus Tracker

This is an educational game which simulates the spread of virus and illustrates the need of vaccination in controlling the outbreak. Players start with an infected state and can return to normal "HUMAN" state once they are vaccinated. Vaccinated players can seek other infected players and vaccinate them in order to earn more points. Virus mutations occurs periodically requiring players to obtain the latest vaccine before they return to the infected state. Players can transit between infected and vaccinated state throughout the game cycle depending on the responsiveness to mutations and their interactions with other players.

Ebola Virus News Alert

This app provides latest news about the Ebola virus disease worldwide. Other apps similar to this are Ebola virus which gives the news about Ebola in French, English, Espanola and Dutch, Ebola news update, Breaking news Ebola virus, Ebola news alert.

Sierra Leone Ebola Trends

This app gives an insight of how Ebola spread in the people of Sierra Leone and West Africa. This app tells all the trends, awareness, alerts about the disease in Sierra Leone.

The Virus Ebola

This is an informative app which includes topics like what is Ebola, its signs and symptoms, information about the previous disease outbreak, transmission, diagnosis, its vaccine and treatments, natural host of the virus, Ebola virus in animals and also the preventive measures one can take to protect themselves from this disease.

MSF Guidance

It is an easily searchable app with latest MSF protocols and essential drug information for the healthcare workers which can be downloaded and accessed anytime. The main aim of this app is to improve the access of clinical information to the healthcare workers to improve the patient care. This is done by constantly taking inputs from the healthcare workers on the field and also asking for any additional features that can be added to the app. This app includes MSF guidelines on Ebola and Marburg.

Novarum Reader

This is a mobile application, which reads and shares the result of Ebola test from a defense diagnostic manufacturer BBI Detection. The main aim of the app is to solve

Table 14.7 Some of the mobile applications related to Ebola virus prevention and education

Application name	Characteristics
ITU Ebola—Info—Sharing	Launched by ITU (International Telecommunication Unit) on December 2014 and focuses on sharing information on Ebola with public and various organizations
Ebola Prevention App	Was created by CloudWare Technologies which has features of showing the user location with respect to affected areas in map, and latest information from all Ebola infected regions
Ebola Guidelines	This app is created by Medical eGuides and helps in improving the patient outcome by updating the healthcare workers with correct information at any time
About Ebola	This app is launched by CODE LLC and mainly focuses on educating people about information, prevention and risks regarding this virus
Stop Ebola WHO official	This WHO Africa official application launched by Sikiwis which has features like preventive measure about Ebola, testimonies from Ebola survivors, FAQS about Ebola. Users will also get WHO alerts for any important information in his/her area
HealthMap	This application is launched by HealthMap and tracks disease outbreaks in real time
#ISurvivedEbola	It houses the stories of Ebola survivors from Sierra Leone, Liberia and Guinea
Ebola Report Center	This application is launched by Mobile Software Solutions and has all the relevant data about Ebola which acts as an information center for health and health care workers and also builds a history for future references
ebola	It contains information pertaining to Ebola like how is it transmitted, signs and symptoms, risk of exposure, diagnosis and etc.
Virus Tracker	This is an educational game which simulates the spread of virus and illustrates the need of vaccination in controlling the outbreaks
Ebola Virus New Alert	This app is launched by qrcodings and has all the latest news about the Ebola virus disease worldwide
Sierra Leone Ebola Trends	This application is launched by Teleficient Communication in September 2014 which tells all the trends, awareness and alerts about the disease in Sierra Leone
MSF Guidance	This app is launched by Open Medicine Project and has latest MSF protocols and essential drug information which can be downloaded and accessed anytime
Novarum Reader	This app reads and shares results of Ebola test from a defense diagnostic manufacturer BBI Detection. The main aim is to solve the problem of human error and also to stop the delay that occurs in sending samples to lab for confirmation

the problem of human error and also to stop the delay which occurs in sending tests to a lab for confirmation [5].

Table 14.7 below gives a brief description of the various mobile application that have been launched for the prevention and information sharing of this Ebola virus along with some of the main features of the application.

Work Done by Government

The U.S. has built, coordinated and led a worldwide response to the Ebola outbreak while strengthening the preparedness in U.S. U.S. has sent more than 3000 DOD, CDC, USAID, and other US officials to Liberia, Sierra Leone, and Guinea to assist with the0020resposne efforts [20]. With the help of these officials they have:

- Constructed 15 Ebola treatment units (ETUs) [20]
- Provided more than 400 metric tons if personal protective equipment and other medical supplies [20]
- Operated more than 190 burial teams in the region. [20]
- Conducted aggressive contact tracing to identify the chain of transmission [20]
- Trained healthcare workers [20]
- Worked with international partners to identify travelers who may have Ebola before they leave the region [20]

CDC and Homeland Security have conducted screening of around 7700 adults and children [20] to detect signs of Ebola among all the passengers arriving in the U.S. Each passenger is asked to forward his/her medical information to the state or local health authorities and they are kept under 21 days compulsory monitoring to ensure any signs of Ebola. If the individual displays any signs or symptoms of the disease they are isolated, diagnosed and are treated. The United States has already committed more than $921 million [20] towards fighting for Ebola in West Africa and will continue to use a strategy that meets the evolving conditions on the ground until there are zero cases of Ebola in the regions.

The Democratic and Republican lawmakers have come with $5.4 billion funding from President's emergency funding which will be used to [20]:

- Prepare US healthcare system for Ebola cases
- Development of Ebola vaccines and treatments.
- Enable detection and prevent the spread of this disease to other countries.

CDC is developing an introductory safety training course for licensed clinicians who wants to work in an Ebola Treatment Unit in Africa.

Innovative Mobile Application for Ebola Spread

Infectious disease spread across population usually follows a well-defined patterns determined by the transmission mechanism that the pathogen can use and the network of relationship that the pathogenic agent can follow to spread throughout the community. Ebola virus is transmissible through contact with vital fluids and secretions and only present in these after the 21 days' incubation period is completed and the patient exhibits the symptoms of the disease.

Identifying and tracking individuals affected by this virus in densely populated areas is a unique and urgent set of challenges in the public health sector. Currently, mapping the spread of the Ebola virus is done manually, however with the help of this Innovative Mobile Application which will use massive amounts of data from various sources like Twitter and Facebook being fed into the decision support system that will model the spread pattern of Ebola virus and create dynamic graphs and predictive diffusion models on the outcome and impact on either a specific person or a specific community. With the help of this model, we can make more precise forward predictions of the disease propagations and to identify possibly infected individuals which will help perform trace—back analysis to locate the possible source of infection for a particular social group. This model will visualize and identify the families and tightly connected social groups who have had contact with an Ebola patient and is a proactive approach to reduce the risk of exposure of Ebola spread within a community or geographic location.

This system will query the movement of a person and possible contacts in the areas affected by Ebola and will send notification notifying the user about the level of alert in their specific areas and caution them as to where there is a disease outbreak within the geographic location. For example, if a person affected with the virus was in movie theatre the previous day, a monitoring alert will be issued to other people who came in contact with that person or who were in the vicinity and advise them to take precautionary measures and watch for any signs of symptoms such as fever or headache.

We describe some of the features that are included in this mobile application like registering a new user, login with help of twitter or Facebook, basic information about the virus, symptoms of the disease, precautionary measures an individual can take, spread map of the disease and alert notification system.

Registering a New User

This is a straight forward method of registering a new as is done in many applications when a new user start using any application. The individual will be asked to put in some basic information like his/her name. email address which will be used for verification purposes. Users need to mention unique username and password which he/she will be using at a later stage to login in the application. On successful registration, a confirmation email will be sent to the registered email address for verification.

All these details are stored in database using SQLite which is a relational database management system. In contrast to many other database management systems, SQLite is not a server–client database engine, rather it is embedded into the end program and it implements most of standard SQL queries.

Login the Application

The user will have various options to login into the application like login through social media such as Twitter and Facebook or he/she can also login via the username and password created during registration process. When user logs in with the username and password created during registration process it queries from the database created and once the username and password matches from the database he/she will be directed to the home page.

For twitter login TwitterCore Kit is used which provides login with Twitter and enables application users to authenticate with twitter. When attempting to obtain an authentication token, this Kit will use the locally installed Twitter application to offer a single sign-on experience. If the kit is unable to access the authentication token through the Twitter app, it falls back to using a web view to finish the OAuth process. Once the login is completed successfully, a TwitterSession is provided in the success result which contain token secret, username and userId of the user and the session becomes active and is automatically persisted.

Facebook SDK for Android is used to allow users to login the application via Facebook. When the users login the app via Facebook they grant permissions to the app to retrieve information. To add this functionality firstly the Facebook SDK has to be initialized and then a call back manager has to be created to handle the login responses and finally the login results are passed to the login manger via the call back manager. If the login succeeds, the login result will have a new Access token with all the new granted and declined permissions.

Basic Information

Once the user has successfully logged in wither via Facebook or Twitter or using his/her credentials they will be redirected to the home screen which contains all the basic information about Ebola virus like the history of the disease, symptoms, how the disease can be transmitted from one individual to another, what are the preventive methods one can adopt. The user can use the slide menu option to go to various links for more information as shown in Fig. 14.15.

Geofencing

This is a feature in a software program that uses the global positioning system (GPS) to define geographical boundaries. With the help of this feature triggers can be set up so when the device enters or exits the boundary defined by the administrator, the user will be notified either via a text message or by email or by push notification.

The interested location in our case the Ebola affected sites are specified using the longitude and latitude and the proximity of these locations are adjusted by adding a radius of 500 m as shown in Fig. 14.17.

PREVENTION

There is no FDA-approved vaccine available for Ebola.

If you travel to or are in an area affected by an Ebola outbreak, make sure to do the following:

• Practice careful hygiene. For example, wash your hands with soap and water or an alcohol-based hand sanitizer and avoid contact with blood and body fluids (such as urine, feces, saliva, sweat, urine, vomit, breast milk, semen, and vaginal fluids).

• Do not handle items that may have come in contact with an infected person's blood or body fluids (such as clothes, bedding, needles, and medical equipment).

• Avoid funeral or burial rituals that require handling the body of someone who has died from Ebola.

• Avoid contact with bats and nonhuman primates or blood, fluids, and raw meat prepared from these animals.

• Avoid facilities in West Africa where Ebola patients are being treated. The U.S. embassy or consulate is often able to provide advice on facilities.

• Avoid contact with semen from a man who has had Ebola until you know Ebola is gone from his semen.

DIAGNOSIS

Diagnosing Ebola in a person who has been infected for only a few days is difficult because the early symptoms, such as fever, are nonspecific to Ebola infection and often are seen in patients with more common diseases, such as malaria and typhoid fever.

However, a person should be isolated and public health authorities notified if they have the early symptoms of Ebola and have had contact with

• blood or body fluids from a person sick with or who has died from Ebola,

• objects that have been contaminated with the blood or body fluids of a person sick with or who has died from Ebola,

• infected fruit bats and primates (apes and monkeys), or

• semen from a man who has recovered from Ebola

Samples from the patient can then be collected and tested to confirm infection.

Ebola virus is detected in blood only after onset of symptoms, most notably fever, which accompany the rise in circulating virus within the patient's body. It

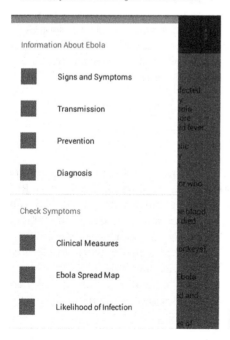

Fig. 14.17 Some features of innovative mobile application

```xml
<?xml version="1.0" encoding="UTF-8"?>
<soap:Envelope xmlns:soap="http://schemas.xmlsoap.org/soap/envelope/"
xmlns:SOAP-ENC="http://schemas.xmlsoap.org/soap/encoding/">
 <soap:Body>
  <roxieoverview1.3Request xmlns="urn:hpccsystems:ecl:roxieoverview1.3">
   <lastname>SMITH</lastname>
  </roxieoverview1.3Request>
 </soap:Body>
</soap:Envelope>
```

Fig. 14.18 Sample query for returning data of users having last name as VED

Firstly, to use geofence feature our mobile application requested to access fine location by adding this permission in the manifest file of the application. To use the intent service for listening the geofence transitions, the IntentService element was added. The geofences are created using the API's class builder which also helps in the setting the desired radius, duration and transition types for the geofence. We have set two triggers one for entering the geofence and the other for exiting. These triggers tell the location services that the respective trigger should be fired if the device is within the geofence. Stopping geofence monitoring when it is no longer needed or desired can help save battery power and CPU cycles on the device. The Geofencing can be stopped by removing the geofence from location service. Figure 14.18 below shows the geofence created around a particular location with a radius of 500 m.

As you can see from the above image the user is currently present in the geofence area and hence will receive a notification alerting him/her that they have entered an Ebola infected area and another notification will be pushed as soon as the user exits the geofence alerting him/her that they have left the Ebola infected area. This feature will help the users by keeping them aware of the infected areas and so to take necessary precautions when entering these areas.

Web Service Through ECL

A SOAP enabled service is created using ECL which is a declarative, data centric programming language designed in 2000 by the organization LexisNexis Risk Solutions to allow developers to process big data across high performance computing cluster. For the current purpose we are using a dummy data with 1000,000 records which has attributes like PersonID, FirstName, LastName, MiddleName, Gender, Street, City, State, Zip. We have designed a Roxie query, Roxie is an HPCC Systems cluster specifically designed to service standard queries, providing a throughput rate of thousand-plus responses per second. This service has been published on hThor which is an R&D platform designed for iterative, interactive development and testing of Roxie queries.

	personid	firstname	lastname	middleinitial	gender	street	city	state	zip	recpos
1	518991	AALIYAH	SMITH	R	M	103 LARKSPUR LN	KIOWA	CO	80117	59164860
2	519991	ABHI	SMITH		F	575 MAIN ST APT 1507	ALLENTOWN	NJ	08501	59278860
3	520991	ABHIJEET	SMITH	Y	F	205 BERKSHIRE WAY	WARREN	MI	48089	59392860
4	521991	ABIGAAIL	SMITH		F	90 GAYNOR AVE	NACHUSA	IL	61057	59506860
5	522991	ABIR	SMITH	U	F	95 LOUVAIN ST	BROOKVILLE	OH	45309	59620860
6	523991	ABRAM	SMITH	R	M	19 OVERLOOK DR	JUNTURA	OR	97911	59734860
7	524991	ABREISHA	SMITH		F	100 LAKESIDE DR	LONE OAK	TX	75453	59848860
8	525991	ABUR	SMITH	W	F	526 EMERALD	BALDWIN CITY	KS	66006	59962860
9	526991	ACHELLE	SMITH	D	M	176 W 87TH ST # 7D	STETSONVILLE	WI	54480	60076860
10	527991	ACHILLE	SMITH	P	M	430 W 34TH ST APT 5G	REPUBLIC	OH	44867	60190860
11	528991	ADAMA	SMITH	I	F	52 FENWAY APT 5	JUNIATA	NE	68955	60304860
12	529991	ADDISWA	SMITH		M	6 DINGLEBROOK LN	HAWLEYVILLE	CT	06440	60418860

Fig. 14.19 Sample output of the service

Image below shows a sample xml query created to return the details of all users having a particular LastName (Fig. 14.18).

This query returned 1000 users having last name as VED in milliseconds. A sample of the result is shown in Fig. 14.19.

This service can be parsed with the mobile application and will help in handling big data with a high throughput rate.

Conclusion

The ongoing epidemic in West Africa offers a unique opportunity to improve our current understanding of the transmission characteristics of the Ebola virus disease in humans, including the duration of immunity among Ebola survivors and the case fatality ratio, as well as the effectiveness of various control interventions. Ending the epidemic requires approximately 70 % of the persons with Ebola to be treated either in an ETU or at home or in community setting such that there is a reduced risk for disease transmission. There are a lot of public health challenges faced during the prediction of number of future cases and if the preventive measures are not scaled-up cases will continue to double in approximately every 20 days. However, this epidemic can be controlled using various models described in the paper. As many consumers are now receiving news from real-time social media platforms, it is important to have quantitative methods like SIR, ISIR, SIS and SEIZ to distinguish news from rumors as misinformation on social platform can some-time resemble as a genuine news.

There was an argument that digital health tech could have played a better role in stopping the Ebola outbreak had there been a quick ground response. With this outbreak, many developers and manufacturers were able to test their apps in the field and had a very positive results. Now the developers have to analyze the data and find better and innovative ideas to bring these technologies together in the fight against Ebola.

Innovative mobile application uses large amount of data from various different sources which are fed into the decision support system that will model the spread pattern of Ebola virus and create dynamic graphs and predictive diffusion models of the impact of virus on either a specific person or a specific community. LexisNexis has provided the big data needed to develop and model this program with the help of their expertise in big data analytics. The data provided by LexisNexis is completely secured and are in compliance with the Centers for Disease Control and Prevention and the National Institute of Standards and Technology for transmission of public health information. The model created leads to more precise predictions of disease propagation, it also helps in identifying the individuals who are possibly infected by the virus and perform a trace back analysis to locate the possible source of infection in a particular social group. All the data is being presented in form of report to the responsible government agencies.

Acknowledgments This work has been funded by the NSF Award No. CNS 1512932 RAPID: Modelling Ebola Spread and Developing Decision Support System Using Big Data Analytics, 2015–2016.

References

1. Browne C, Huo X, Magal P, Seydi M, Seydi O, Webb G. Model of 2014 Ebola Epidemic in West Africa with contact tracing.
2. Kouadio KI, Clement P, Bolongei J, Tamba A, Gasasira AN, Warsame A, Okeibunor JC, Ota MO, Tamba B, Gumede N, Shaba K, Poy A, Salla M, Mihigo R, Nshimirimana D. Epidemiological and surveillance response to Ebola virus disease outbreak in Lofa County, Liberia.
3. Lutwama JJ, Kamugisha J, Opio A, Nambooze J, Ndayimirije N, Okware S. Containing Hemorrhagic Fever Epidemic. The Ebola experience in Uganda.
4. https://en.wikipedia[9].org/wiki/Ebola_virus_disease#Onset.
5. http://www.imedicalapps.com/2015/04/lessons-ebola-outbreak-using-apps-fight-infectious-diseases/.
6. https://www.whitehouse.gov/ebola-response.
7. https://en.wikipedia.org/wiki/Basic_reproduction_number.
8. http://www.healthmap.org/site/diseasedaily/article/estimating-fatality-2014-west-african-ebola-outbreak-91014.
9. http://epidemic.bio.ed.ac.uk/ebolavirus_fatality_rate.
10. Jin F, Dougherty E, Saraf P, Cao Y, Ramakrishnan N. Epidemiological modeling of news and rumors on Twitter.
11. Anastassopoulou SC, Russo L, Grigoras C, Mylonakis E. Modelling the 2014 Ebola virus epidemic—agent based simulations, temporal analysis and future predictions for Liberia and Sierra Leone.
12. Merler S, Ajeli M, Fumanelli L, Gomes MFC, Pastore y Piontti A, Rossi L, Chao DL, Longini IM, Halloran ME, Vespignani A. Spatiotemporal spread of the 2014 outbreak of Ebola virus disease in Liberia and the effectiveness of non-pharmaceutical interventions: a computational modelling analysis.
13. http://ebolaresources.org/ebola-mobile-apps.htm.
14. Meltzer MI, Atkins CY, Santibanez S, Knust B, Petersen BW, Ervin ED, Nichol ST, Damon IK, Washington ML. Estimating the future Number of cases in the Ebola Epidemic—Liberia and Sierra Leone, 2014–2015.

15. Zhang Z, Wang H, Wang C, Fang H. Modeling epidemics spreading on social contact networks.
16. http://www.who.int/csr/disease/ebola/one-year-report/factors/en/.
17. Haas CN. On the quarantine period for Ebola virus. PLoS Curr. 2014; Edition 1. doi:10.1371/currents.outbreaks.2ab4b76ba7263ff0f084766e43abbd89.
18. https://data.hdx.rwlabs.org/dataset/number-of-health-care-workers-infected-with-edv.
19. http://jmsc.hku.hk/ebola/2015/03/23/fighting-ebola-does-the-mobile-application-help/.
20. http://who.int/hrh/news/2015/ebola_report_hw_infections/en/.
21. http://apps.who.int/ebola/ebola-situation-reports.
22. http://www.who.int/mediacentre/factsheets/fs104/en/.
23. http://www.who.int/features/factfiles/malaria/en/.
24. http://www.gleamviz.org/model/.
25. http://www.math.washington.edu/ ∼ morrow/mcm/mcm15/38725paper.pdf.
26. Chowell G, Nishiura H. Transmission dynamics and control of Ebola virus disease (EVD): a review.
27. http://www.cdc.gov/vhf/ebola/transmission/index.html.
28. http://www.who.int/mediacentre/factsheets/fs103/en/.
29. Centola D. The social origins of networks and diffusion. Am J Sociol. 2015;120(5):1295–1338. http://doi.org/10.1086/681275.

Chapter 15
Unsupervised Learning and Image Classification in High Performance Computing Cluster

I. Itauma, M.S. Aslan, X.W. Chen and Flavio Villanustre

Introduction

Representing objects using lower dimensional, representative, and discriminative features is an ongoing research topic that has many important rami cations. This concept leads to important questions that need to be answered. For example:

How to design an optimal and fast optimization method that can avoid local minimums and converge?

- How to speed up the computational process using hardware systems?
- How to choose system/network parameters?
- How to use unlabeled data more efficiently?
- How to normalize the learned features before classification?
- How to avoid over-fitting?
- How to extract features from labeled data?

In this chapter, some of these questions are answered. Hand-engineering approaches have been proposed to extract good features from data to be used in the classification stages. In addition to the labor-intensive techniques that do not scale well to new problems, there have been many methods proposed (such as sparse coding [2] and sparse auto-encoders [3]) that can automatically learn better feature representations compared to the hand-engineered ones. Although those unsupervised methods achieve good performance if required settings are satisfied, one of the major drawbacks is their complexity. Many of those methods also require careful selection of multiple hyper parameters like learning rates, momentum, sparsity penalties, weight decay, and many other parameters that must be chosen through cross-validation, thus increasing running times dramatically. Coates et al. [1] compared sparse auto-encoders, sparse restricted Boltzmann ma-chines,

© Springer International Publishing Switzerland 2016
B. Furht and F. Villanustre, *Big Data Technologies and Applications*,
DOI 10.1007/978-3-319-44550-2_15

Gaussian mixture models, and K-means learning methods. Surprisingly, the best results were achieved using the K-means method that has been used in image processing, but that has not been widely practiced for deep unsupervised feature learning. To obtain the best results of the K-means method, a selection of the best number of centroids from the data is needed. In this study, we extend the use of the K-means algorithm with multimodal learning and recognition framework in the High Performance Computing Cluster environment for any dimensional data.

There is a high demand for new ideas to deal with the feature learning and classification stages on high dimensional data. The high dimensionality of un-labeled data requires new developments in learning methods. In spite of recent advances in representation learning, most of the current methods are limited when dealing with large scale unlabeled data. Complex deep architecture and expensive training time are mostly responsible for lack of good feature representations for large scale data. As a solution to dealing with high dimensional data, researchers in the machine learning community have adopted the use of GPUs and parallel programming techniques to speed up computationally intensive algorithms. Furthermore, important studies have been carried out to propose more efficient optimization methods to speed up the convergence (such as [4]). In response to these various ideas and platforms, we investigate High Performance Computing Cluster (HPCC Systems^R) as a new environment to assess our framework's effectiveness in terms of computation time and classification accuracy.

Background and Advantages of HPCC Systems^R

HPCC Systems^R is a massively parallel processing computing platform used for solving Big Data problems. A multi-node system leverages the full power of massively parallel processing (MPP). While the single-node system is fully functional, it does not take advantage of the true power of an HPCC Systems^R platform which has the ability to perform operations using MPP. Algorithms are implemented in HPCC Systems^R with a language called Enterprise Control Language (ECL). ECL compiler generates highly optimized C ++ for execution. It is open source and easy to setup. Figure 15.1 shows an HPCC Systems^R multi-cluster setup. The figure shows a THOR processing cluster which is similar to Google and Hadoop MapReduce platforms with respect to its function, filesystem, execution, and capabilities but o ers higher performance [5].

In [6], Payne et al. discussed the challenges of academic data in heterogeneous formats and diverse data sources. They assessed HPCC Systems^R in the analysis of academic big data. Based on their evaluation, HPCC Systems^R pro-vides mechanisms for ingesting and managing simple data such as CSV data as well as complex data.

Fig. 15.1 HPCC systems
THOR cluster

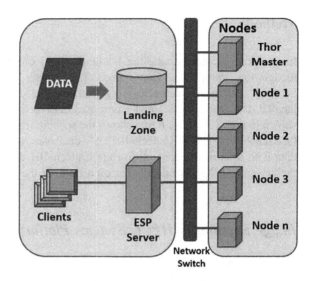

We chose HPCC Systems[R] because of its scalability with respect to code reuse irrespective of the size of the dataset and number of clusters. It provides programming abstraction and parallel runtime to hide complexities of fault tolerance and data parallelism.

One of the goals of this study is to show that researchers are able to run their proposed methods on HPCC Systems[R] even using a single core computer. We expect a faster training time if the algorithms are tested on a multinode HPCC Systems[R] platform. We leave the use of a system combining multiple computers for our future studies.

Contributions

The use of HPCC Systems[R] is adopted in the implementation of the feature learning and object classification tasks. We show that (i) HPCC Systems[R] enables researchers to leverage a multi-cluster environment to speed up the running time of any computationally intensive algorithm; (ii) it lowers the budget costs by using existing computers instead of designing an expensive system with GPUs; and (iii) it is scalable with respect to code reuse irrespective of the size of the dataset and number of clusters.

We implement a new feature learning and recognition framework using a multimodal strategy. Our novel idea is to use the HPCC Systems[R] platform that can handle identity recognition with high recognition accuracy rates. For instance, by dividing a face image into several subunits, we can extract intra-region information more precisely. We will discuss this in the next sections.

Methods

In this section, we describe the learning of object representations as well as the recognition framework. Our framework consists of image reading in HPCC Systems[R] platform, feature learning from unlabeled data, feature extraction from labeled data using the learned bases, and classification stages. Our framework is shown in Fig. 15.2. This figure shows the specific framework that we follow for the face databases such as Caltech-101, AR databases, and a subset of wild PubFig83 data with multimedia content. For the Caltech-101 data, we use patches instead of facial regions. We give details for each stage in the following sections.

Image Reading in HPCC Systems Platform

This work is the first study on image classification using HPCC Systems[R], to the best of our knowledge. First, we explain how we integrate databases into the HPCC Systems[R] in which images are represented as Binary Large OBject (BLOB). BLOB support in ECL begins with the DATA value type which makes it perfect for housing BLOB data. There are essentially three issues around working with this type of data: (i) How to get the data into the HPCC Systems THOR Cluster (Spraying). (ii) How to work with the data, once it is in HPCC. (iii) How to get the data back out of HPCC Systems[R] THOR Cluster (Despraying).

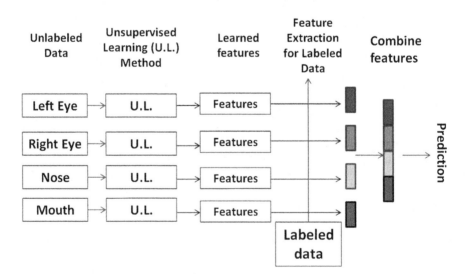

Fig. 15.2 The framework that we follow for the classification of AR data

The BLOB spray is described in [7, 8]. The image dataset should be sprayed in BLOB format. There are different formats for spraying data such as delimited for Comma Separable Value (CSV), fixed for texts and blob for images. We explored the BLOB spray option which will result in a dataset on the cluster where each record is one of the image datasets. Typically, we use a pre x of both the name and length to de ne the record structure of the image dataset. Since we use grayscale images, we convert all images to Comma Separable Value.

The following steps are followed to use the image database: (i) Extract patches or regions from images. (ii) Normalize the patches. (iii) Convert these patches to CSV. (iv) Spray the CSV to HPCC Systems^R platform. The next section describes how we learn the features from the data.

Feature Learning

Most applications in image processing involve the use of high dimensional data. The goal of unsupervised learning is to find a lower dimensional projection of the unlabeled data that preserves all the information in the data while reducing redundant dimension. The problem in unsupervised learning is to find hidden structures in unlabeled data.

We implement a multimodal feature learning framework that runs the K-means learning method for each region of the data. Using the multimodal learning, we are able to extract representations that capture intra-region changes more precisely. The K-means clustering method obtains specialized bases for the corresponding region of the data. Instead of estimating a single centroid of an image/data, feature learning for each divided region increases the deep representation that learns more representative information as we assess this point in our experimental results.

Coates et al. [1] proved that the K-means method can achieve comparative or better results than other possible unsupervised learning methods. In view of this, one objective was to extend the K-means method for multimodal learning and classification framework in HPCC Systems. The algorithm takes the dataset X and outputs a function f: $R^n!R^k$ that maps an input vector $x^{(i)}$ to a new feature vector of k features. To extract high quality features in order to obtain a high classification accuracy, we ran the methods with respect to the key points of this stage such as using: (i) a good number of samples, (ii) choice of parameters, and (iii) number of bases.

K-means is a partitioning algorithm in which we construct various bases or centroids and evaluate based on specific criteria. It is an unsupervised clustering method and partitioning algorithm where data are assigned in clusters defined by their centroid, based on their features and distance from the centroids. The goal is to minimize the sum of the square errors (SSE) as can be seen in Eq. (15.1). The SSE is used to make partitions and it is the sum of squared differences between each observation in its cluster as a centroid over all the k clusters. The SSE strictly decreases after re-computing new centers in the K-means algorithm. The new center of a cluster comes from the average of all data points in this cluster, which minimizes the SSE [9]; as follows:

$$SSE = \sum_{i=1}^{k} \sum_{i \in C_i} ||x - m_i||^2 \qquad (15.1)$$

where m_i is the mean of points in C_i and x is the data point in cluster C_i. Given two partitions, we choose the one with the smallest error. Each cluster is represented by the center of the cluster which is the centroid. Points are assigned to the cluster with the nearest centroid. The distance between clusters is based on their centroids:

$$\text{dis}(K_i;\ K_j) = \text{dis}(C_i;\ C_j); \qquad (15.2)$$

where K_i and K_j are two groups of points or information, and C_i and C_j are the corresponding centroids. Given k, the number of clusters, the K-means clustering algorithm is outlined as:

- Select k points as initial centroids repeat

  - Form k clusters by assigning each point to its closest centroid
  - Re-compute the centroids of each cluster until convergence criterion is satisfied

In order to specify the best k, we run a range of values. The computational complexity is O(tkn) where n is the number of data points, k is the number of clusters and t is the number of iterations. It is an efficient method since usually, k; t ≪ n. The work [10] summarizes recent results and technical points that are needed to make elective use of K-means clustering for learning large-scale representations of images. Figure 15.3 shows the centroids learned by K-means implemented in ECL from the AR dataset without whitening as an example.

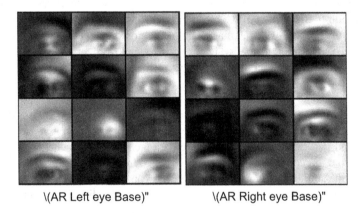

\(AR Left eye Base)"   \(AR Right eye Base)"

Fig. 15.3 Selected bases (or centroids) trained on AR images using K-means in HPCC systems

Feature Extraction

For each region, we train one K-means algorithm. The learned and specialized bases are able to capture nonlinear structure of the corresponding image regions. We use these bases for the feature extraction and dimensionality reduction of the labeled data. The new projected data is calculated using the correlation information between the labeled data and estimated bases or centroid.

Let X_i be any image region and C_i is the corresponding learned bases using the K-means method. The feature of labeled data corresponding to image regions is calculated as $Y_i = X_i C_i^t$. Then, these extracted features are fused together through concatenating one by one to get the multimodal representation as

$$Y = [Y_1; Y_2; \ldots; Y_M]; \tag{15.3}$$

where M equals to the number of image region (and sometimes equals to the number of image region and multimedia data such as speech in addition to image information). The multimodal learning and classification idea improves the recognition rates as seen in the results. A reason for this is that the extraction of intra-region information is estimated more precisely when the learning method focuses on a specific image regions separately.

Classification

Classification is a supervised learning process that aims at accurately predicting some value or attribute of an object based on known facts about the object. It involves deriving a rule or model from a training set which is then used to predict a test set. In machine learning, all classification algorithms follow three logical steps. Learning the model from a training set, testing with respect to obtaining measures of how well the classifier fits, and classifying which involves testing the model on new data in order to compute a classification accuracy.

We apply our multimodal object representation learning method to the object classification task. To do this, we train classification methods and configurations in our experiments. Once the framework is trained, it can be used to identify a testing object. The testing data should undergo the same procedure that the training data goes through.

Experiments and Results

In this study, we assess our design on a subset of the Caltech-101 [11], AR [12], and a subset of PubFig83 database [13] that we add speech content in addition to face images. Our goal is to assess our feature learning and classification framework in the

HPCC Systems platform. Note that all data in our experiments are locally normalized to have the Gaussian distribution.

Evaluation on Caltech-101. The Caltech-101 database consists of 102 categories. As a subset of Caltech-101, we use 10 classes (which have more than 60 images per class) for both unsupervised and supervised learning steps. We randomly select up to 60 images per class and pre-process them as in [14]: The images are converted to gray-scale, then down-sampled and zero padded to 143×143 pixels. Finally, we normalize the images to have the standard Gaussian distribution.

In our study, we assess the performance of our method in HPCC Systems and compare directly with [1]. We run methods using $3; 000$ randomly selected patches of 16×16 dimensional pixels. In the unsupervised learning part, we train the entire unlabeled training set of images before the classification step. We learn 32 bases in the unsupervised learning of all methods in two platforms.

For the supervised learning, we use 30 training and 30 testing images for each category. To extract features from the labeled training samples, we follow the convolutional extraction process of Coates et al. [1]. We use stride 1 with 16×16 patches to obtain a dense feature extraction. The non-linear mapping transforms the input patches into a new representation with 32 features using the learned bases. Then we use pooling for dimensionality reduction. 132 pooled features are used to train the classifiers.

We show the visualizations of the bases (or centroids) learned by K-means in Fig. 15.4. We achieve $83:5$ % identification accuracy using the C4.5 Decision tree classification method; whereas Coates et al. [1] achieves only $80:7$ % rate using the linear SVM.

Evaluation on AR Face Database We also test our proposed idea on AR [12] face database. The aligned AR database contains 100 subjects (50 men and 50 women), with 26 different images per subject which totals $2; 600$ images taken in two sessions. In this database, there are facial expression (neutral, smile, anger, scream), illumination and occlusion (sunglass, scarf) challenges. In our study, we use images without the occlusion challenges which totals to $1; 400$ images for both the unsupervised learning and classification steps. Figure 15.5 shows some example images from a subject.

First, we learn the base for each facial region separately. We segment four essential facial regions with sizes of 39×51 (left eye and right eye), 30×60 (mouth), and 45×42 (nose). We believe that better representations are obtained by running unsupervised learning for each region. We also obtain the features of the labeled facial regions using the corresponding learned bases separately. To do this, we calculate the correlation between each labeled sample and each center vector (base) to get a vector of features. We combine the features extracted from the four facial regions (and other possible modalities), and train the classifiers.

For the AR database, we follow a scenario described in [15] which reported one of the state-of-the-art recognition rates. Each subject has 14 images with facial expression and illumination changes. Various train-test image partitions are tested.

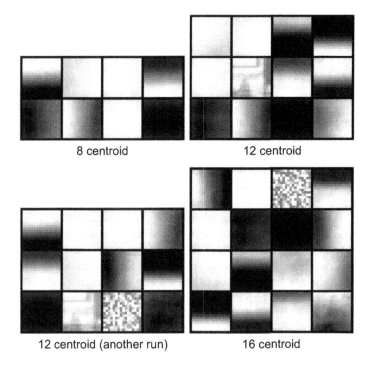

Fig. 15.4 Selected bases (or centroids) trained on Caltech101 images using K-means in HPCC Systems

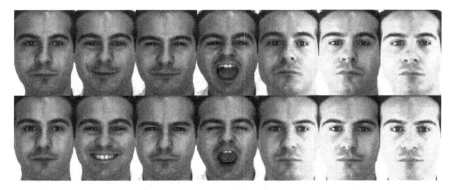

Fig. 15.5 Example images from one subject in AR database with various facial expressions and illumination

We conduct 10 runs for train-test procedure to get the average recognition rate for each partition. Table 15.1 shows the face classification results of our proposed framework (using K-means and C4.5 Decision tree methods) in HPCC Systems and [1]. We improve the classification results of [1] by 4:6 %. From this result, we can

Table 15.1 Comparison of face recognition rates on AR database

Methods	Acc. (%) with train
	5 train
Coates et al. [1]	74.3
Ours	78.9

infer that we learn more representative features and our classification method is better than [1]. This proves that our framework developed in HPCC Systems achieves at least comparative or better results than its alternative.

Our preliminary results show that more research studies on HPCC Systems are beneficial to the machine learning society.

Identity Recognition on the Wild and Multimedia Database In recent years, several unconstrained databases have emerged in the literature for face identification or verification. Unlike the traditional face databases which are com-posed of images taken in controlled environments, face images in unconstrained databases are generally collected from Internet sources. In particular, these images contain unrestricted varieties of expression, pose, lighting, occlusion, resolution, etc. Therefore, unconstrained face recognition is a very challenging task.

We prepare a data set from aligned version of wild PubFig83 database [13]. We select 10 subjects which totals to 1; 000 face images. Some example images are shown in Fig. 15.6. For the images, we randomly select 50 images per subject as the training set, and the rest of the images are used as the testing set in the supervised learning step. Four essential facial regions are used for facial representation learning. We segment four essential facial regions with sizes of 32×52 (left eye and right eye), 48×76 (mouth), and 60×48 (nose), which are further reduced by half, using bicubic interpolation.

Fig. 15.6 Example images of 10 celebrities with various real-world changes on facial expression, pose, illumination, occlusion, resolution, etc

Table 15.2 Identity recognition results using visual and/or speech contents on multimedia database. Note that we use the K-means in the HPCC Systems platform for all cases

Methods	Acc. (%)
	50 train
Only visual content	
Naive Bayes (HPCC)	55.5
Linear SVM (MATLAB) [1]	71.6
Random forest (HPCC)—maxLevel = 10	74.2
Random forest (HPCC)—maxLevel = 15	75.5
Softmax classification (HPCC)	83.0
C4.5 decision tree (HPCC)—maxLevel = 25	91.5
Only speech content	
C4.5 decision tree (HPCC)	91.6
Multimedia (visual + speech)	
C4.5 decision tree (HPCC)	94.0

Table 15.2 shows the classification results across 10 runs using various classification configurations. We run the K-means method in HPCC Systems plat-form, then the learned bases are used to extract features from labeled data. Our method achieves 91:5 %. As we observe from the two databases that we reported above, the C4.5 decision tree achieves the best and most superior classification rate.

To assess the feasibility of the proposed framework on the multimedia data, we downloaded several videos for 10 subjects from YouTube to extract 5 min speech information. There have been research e orts to show that it is beneficial to leverage the knowledge from multimedia data [16]. For instance, multimedia entertainment companies have started to order information on cast and characters for movies and shows during playback, presumably via a combination of visual and sound contents [17]. It is also beneficial to borrow knowledge from some other related tasks for feature extraction.

We employ the Mel frequency cepstral coefficients (MFCCs) [18], and their first and second derivatives to represent the acoustic features. Note that the content and quality of the speech data are heterogeneous. These features are calculated every 10 ms using 25 ms Hamming-window,[1] and their first 12 elements are selected to form a 36-dimensional feature vector for each frame. In this study, features extracted from every 40 consecutive frames are concatenated to be a 1440-dimensional feature vector which is considered as one training/test example. The unsupervised generic features are learned over 5000 examples (500 per subject). Half of the samples are randomly selected to train the classifiers and the rest are used for the testing.

In addition to the experiments on the visual content, Table 15.2 shows the identity recognition across 10 runs using speech only and multimedia representation with visual and speech contents. For these two cases, we run the C4.5 decision tree

---

[1]We used Dan Ellis' implementation for MFCC which is available at http://www.ee.columbia.edu/dpwe/resources/matlab/rastamat/.

classification method. We achieved 91:6 % recognition rate only using the speech information. When we combine the visual and speech information together in a multimodal way, we achieved 94:0 % recognition accuracy. By combining different features of various facial regions and corresponding speech content, we increase the classification accuracy by 2:5 %.

Discussion

We successfully performed feature learning and object classification in HPCC Systems platform. Although we used considerably smaller dimensional features such as 16 for AR and 32 for Caltech-101 databases, we achieved very promising classification results. A faster training time is expected when a multinode HPCC Systems platform cluster is used. We leave the use of a system combining multiple computers for our future studies. Further evaluations with higher dimensional features would be straightforward using an HPCC Systems platform with multiple CPUs.

We observed that the C4.5 decision tree classification method achieves higher accuracy compared to other alternatives. When we increased the depth of the C4.5 decision tree algorithm, we obtained a higher classification accuracy. Hence, the deeper the tree, the more complex the decision rules and the fitter the model. The C4.5 decision tree is also less variant to the parameter and structure selection than some of the classification methods such as deep neural networks.

In addition to visual information, our multimodal learning framework is also capable of integrating other multimedia content, e.g., speech, by treating individual sources as a unique modality. Representations of multimedia data are learned in a similar way to face images and then fused together with face descriptors to feed the identity recognition step.

The K-means learning can be used for engineering complex features. A potential application of this project would be in the area of medical imaging. It can be used in finding distinct features that could lead to improved diagnostic accuracy. Considering medical databases, patients may have a unique real-value measure for certain tests such as glucose or cholesterol. Clustering patients first would help us understand how binning should be done on real-value features to improve accuracy on classification.

Conclusion

We have presented an interesting and novel idea to analyze feature learning and object classification problems in a considerably new platform (HPCC Systems[R]) that can lead to faster optimization/calculation of algorithms and low cost of hardware designs. We have assessed our idea on several databases such as

CALTECH-101, AR databases, and a subset of wild PubFig83 data with multimedia content. Our framework developed in ECL programming language in HPCC Systems is compared with a well cited method [1] developed using MAT-LAB. We observed that the results obtained using the HPCC Systems plat-form are at least comparable with the outcome from this alternative method. Our novel identity recognition algorithm can lead to further exploration of face recognition problems. We increased the classification accuracy of AR database to 78:6 % when compared with the method described in Coates et al. [1]. We analyzed several supervised learning methods and observed that the C4.5 decision tree classification method boosts the recognition rates. We also prove that our learning framework leverages new representations that are learned over multi-media data automatically.

We plan to integrate an improved method to select the feature or cluster number automatically to obtain a higher accuracy. We also plan to integrate multiple CPUs in the HPCC system to speed up the process, compete with GPU hardware and handle larger dimensional calculations.

Acknowledgments The authors would like to acknowledge the support from National Science Foundation awards OIA-1028098 and LexisNexis Risk Solutions.

References

1. Coates A, Ng AY, Lee H. An analysis of single-layer networks in unsupervised feature learning. In: International conference on artificial intelligence and statistics; 2011, p. 215–23.
2. Ng A. Sparse autoencoder. CS294A Lecture notes, 72; 2011.
3. Vincent P, Larochelle H, Bengio Y, Manzagol P. Extracting and composing robust features with denoising autoencoders. In: ICML; 2008.
4. Bristow H, Eriksson A, Lucey S. Fast convolutional sparse coding. In: 2013 IEEE conference on computer vision and pattern recognition (CVPR); 2013. p. 391–98.
5. Middleton A. HPCC systems: introduction to HPCC (high performance computing cluster); 2011.
6. Payne ME, Ngo LB Villanustre F, Apon AW. Managing the academic data lifecycle: a case study of HPCC. In: 2014 IEEE international conference on big data; 2008. p. 22–30.
7. HPCC systems: ECL programmers guide. Boca Raton Documentation Team; 2015.
8. HPCC systems: HPCC client tools. Boca Raton Documentation Team; 2014.
9. K-Means Clustering Algorithm. http://en.wikipedia.org/wiki/K-means_clustering.
10. Coates A, Ng AY. Learning feature representations with k-means. In: Montavon G, Orr GB, Mller K-R, editors. Neural networks: tricks of the trade. Springer; 2012.
11. Fei-Fei L, Fergus R, Perona P. Learning generative visual models from few training examples: an incremental Bayesian approach tested on 101 object categories. Comput Vis Image Underst. 2007;106(1):59–70.
12. Martinez AM, Benavente R. The AR face database. Computer Vision Center, Technical Report, vol 24; 1998.
13. Becker BC, Ortiz EG. Evaluating open-universe face identi cation on the Web. In: Proceedings of the IEEE conference computer vision and pattern recognition workshops; 2013. p. 904–11.
14. Kavukcuoglu K, Ranzato MA, Fergus R, Le-Cun Y. Learning invariant features through topographic filter maps. In: IEEE conference on computer vision and pattern recognition (CVPR); 2009. p. 1605–12.

15. Zang F, Zhang J, Pan J. Face recognition using elastic faces. Pattern Recogn. 2012;45 (11):3866–76.
16. Yang Y, Ma Z, Hauptmann AG, Sebe N. Feature selection for multimedia analysis by sharing information among multiple tasks. IEEE Trans Multimed. 2013;15:661–9.
17. Bauml M, Tapaswi M, Stiefelhagen R. Semi-supervised learning with constraints for person identification in multimedia data. In: IEEE conference on computer vision and pattern recognition; 2013. p. 3602–09.
18. Davis S, Mermelstein P. Comparison of parametric representations for monosyllabic word recognition in continuously spoken sentences. In: IEEE transactions on acoustics, speech, and signal processing; 1980. p. 357–66.

Printed in the United States
By Bookmasters